Development and Prospective Applications of Nanoscience and Nanotechnology

(*Volume 2*)

(Nanomaterials for Environmental Applications and their Fascinating Attributes)

Edited by

Prof. Sher Bahadar Khan

*Center of Excellence for Advanced Materials Research and Chemistry
Department, Faculty of Science,King Abdulaziz University,
P. O. Box 80203, Jeddah 21589, Saudi Arabia*

Prof. Abdullah M. Asiri

*Center of Excellence for Advanced Materials Research and Chemistry
Department, Faculty of Science, King Abdulaziz University,
P. O. Box 80203, Jeddah 21589, Saudi Arabia*

&

Dr. Kalsoom Akhtar

*Division of Nano Sciences and Department of Chemistry, Ewha Womans
University, Seoul 120-750, Korea*

Development and Prospective Applications of Nanoscience and Nanotechnology

Volume # 2

Nanomaterials for Environmental Applications and their Fascinating Attributes

Editors: Sher Bahadar Khan, Abdullah M. Asiri and Kalsoom Akhtar

ISSN (Online): 2452-4085

ISSN (Print): 2452-4077

ISBN (Online): 978-1-68108-645-3

ISBN (Print): 978-1-68108-646-0

General:

1. Any dispute or claim arising out of or in connection with this License Agreement or the Work (including non-contractual disputes or claims) will be governed by and construed in accordance with the laws of the U.A.E. as applied in the Emirate of Dubai. Each party agrees that the courts of the Emirate of Dubai shall have exclusive jurisdiction to settle any dispute or claim arising out of or in connection with this License Agreement or the Work (including non-contractual disputes or claims).
2. Your rights under this License Agreement will automatically terminate without notice and without the need for a court order if at any point you breach any terms of this License Agreement. In no event will any delay or failure by Bentham Science Publishers in enforcing your compliance with this License Agreement constitute a waiver of any of its rights.
3. You acknowledge that you have read this License Agreement, and agree to be bound by its terms and conditions. To the extent that any other terms and conditions presented on any website of Bentham Science Publishers conflict with, or are inconsistent with, the terms and conditions set out in this License Agreement, you acknowledge that the terms and conditions set out in this License Agreement shall prevail.

Bentham Science Publishers Ltd.
Executive Suite Y - 2
PO Box 7917, Saif Zone
Sharjah, U.A.E.
Email: subscriptions@benthamscience.org

BENTHAM SCIENCE

CONTENTS

Preface

Nano is assuredly of incredible small size but its beauty is its real perfection, potency and wide range of applications. Therefore, nanomaterials and nanotechnology have become a fundamental arena of scientific activities because of their enormous applications especially their roles in environmental monitoring and remediation. The inimitable properties of nanomaterials make them suitable for the removal of pollutants from the environment and ultimately cleaning up the environment. The unique properties of nanomaterials are mainly due to their extremely small size, typically in the range between 1 and 100 nanometres, creating a large surface area in relation to their volume and these properties makes them highly reactive with high capacity and better recyclability. Thus nanomaterials provide high surface area, high capacity, well defined structure, high reactivity, insolubility, good chemical and thermal stability, can be easily recycled, with fast sorption kinetics and readily tailored for application in different environments and these properties make them unique for developing a new generation of efficient, cost effective and environmentally acceptable functional materials for water treatment processes compared to non-nano forms of the same materials.

Environmental pollution treatment by nanomaterials is one of the emerging fields which is becoming important day-by-day because of the current and increasingly establishment of the industries. Majority of the industries uses various toxic organic and inorganic chemicals. These industries discharge their toxic and used chemicals to the nearby water streams which pollutes the aquatic world as well as indirectly influences human's life. Therefore, scientific awareness and methods are needed to overcome the mentioned challenges. Being a subject of key interest, it was thought to summarized the cutting edge research on nanomaterials utilization for environmental challenges in the form of a book.

The book is composed of eight chapters. The first chapter is related to different treatment techniques of environmental pollutions using nanomaterials. It highlighted how to resolve the old challenges with new solutions, reviewed different methods used for environmental remediation and highlighted the importance of nanomaterials in environmental remediation. The second chapter is related to nanotechnology for safe and sustainable environment. Nanotechnology is one such revolutionary and state-of-the-art for environmental protection, remediation and pollution prevention. This chapter is set to explore the role of nanotechnology in regard to safe and sustainable environment, which can truly be regarded as a "Realm of Wonders". In the third chapter, basic concepts of photocatalysis are explored. Various parameters which control and influence the photocatalytic process are studied in relation to the mechanistic approach. Nanomaterial such as metal oxides and some new types of materials, like perovskite and metal organic framework, (MOF) are used as efficient photocatalyst. The role and mechanism of these materials have been discussed. All these nanomaterials are used for the environmental remediation, dye sensitized solar cells, air purifications, hydrogen production and self-cleaning process. In the fourth chapter, the role of clay based nanocomposites for environment protection is presented. In particular, the removal of heavy metal ions, toxic organic compounds, hazardous dyes and antibiotics from aqueous environment has been discussed and recent studies are summarized. Purification and remediation of contaminated soil and air with the help of clay based nanocomposites are also discussed. The fifth chapter deals with the introduction of cation exchange materials especially nanocomposite cation-exchange materials, the drivers for green technology. This chapter also describes nanocomposite cation-exchange materials with their technological improvement from old era to the latest age of nano because green chemistry can be applied to real processes. The sixth chapter describes the synthesis of iron oxide and its derivative nanoparticles and their wide scale applications. This chapter summarizes comparative and brief study of the methods for the preparation of iron oxide magnetic nanoparticles with a control over the size, morphology and the magnetic properties. Some future applications of microwave irradiation for magnetic particle synthesis are also addressed. The seventh chapter reviews relative and comprehensive techniques for the preparation of polymeric membrane in cooperated with typical additives and their influence on membrane significant in terms of permeability and selectively. In addition, the recent development in polymeric membranes loaded with nanoparticles for evaluating their properties against biofouling. The eighth chapter offers a brief knowledge about the nanocatalyst for the removal of organic toxins, for instance nitrophenols and dyes, which are at alarming condition. This chapter also deals with the use of metal oxides and layered double hydroxide for the removal of these organic pollutants. Metal oxides and layered double hydroxide worked as a solar catalyst for the removal of contaminant and also how various support work enhance its catalytic performance of these materials. Nanocatalyst on a solid supported materials are also explained in this chapter, which avoids them from aggregation and ease separation after the reaction, which are highly demanded at the industrial level.

We believe that this book will properly convey the savor of the nanotechnology and their approaches toward environmental challenges. We intently anticipate that this book will be beneficial for students, teachers and practitioners.

Sher Bahadar Khan
&
Abdullah M. Asiri
Center of Excellence for Advanced Materials Research &
Chemistry Department
Faculty of Science
King Abdulaziz University
Jeddah 21589
Saudi Arabia

&

Kalsoom Akhtar
Division of Nano Sciences and Department of Chemistry
Seoul 120-750
Korea

List of Contributors

Abdullah M. Asiri	Center of Excellence for Advanced Materials Research, King Abdulaziz University, P. O. Box 80203, Jeddah, 21589, Saudi Arabia Chemistry Department, Faculty of Science, King Abdulaziz University, P. O. Box 80203, Jeddah, 21589, Saudi Arabia
Aftab Aslam Parwaz Khan	Center of Excellence for Advanced Materials Research (CEAMR), King Abdulaziz University, P. O. Box 80203, Jeddah, 21589, Saudi Arabia Chemistry Department, Faculty of Science, King Abdulaziz University, P. O. Box 80203, Jeddah 21589, Saudi Arabia
Anish Khan	Center of Excellence for Advanced Materials Research (CEAMR), King Abdulaziz University, P. O. Box 80203, Jeddah, 21589, Saudi Arabia Chemistry Department, Faculty of Science, King Abdulaziz University, P. O. Box 80203, Jeddah 21589, Saudi Arabia
A. Iqbal	International Islamic University, Islamabad, Pakistan
Enrico Drioli	Institute on Membrane Technology (ITM-CNR), University of Calabria, *Via* P. Bucci 17/C, 87030 Rende, (CS), Italy
Farman Ali	Department of Chemistry, Hazara University, Mansehra, 21300, Pakistan
Fazal Rahim	Department of Chemistry, Hazara University, Mansehra, 21300, Pakistan
Hafiz Nidaullah	School of Industrial Technology, Universiti Sains Malaysia, 11800 Pulau Penang, Malaysia
Iftikhar Ahmad	Pakistan Institute of Engineering and Applied Science (PIEAS), Nilore 45650, Islamabad, Pakistan Department of Center for Nuclear Medicine and Radiotherapy (CENAR), Quetta, 28300, Pakistan
Imtiaz Ahmad	Institute of Chemical Sciences, University of Peshawar, Peshawar, Pakistan
Iqbal Ahmed	Center of Excellence in Desalination Technology, King Abdulaziz University, P. O. Box 80203, Jeddah 21589, Saudi Arabia
Jehanzeb Qureshi	Skudai Johor Unit for Ain Zubaida Rehabilitation and Groundwater Research, King Abdulaziz University, P. O. Box 80203, Jeddah 21589, Saudi Arabia
Kalsoom Akhtar	Division of Nano Sciences and Department of Chemistry, Ewha Womans University, Seoul 120-750, Korea
Kamisah Kormin	Faculty of Management, University Technology Malaysia, Johor, Malaysia
Kashif Gul	Institute of Chemical Sciences, University of Peshawar, Peshawar, Pakistan
L. Gzara	Center of Excellence in Desalination Technology, King Abdulaziz University, P. O. Box 80203, Jeddah 21589, Saudi Arabia
Mohd Omar A.K	School of Industrial Technology, University Sains Malaysia, 11800 Pulau Penang, Malaysia
Muhammed H. Albeiruttye	Center of Excellence in Desalination Technology, King Abdulaziz University, P. O. Box 80203, Jeddah 21589, Saudi Arabia
M. Nadeem	Allama Iqbal Open University, Islamabad, Pakistan

Rizwan Rajput   Department of Chemistry, Government (MPL) Higher School Nawabshah, Nawabshah, Sindh, Pakistan

Saima Sohni   School of Industrial Technology, University Sains Malaysia, 11800 Pulau Penang, Malaysia
Institute of Chemical Sciences, University of Peshawar, Peshawar, Pakistan

S. Sajjad   International Islamic University, Islamabad, Pakistan

S.A.K Leghari   Pakistan Institute of Engineering and Applied Sciences, Islamabad, Pakistan

Shahid Ali Khan   Center of Excellence for Advanced Materials Research (CEAMR), King Abdulaziz University, P. O. Box 80203, Jeddah, 21589, Saudi Arabia
Chemistry Department, Faculty of Science, King Abdulaziz University, P. O. Box 80203, Jeddah, 21589, Saudi Arabia

Sher Bahadar Khan   Center of Excellence for Advanced Materials Research, King Abdulaziz University, P. O. Box 80203, Jeddah, 21589, Saudi Arabia
Chemistry Department, Faculty of Science, King Abdulaziz University, P. O. Box 80203, Jeddah, 21589, Saudi Arabia

Zulfiqar Ahmad Rehan   Department of Polymer Engineering, National Textile University Faisalabad, Faisalabad 37610, Pakistan

CHAPTER 1

Nanomaterials and Environmental Remediation: A Fundamental Overview

Kalsoom Akhtar[1], Shahid Ali Khan[2,3,4], Sher Bahadar Khan[2,3,*] and Abdullah M. Asiri[2,3]

[1] *Division of Nano Sciences and Department of Chemistry, Ewha Womans University, Seoul120-750, Korea*

[2] *Center of Excellence for Advanced Materials Research, King Abdulaziz University, Jeddah 21589, P.O. Box 80203, Saudi Arabia*

[3] *Chemistry Department, Faculty of Science, King Abdulaziz University, P. O. Box 80203, Jeddah21589, Saudi Arabia*

[4] *Department of Chemistry, University of Swabi, Anbar-23561, Khyber Pakhtunkhwa, Pakistan*

Abstract: In this chapter, we have made an overview of the whole book and summarized the environmental pollutions and their treatment with new materials and technology. We highlighted how to resolve the old challenges with new solutions. We reviewed different methods used for environmental remediation and highlighted the importance of nanomaterials in environmental remediation. Different processes related to the management of waste water polluted by bacteria, organic, inorganic pollutants, and toxic metal ions, *etc.* have been discussed. We discussed how nanomaterials are economical solutions for the resolution of the old challenges related to waste water treatment. We also deliberated that the waste water containing harmful metal ions, organic pollutants, bacteria *etc.* can be treated with nanomaterials and for this purpose, development of novel nanomaterials is paramount because nanomaterials have revolutionized the scenario of emerging catalytic and adsorption technologies with recently certified efficient removal of pollutants along with the low cost and high stability. Therefore, the molecular engineering of nanomaterials to use them to reach stable state-of-the-art efficiency for the removal of pollutants as adsorbent and catalysts is vital and the high efficiency coupled with low cost and easy treatment process of the developed nanomaterials has probability to compete and replace the established technologies.

Keywords: Adsorption, Chemical degradation, Environmental pollution, Nanofiltration, Nanomaterials, Photodegradation.

* Corresponding author Sher Bahadar Khan: Chemistry Department and Center of Excellence for Advanced Materials Research, King Abdulaziz University, Jeddah 21589, P.O. Box 80203, Saudi Arabia; Tel: +966-59-3541984; Email: sbkhan@kau.edu.sa

INTRODUCTION

Water is a fundamental need of all living beings, and its contamination affects them to a great extent. Sea water is mainly polluted containing a lot of waste and metal ions. Some of the water resources are contaminated mainly by the mineral waste products, colored materials, organic byproducts released from the industries and to some extent by microorganisms. The wastes released from the industries particularly textile industries pollute the water to a large extent as it contains colored materials that are carcinogenic and toxic in nature, thereby affecting the living beings and the environment [1 - 5]. It is very important to protect the environment from pollutants because agriculture and industries wastes cause serious problem and big threat to the environment. Thus, the wastewater from the sea and industries needs to be detoxified and must be treated before use for drinking and agricultural purposes. Therefore, developing new resourceful methods for curing and purification of contaminated water is very much preferred these days [6, 7].

Nanomaterials are considered as effective purification substances regarding elimination of toxic contaminants from waste water. Nanomaterials function as adsorbents and catalysts for the removal heavy metals, SO_2, CO, NOx, manganese, iron, arsenic, nitrate, heavy metals, dyes, nitrophenols, aliphatic and aromatic hydrocarbons, viruses, bacteria, parasites, antibiotics *etc*. Among different materials used for environmental remediation, nanomaterials exhibited excellent performance as compared to micro and macro-materials [8, 9]. The main reason for good performance of nanomaterials is their high capacity, high reactivity, high surface area, well defined structure, easy dispersability, high chemical and thermal stability, and easy regeneration and recyclability. Another advantage of nanomaterials is that they can be easily designed for use in different environment and can be readily modified for a specific new target species. Nanomaterials have generally rigid structure with open pore assembly which usually offers fast sorption kinetics. Due to large surface areas of particles as compared to their volumes, nanomaterials are more suitable for environmental applications. Thus their reactivity in specific surface mediated reactions can be greatly increased in contrast to the similar material having much bigger sizes. The presence of a comparatively larger number of reactive sites is responsible for nanomaterials' high reactivity along with large surface area to volume ratio; but may also show different reaction rates that surface-area alone cannot rationalize [10, 11]. These properties mark the possibility for increased interaction with contaminants, thus subsequently decreasing contaminant concentrations.

Nanomaterials have been utilized as adsorbent for elimination of metal ions and catalysts for the decontamination of organic pollutants [11]. The adsorption method is extremely valuable toward purifying water. Various adsorbents are developed and applied recently for waste water treatment. However, the nanomaterials are more efficient, cheap, and stable adsorbent and their practical applicability and cost-effectiveness are responsible for their selection toward treatment of waste water [12, 13]. A huge number of metal oxides nanomaterials have been utilized for discarding various environmental contaminants [14, 15]. To increase adsorption ability, the modification of nanomaterials will be carried out.

Nanomaterials, especially nanoparticles [16, 17] and metal oxides have also played important role in the catalysis of organic pollutants [10]. Metal oxides work as photocatalyst for the elimination of different organic impurities and waste water treatment. TiO_2, ZnO, Fe_2O_3, CdO, CeO_2, CdS, WO_3, SnO_2, *etc.* are widely used as catalysts. TiO_2 and ZnO have shown their self as excellent photocatalyst. However, these photocatalyst only encourage photocatalysis during irradiation using UV light as it absorb only in UV region of round about 375 nm with the band gap (~3.2 ev) in UV region. For solar photocatalysis, a photocatalyst must promote photocatalysis through irradiation using visible light. The visible light is almost 46% in solar spectrum which is much more as compared to UV light (5-7%). This least coverage of UV light in the solar spectrum, the high band gap energy (3.2 eV), and fast charge carrier recombination (within nanoseconds) of ZnO confines its extensive application in the solar light [18, 19]. Hence preparation of solar active photocatalyst is vital. For this purpose, several attempt has been made to tune the absorption range of TiO_2 and ZnO to visible region of the solar spectrum by doping with various materials. Similarly, LDH have also been largely investigated as solar photocatalyst. We have recently developed different LDHs and have shown high efficiency toward catalytic degradation of organic pollutants under sun light [3, 4]. Dom *et al.* synthesized $MgFe_2O_4$, $ZnFe_2O_4$ and $CaFe_2O_4$ by low temperature microwave sintering and applied for organic pollutant removal using solar light. They found high photocatalytic activity of these oxides by mineralization of methylene blue under visible light [20]. Raja *et al.* reported a solar photocatalyst based on cobalt oxide and found as a good solar photocatalyst by degrading azo-dye orange II [21]. Wawrzyniak *et al.* have synthesized a solar photocatalyst based on TiO_2 containing nitrogen and applied for the degradation of azo-dye which completely degraded under solar light [22]. Wang *et al.* degraded L-acid up to 83% by using S-doped TiO_2 under solar light [23, 24]. Mohapatra and Parida have synthesized Zn based layered double hydroxide and applied for the degradation and found that layered double hydroxide will be a prominent solar photocatalyst for the detoxification of toxic chemicals [24]. Zhu *et al.* have developed several solar photocatalyst based on Sm^{3+}, Nd^{3+}, Ce^{3+} and Pr^{3+} doped titania-silica and found as good applicants for industrial applications [25]. Parida and Mohapatra reported Zn/Fe layered double hydroxides as an efficient solar photocatalyst for decolorization of hazardous chemicals [26]. Zhao *et al.* synthesized TiO$_2$ modified solar photocatalyst and reported as good candidate for the detoxification of plastic contaminants under solar light [27]. Im *et al.* have synthesized hydrogel/TiO$_2$ photocatalyst for removal of hazardous pollutants under solar light [28]. Pelentridou *et al.* treated aqueous solutions of the herbicide azimsulfuron with titania nanocrystalline films under solar light and found photodegradation of herbicide in few hours demonstrated titania as best candidate for purification of water containing herbicide [29].

This chapter aims to summarize nanomaterials based adsorbents and catalysts for polluted water purification. Different novel and innovative nanomaterials were utilized as adsorbents as well as catalysts for the exclusion of pollutants from sea and waste water. We have also discussed which were used as adsorbent for the elimination of metal ion, metal oxides as solar and UV-vis catalysts while zero valent nanoparticles as catalysts for the removal of organic pollutants.

NANOMATERIALS

Nanometer scale materials are extremely small size and it is estimated that this scale is one ten-thousandth small then the width of human hair. So we can say that this nano-scale constituents includes the sub-microns particle, while the technology which govern, understand and generate substances with dimensions 1-100 nanometers is called nanotechnology. The incredible characteristics of these small scale materials are due to their nano-scale dimensions. Incidental, natural, and engineered materials are the three classes of nano-scale materials. For instance, organic matter, clays, and oxide of Fe inside the soil are comprising in natural nano-scale materials, which playing a key character in biogeochemical practices, while when these small scale substance if come into the environment through waste streams of liquid or solid, atmospheric emissions, fuel combustion, agricultural operations, and weathering process, then it is called incidental nano-scale. After industrial and environmental processing of these materials, if it is applied on industrial or environmental scale, if these are enter environmental through these process it is known as engineered nano-scale materials. There are two methods for nano-scale materials, top down and bottom up methodologies, in the former grinding or milling the macro to nano or by reduction process such as borohydride, creating nano materials by aggregating or combining atoms or molecules [4, 7].

Nanosized materials, nanoparticles, nanomaterials, nanosized particles, nanostructured and nano-objects, are some terms used for nano-scale materials. However, all these materials must have one dimension less than 100 nm. These materials have diverse uses and are highly demanded in the environmental, industrial, biological, chemical, physical and medical fields [14, 24].

The naturally occurring or engineered materials are:

- The carbon containing materials carbon nanotubes or fullerenes, existing in the form of ellipsoids, tubes or hollow spheres. They are highly stable, less reactive, with exceptional electrical and thermal conductivity, and are largely used in photovoltaic cells, sensing, super-capacitor and biomedical applications.
- Metal oxide nanomaterials such as TiO_2, ZnO, CeO_2, Fe_3O_4, *etc.* are to block and absorb ultraviolet light. These nanomaterials are comprise of closely and strictly packed semiconductor crystals, which are composed of hundreds or thousands of atoms. These metal oxide nanomaterials have applications in environmental remediation as photocatalysts and adsorbent for the removal of pollutants.
- Zero-valent metal nanoparticles such as zero-valent iron, copper, silver, nickle, *etc.* are the example of engineered nanomaterials. These nanomaterials have high surface area which cause increase in surface reactivity. Zero-valent metal nanoparticles have been utilized in waste water treatment.
- Excitons are the bounded electron hole pairs, which have three-dimensional arrangement in the so-called quantum dots. Quantum dots are semiconductors having 10 to 50 nanometers. And are largely applied in telecommunications, medical imaging, photovoltaics, and sensing technology.
- Incorporating various functional groups in dendrimers; exceedingly branched polymers, which are manufactured and designed with different contours like discs, spheres, and cones like structures. It has potential application in chemical sensing, drug delivery, modified electrodes, and DNA transferring agents.
- Composite is another important materials of this class that are made from two nano or one nano with macro-materials. Such materials can be combined with synthetic and biological molecules, with novel catalytic, electrical, magnetic, thermal, mechanical, and imaging practices. This class of materials is also potentially applied in cancer detection and drugs delivery. It is also used in packaging and auto-parts materials to improve its flame-retardant and mechanical characteristics.

ENVIRONMENTAL APPLICATIONS OF NANOMATERIALS

Nanomaterials efficiently removed various chemical and biological pollutants from waste water. Nanomaterials function as adsorbents and catalysts for the removal heavy metals, SO_2, CO, NOx, arsenic, iron, manganese, nitrate, heavy metals, dyes, nitrophenols, aliphatic hydrocarbons, aromatic hydrocarbons, viruses, bacteria, parasites, antibiotics *etc.* [30 - 32]. Different materials have been used for environmental remediation. However, the role of nanomaterials is highly appreciated for the water purification as compared to micro and macro-materials.

The main reason for better performance of nanomaterials is their high capacity, high reactivity, high surface area, well defined structure, easy dispersability, high chemical and thermal stability, and easy regeneration and recyclability. Another advantage of nanomaterials is that they can be easily designed for use in different environment and can be readily modified for a specific new target species. Nanomaterials, especially mesoporous nanomaterials have generally rigid structure with open pore assembly which usually offers fast sorption kinetics [33].

Nanomaterials are used for the de-contamination approach of soil, water and environment, as well places contaminated by oil spills dyes, chlorinated solvents and heavy metals. Nano-scale materials got much interest in various scientific sectors due to its large surface/volume ration as compared to their bulk materials. The constituents of the macro, or micro and nano-scale materials are the same, only the difference in the particle size. These materials can be tuned for specific application as compared to their bulk counterparts. Due to their large reactive sites and surface to volume ratio, accelerate the reaction rate with a high rate constant. These characteristics make them their facile contact to chemicals, thus a quick decrease the concentrations of pollutant are achieved. By using a suitable coating around these materials it remains suspended in groundwater due to its small size. These materials achieved a broader dissemination and large traveling than macroscopic particles by using appropriate coatings, which enhanced the reaction rate [32, 33].

Most of these materials are already find its jobs in various sectors, while others are in the process for their full implementations. The ongoing small scale and industrial research are at play as to explore the particles like TiO_2, dendrimers, self-assembled monolayers on mesoporous supports, carbon nanotubes, swellable organically modified silica and metalloporphyrinogens. Many un-un-answered questions should be addressed in this field. For instance, much knowledge is needed about the fate and passage of free nanomaterials in the environment, there staying, toxicology, and their broad marketable benefit.

The nanomaterials are highly ambitious and it is important to optimize and characterize adsorbents as well as solar photocatalysts in order to understand their morphology, efficiency and stability relationships in waste water treatment applications. It is also important to develop innovative nanomaterials which are expected to show high record efficiency and lead to a maximum removal of pollutants under full-sun illumination. Further the performance of nanomaterials should be optimized under different conditions. The demanding task of nanomaterials is to address the stability under heat and light soaking conditions. Since, the metal oxides nanomaterials are powder, it is also important to address questions regarding the tolerance limit for recyclability and methods to overcome the problem of recyclability. One possible solution is development and incorporating of metal oxides into the polymer hosts. Given the simple preparation, easy implementation and high efficiency, it is in the interests to develop tailored nanomaterials that show enhanced stability and very high efficiency toward the removal of different pollutants [34].

NANOMATERIALS BASED PROCESSES FOR WASTE WATER TREATMENT

Adsorption Based Environmental Remediation

Adsorption is the most interesting and most used method for water purification [27, 34]. Sorption is a phenomenon where the ions deposited on the solid surface, and this deposition is due to the transfer of the ions from the liquid to solid phase. Fundamentally, mass transfer are occurred in adsorption phenomenon, and attached to the solid surface by chemical or physical processes. Numerous low cost adsorbents, agricultural waste, by-products from industries, biopolymers and its modified form, natural materials and silica have been employed for the heavy metals removal and waste water purification. Beside, the beneficial application of the aforementioned materials as adsorbent, nano-scale materials are preferred as adsorbent materials due to it high surface area, stable nature practical application and low-cost for the treatment of waste water. Usually, the sorbents phenomena occurred in three main process: (i) transportation of the pollutant from the bulk solution to the sorbent surface; (ii) adsorption of the pollutants onto particle surface; and (iii) inward movements inside the sorbent particles [35].

Different Nanomaterials Used as Adsorbent

Metal Oxides

Metal oxides with diverse morphological and textural characteristics has been successfully employed as adsorbent for environmental remediations [35 - 37]. Fe^{3+} has been selectively removed through ZnO-CdO nanoblock from iron ions contaminated water [38]. Nickel ions were largely removed by using Cs doped ZnO [32]. SnO_2-TiO_2 nanocomposite and silver oxide nanoparticles removed La^{3+} selectively and efficiently [39] while Ag_2O_3-ZnO nanocones were selective for the adsorption of cobalt ions. Inorganic based nanomaterials and mesoporous structures removed organic contaminants by two different methods (1) static force (containing Lewis adsorption); and (2) weak chemical interaction due to the functional groups located on the surface which facilitate hydrogen bonding. In order to improve the desired adsorption performance, it is utmost to tune the chemical nature of the nanomaterials. For instance, As^{3+} and As^{4+} was largely removed by using nanocrystalline TiO_2 as compared to TiO_2 [40, 41].

The pH of the system in adsorption play a significant role in metal ions removal as it effect the surface charge. For instance, at the zero point charge pH (ZPC), charge on the surface is neutral, where ZPC pH of maghemite nanoparticles is 6.3. In case of metal oxides, the hydroxyl groups usually covers the surface and different pH effect the hydroxyl groups and thus hydroxyl groups on the surface of metal oxide can vary with pH variation [11]. Below the ZPC pH, the surface of adsorbent materials become positive leading to anion adsorption. Above pH 4, the adsorption of MnO_4^{2-} ions upsurges and became persistent at 4–6 pH [11]. Similarly, the adsorption of MoO_4^{2-} ions decreases as the pH raise above pH 6 because at pH > pHzpc, the surface becomes negatively charged, thus the electrostatic repulsion increased amongst the negatively charged MoO_4^{2-} ions and the negatively charged adsorbent, which releases the adsorbed MoO_4^{2-} ions [11, 42].

Magnetic Sorbents

Magnetic nanomaterials find its fascinating application in the elimination of organic contaminants, inorganic ions and dissolved carbon. Organic dyes, polycyclic dyes, crystal white and malachite green have been successfully removed from waste water by using various magnetic materials like magnetic chitosan gel particles, magnetic alginates, and immobilization of copper phthalocyanine dye covalently and magnetic charcoal [11]. Similarly, magnetite nanoparticles with surfactant-coated have been used for the removal of 2-hydroxyphenol [43]. Polymer-coated vermiculite iron oxide composites acting as floating magnetic sorbents. These materials lift on water surface, which removed spilled oils from contaminated water [11, 44]. Fe^0 nanoparticles removed As^{3+} and As^{4+} by rapidly adsorbing than precipitated by weak electrostatic attraction amongst the binding sites of the catalyst and adsorbed materials [11, 45, 46]. Similarly, polyvinyl alcohol-co-vinyl acetate-co-itaconic acid stabilized zero valent iron nanoparticles was used for the uptake of heavy metal ions.

The Zn^{2+}, Cd^{2+}, Cu^{2+} and Pb^{2+} are efficiently removed through modified nanoparticles of Fe oxide with 3-aminopropyltriethoxysilane and copolymers of acrylic acid and crotonic acid [11, 47]. Similarly, the heavy metal ions are also removed by reporting the carboxylated chitosan changed magnetic nanoparticles [48]. Qi and Xu altered the chitosan nucleus by ionic gelation with tripolyphosphate used as ionic cross linker for Pb^{2+} [11, 49]. However, modified chitosan nanoparticles are breakup in the aqueous solution or aggregated in alkaline solution at pH 9, due to weak force attractions between tripolyphosphate molecules and chitosan. Dissolved organic colloids and organic carbon in small size have been identifies as a distinctive non-aqueous organic phase, which adsorbed the pollutants and thus decreases its bioavailability.

Zeolites

Zeolites, have microporous materials consisting of a 3D configuration of $[SiO_4]^{4-}$ and $[AlO_4]^{5-}$ polyhedra linked with oxygen atoms forming negative lattice providing Lewis and Bronsted acid sites. It has pore size less than 2 nm. Zeolite received significant Bronsted acidic characteristics by cations and protons exchange. Zeolite materials are largely applied in environmental studies due to their vast availability, non-toxic nature, selective adsorbent properties and low cost. It is extensively applied in the deletion of heavy metals, for instance Cu^{2+}, Ni^{2+}, Cr^{3+}, Zn^{2+}, Fe^{2+}, Cd^{2+} and Pb^{2+} [11, 50, 51]. Zeolites are stable however below pH 2, it is unstable and collapse. Zeolites are documented to better as compared to activated carbon for the retention of chloroform, methyl-tert-butyl ether, and TCE in water.

Silica and Silica Based Nanomaterials

Another important materials is the mesoporous silica and its derivatives materials having approximately 2–50 nm pore size which successfully adsorbed heavy metals from contaminated water. Mesoporous silica with functionalized monolayers have been valuable for mercury and others heavy metals removal.

Amino-functionalized silica showed better performance for the removal of Cu^{2+}, Ni^{2+}, Zn^{2+} and Cr^{2+}, whereas, mercuric ions efficiently adsorbed on thio-functionalized silica [11, 52]. Activated alumina, having high porosity, is widely used in filtering apparatuses which are used for the purification of drinking water. Aminated and mesoporous alumina, and alumina-supported MnO are able to take out As^{3+}, As^{4+}, Cu^{2+} and TCE from the polluted water [11, 53, 54].

Polymer Based Nanomaterials

Various polymers have been assembled in the form of nanoparticles and were applied as adsorbent for the removal of various pollutants. For instance, poly *N*-isopropylacrylamide nanoparticles eradicate Pb^{2+} and Cd^{2+} from contaminated water by Coulombic interaction between the polymer carboxylate group and metal species with positive charge [55]. Nevertheless, the use of poly N-isopropylacrylamide in treatment of polluted water is not extensive because the isopropylacrylamide is not showing promising capability toward metal ions removal. The copolymerization of pyridyl monomer to form polymeric nanoparticles with styrene have been used for metal ions removal. Bipyridine groups on developed nanoparticles surface enhanced the metal removal performance from wastewater due to their high interacting capability with metal ions [56]. Poly(vinylpyridine) nanoparticles specifically removed Cu^{2+} as compared to other metal ions [11]. Azo-chromophore modified polystyrene nanoparticles showed high adsorption capacity toward Pb^{+2} Bell *et al.* [57]. proposed that the polymeric nanoparticles grafted with macrocyclic ligand having core-shell structure are highly active for the selective removal of heavy metals. Without grafting, this unique core–shell morphology removed Hg ion selectively, while with grafting technology it removed Co^{2+} selectively from other metal ions [58]. Thus, polymeric nanoparticles may be easily structured for the selective removal of heavy metals. Tungittiplakorn *et al.* suggested that polyurethane-based nanoparticles could be practical in the transportation and desorption of organic contaminants with the hydrophobic core of the nanoparticles [11, 59, 60].

Dendrimers Based Nanomaterials

Dendrimers based metal nanoparticles (NPs) have got much interest in the scientific research and technology in ecological treatment owing to their unique crystal shapes, size, and lattice morphology [59]. Synthesis of functionalized nanoparticles by different approaches is remain a great deal. One of the exceptional methods used to make inorganic NPs is the dendrimers approach. Dendritic nanopolymers are extremely 3D branched globular nanoparticles with precise shapes and designs. Their sizes are 1–100 nm in range, and are built from a preliminary atoms, for instance, attachment of nitrogen to carbon as well as other elements, which built their self through a series of chemical reactions, that produces a sphere-shaped branching structure by involving hierarchical assembly of divergent or convergent methods. As the development progresses, layers are supplemented succeedingly and the sphere can be extended to the targeted size. It contains three constituents: (1) core, (2) internal branched cells and (3) end branched cells [61, 62]. It is an innovative polymers class with a thick sphere-shaped structure and exceptional performance with narrow size distribution, used as templates or stabilizers to form comparatively monodispersed organic/inorganic hybrid nanoparticles. During the synthesis of dendrimer-stabilized nanoparticles, dendrimer played an extremely important role by coordinating with the metal ions through electrostatic contact, *etc.*, followed by reduction of the nanoparticles to make inorganic nanoparticles. Dendrimer established nanomaterials can sum up a wide range of solutes in water, comprising cations (*i.e.*, iron, silver, gold, zinc, copper, nickel, uranium, cobalt and lead) by connecting to the functional groups of dendrimers, such as carboxylates anions, hydroxymates and primary amines [63]. The removal aptitude and selectivity can be upgraded by varying functional group of dendrimers. Dendrimers-based nanomaterials might be useful in the regaining of uranium metals and perchlorate anions from unclean groundwater.

Carbon Nanomaterials

These materials are largely used for sorption. Carbon nanomaterials are existing in different morphologies, such as activated carbon, carbon fibers, carbon beads, single-walled carbon nanotubes (SWCNTs), multi-walled carbon nanotubes (MWCNTs), nanoporous carbon, graphene, graphene oxide, carbon nanotubes. These carbon nanomaterials have been largely used to remove different pollutants from waste water because they have many benefits as compared to the old-fashioned materials, because of their high surface/volume ratio. High electronic, mechanical and optical properties make them potent adsorbents [64]. The high degree of successful ecofriendly process is because of high adsorb capacity for diverse pollutants, favorable kinetics, high surface area, and selective removal of aromatic solutes [65].

Carbon Black and Activated Carbon

Activated carbon is a well known adsorbent and has been widely used for the waste water treatment. Usually the carbon black is activated by oxidation in the presence of HNO_3 and thus make it functionalize by creating functional groups on the surface of carbon black. This functionalized carbon showed high adsorption of Cu^{2+} and Cd^{2+} because of the increased amount of functional groups created by oxidation on the surface of carbon black. The Adsorption behavior of modified carbon black (CB) toward Cu^{2+} or Cd^{2+} is highly dependent on pH of the solution. Adsorption is directly related to solution pH [11]. The functionalized carbon black uptake most of the metal when pH raise above 5.5 and the reasons might be due to the development of charge on the surface of carbon black (modified) and the concentration distribution of Cd^{2+} or Cu^{2+} that are pH dependent. At lower pH, Cd^{2+} or Cu^{2+} adsorption on modified CB is lower due H^+ and Cd^{2+} or Cu^{2+} competition for the adsorption sites. There are negative charges on the modified CB surface for wide pH range and Cd^{2+} or Cu^{2+} carrying positive charge, remaining as either Cd^{2+} or Cu^{2+}. When the pH level of the solution increases, the concentration of competitor H^+ ions decreases and Cu^{2+} or Cd^{2+} adsorption increases [11]. Nano-scale hydroxyapatite and carbon black have shown strong adsorption of Cu^{2+}, Zn^{2+}, Pd^{2+} and Cd^{2+}. CB and activated carbon have different adsorption affinities for different metal ions. The adsorption capacity also depend upon the size of carbon particles, smaller the particle size, higher will be the adsorption capacity while carbon with larger particles result in lower adsorption because the micropores at the internal surface of the activated carbon are not accessible to pollutant whereas the nanosized pores of carbon black are more accessible to pollutant [11, 66].

Carbon Nanotubes

CNTs are like cylindrical hollow micro-crystals of graphite. Due to its high specific surface area, CNTs have attracted the interest of researchers as a new type of adsorbent. It is the graphitic carbon needles which have 4–30 nm external diameter and 1 mm a length [67]. MWCNTs are made of concentric cylinders with spacing between the adjacent layers of about 3.4 angstrom [68]. Iijima was first discovered SWCNTs [69]. The modified CNTs which was functionalized by oxidation was found to be a good adsorbent of Cd^{2+} and it was found that the specific area and pore specific volume of CNTs were increased after oxidation. Due to the large specific area, The modified CNTs have displayed extraordinary adsorption capabilities and adsorption efficiencies toward several organic pollutants [70]. Further it was observed that CNTs require less time to adsorb organic pollutants as compared to activated carbon [44]. The reason behind the fast adsorption is that CNTs are lack of porosity while the activated carbons have porous structure. In order to achieve the equilibrium, adsorbate move from external to internal pores surface [44]. The fastest response of CNTs indicated its high removal potential for dichlorobenzene from water. Even CNTs were found superior adsorbent of dichlorobenzene as compared to graphitized CNTs which is due to the rough surface of CNTs that makes the adsorption of dichlorobenzene much easier for CNTs while the smooth surface of graphitized CNTs reduces the adsorption of dichlorobenzene. It is also adsorbed that SWCNTs exhibit higher adsorption than hybrid carbon nanotubes (HCNTs) and MWCNTs. It is reported that the ethylbenzene adsorption on the surface of CNTs rely on their porosity and chemical nature. HCNT hybride might created more absorbent structure for MWCNTs and exposing larger surface area as compared to MWCNTs for ethylbenzene adsorption and thus predominantly removed ethylebenzene. SWCNT efficiently adsorbed ethylbenzene as compared to MWCNT due to the electrostatic force of interactions between ethylebenzene and SWCNT [44] due to the positively charged ethylbenzene with negatively charged SWCNT, making more electrostatic interactions and thus high adsorption capacity. Zn^{2+} sorption from the aqueous solutions was studied by SWCNTs and MWCNTs [71], where it was observed that the Zn^{2+} adsorption on the CNTs was directly related to temperature. Under the same experimental conditions, the sorption capacity of Zn^{2+} by CNTs was higher compared to commercially powdered activated carbon, indicating that MWCNTs and SWCNTs are effective sorbents. All these results indicated the high reusability of CNTs in wastewater treatment. CNTs activation under oxidizing conditions displaying a key role by increasing the sorption capacity, because it brought changes in surface functional and morphology which helped amorphous carbon removal [72]. The activation alter the surface characteristics of functional with many defects on the surfaces. Additionally they had higher lead adsorption capacity and become a highly adsorbents materials for waste water purifications. The Cu^{2+}, Cd^{2+} and Ni^{2+} removal efficiency was documented in the literature. The dyes removal are pH dependent due to the electrostatic force of attraction amongst the negatively charged CNT and positively charged cationic dyes, such as, methylene blue and methyl violet [73, 74].

Nanocomposites

Nanocomposites play an important role in water purification. Nanocomposites are either carbon nanocomposites or polymer nanocomposites which shows applications in environmental remediation.

Carbon Based Nanocomposites

Carbon based nanocomposites have shown excellent performance for adsorption as compared to pure carbon based materials, such carbon black, carbon nanotube, graphene and graphene oxides. Carbon based nanomaterials are used widely in the field of removal heavy metals in recent decades, due to its nontoxicity and high sorption capacities. Activated carbon is used firstly as sorbents, but it is difficult to remove heavy metals at ppb levels. Then, with the development of nanotechnology, carbon nanotubes, fullerene, and graphene are synthesized and used as nanosorbents. These carbon nanomaterials have shown significantly higher sorption efficiency comparing with activated carbons. Therefore CNT/MO nanocomposites sorption capacity are highly demanded for environmental remediation. It was reported that for the physical adsorption of contaminants, nanocomposites being the highly demanded materials due to its inert and specific surface areas, with comparatively uniform assemblies, leading to high adsorption sites. For instance, the hybride nanocomposite (MWCNT/alumina) has been described an effective sorbent for lead ions removal 3-7 pH [75]. For instance, magnetic composites adsorbent (CNT-iron oxides) has been effectively practiced for different metals removal from wastewater. The europium adsorption was achieved by incorporating iron oxide magnetite with CNTs. This composite is potentially a promising to facilitate the separation and recovery of CNTs from solution with magnetic separation technique. CNTs become not the part of pollutants and can be easily be recovers [76]. For instance, different nanostructure materials was water purification including magnetic MMWCNT nanocomposite for removal of cationic dyes removal [77] and manganese oxide-coated carbon nanotubes for Pb(II) removal, the Pb(II) is enhanced with manganese oxide loading, as manganese oxide provide high adsorption sites [78]. Graphene and reduced graphene oxide (rGO) are kinds of novel and interesting carbon materials and have attracted tremendous attentions as adsorbents for the removal of different pollutants. Because of very high specific surface area, graphene and reduced graphene oxide are good candidates as an adsorbent [79]. Graphene-based manganese oxide composites were applied as adsorbent materials for water purification [79]. Graphene-based iron oxide nanocomposites have been used as adsorbents for the removal of tetracycline, dyes, As(III), As(V), chromium, lead, cobalt, and so on [79]. More recently, a ternary composite of highly reduced graphene oxide/Fe_3O_4/TiO_2 has been reported, which exhibited high selectivity and capacity for the removal of phosphopeptides. In another study, graphene/zinc hydroxide nanocomposites were shown high removal efficiency toward hydrogen sulfide. Recently, graphene-based composites have been applied for the extraction of polycyclic aromatic hydrocarbons and parathyroid pesticides with excellent removal efficiency. The preparations of graphene-based magnetic nanocomposites have reported for the removal of arsenic. Similarly, the magnetic reduced graphene oxide nanocomposite have been reported for the removal of dyes and heavy metals such as arsenate, nickel, and lead [80].

Polymer Based Nanocomposites

The difficult separation process limit the reuse of metallic oxides nanoparticles and other nano-scale materials and possible risk to ecosystems and human health caused by the potential release of nanoparticles into the environment. In addition, the use of aqueous suspensions limits their wide applications because of the problems for the separation of the fine particles and the recycling of the catalyst. Immobilization of these nanoparticles in polymer matrix has been available to solve the problems to considerable extent, serving for the reduction of particle loss, prevention of particles agglomeration and potential application of convective flow occurring by free-standing particles [81 - 85]. The widely used host materials for nanocomposite fabrication are polymers and polymeric host materials must possess excellent mechanical strength for long term use. The choice of the polymeric support is influenced by their mechanical and thermal behavior, hydrophobic/hydrophilic balance, chemical stability, bio-compatibility, optical and/ or electronic properties and their chemical functionalities (*i.e.* solvation, wettability, templating effect, *etc.*) [86]. Nanocomposites based on magnetite (Fe_3O_4), maghemite (Fe_2O_3) and jacobsite ($MnFe_2O_4$) nanoparticles loaded alginate beads have shown high ability to remove heavy metal ions (Co(II), Cr(VI), Ni(II), Pb(II), Cu(II), Mn(II)) and organic dyes (methylene blue and methyl orange) from aqueous solutions [87]. Magnetic particles in the nanocomposites allowed easy isolation of the beads from the aqueous solutions after the sorption process. Cellulose/Mn_3O_4 have shown high uptake capacity for the removal of chromium. It has been studied for the removal of Cu^{2+}, Cd^{2+}, Co^{2+}, Cr^{3+}, Fe^{3+}, Ni^{2+}, Zn^{2+} and Zr^{4+}. The nanocomposite has followed the uptake capacity order $Cr^{3+} > Zn^{2+} > Fe^{3+} > Cd^{2+} > Zr^{4+} > Ni^{2+} > Co^{2+} > Cu^{2+}$ [36]. Similarly, ZrO_2 embedded cellulose adsorb Ni^{2+} selectively [34]. Cellulose acetate/ZnO nanocomposite has been utilized for the removal of heavy metals [8]. Polymer based nanocomposites were used as adsorbent for the uptake of cadmium and yttrium [33] while poly(propylene carbonate)/exfoliated graphite nanocomposites extracted Au^{+3} selectively [88]. Polybenzimidazole hybrid membranes were found to be selective for the adsorption of mercury [13]. Cellulose-lanthanum hydroxide nanocomposite selectively removed copper ions [30]. Various chitosan based nanocomposites have showed excellent adsorption capability for different dyes [36].

Chemical Degradation Based Environmental Remediation

There are different methods for water treatment. Among them, chemical degradation is the most widely used methods [11]. Chemical degradation methods include the following:

Ozone/UV Radiation/H_2O_2 Oxidation
Supercritical Water Oxidation
Fenton Method
Sonochemical Degradation
Electrochemical Method
Electron Beam Process
Solvated Electron Reduction
Enzymatic Treatment Methods
Photocatalytic Degradation

The word "photocatalytic degradation" means the degradation in the presence of photon and catalyst while the word "photocatalyst" is a combination of photons and catalyst. Thus photocatalyst is a chemical substance which can accelerate the rate of chemical reaction under illumination of light while the photocatalysis is photon induced redox processes at the surface of a photocatalyst in suspension [89]. Photocatalysis is a type of heterogeneous catalysis for which the basic requirements are as follow.

Photocatlyst: Metal oxides, Metal sulfides, Metal selenides, LDH, Nanocomposite

Redox/Donor or acceptor medium: Water or any other polar solvent (e- donar)

Adsorbed or dissolved O_2 (e- acceptor)

Photon source: Halogen lamps, Mercury vapor lamps and sunlight

A photocatalyst must have the following properties.

• *It must be active under illumination of light and have band gap energy (Eg)*
• *It should be inert both biologically and chemically.*

- *It should be stable toward photocorrosion.*
- *It should be economical and non-toxic.*
- *It must have suitable valance and conduction band potentials to establish a redox reaction.*

The valance and conduction band potential means that a photocatalyst must have band gap. Electrons of an atom occupy discrete energy levels. In a crystal, each of these energy levels is split into many energy levels. Consequently, the resulting energy levels are very close and can be regarded as forming a continuous band of energies. For a metal (or conductor), the highest energy band is half-filled and the corresponding electrons need only a small amount of energy to be raised into the empty part of the band, which is the origin of the electrical conductivity at room temperature. Thus it means that metal has no band gap. Only reduction or oxidation takes place which depends on the band. In contrast, in semiconductors and insulators, valence electrons completely fill a band, which is thus called the valence band, whereas the next highest energy band (termed the conduction band) is empty, at least at 0 K. Thus both semiconductors and insulators has bandgap. Semiconductors have low bandgap (bandgap < 5) while insulators have high bandgap (bandgap > 5). Thus semiconductors are much more important because semiconductors can easily excite valance band electrons to generate e^{-}-h^{+} pair [3, 4, 90]. Generally, a potential difference is established across valance band and conduction band of semiconductor when it in contact with solvent. When semiconductor is illuminated by light and get energy equal or more then bandgap energy, it excite valence band electron and generate e^{-}-h^{+} pair. This e^{-}-h^{+} pair cause redox reaction when a donor and acceptor get adsorbed on the surface of the semiconductor. The photo-generated holes at the surface of the irradiated semiconductor can oxidize a variety of hazardous species or produce ˙OH radicals [91 - 94]. However, the main problem is the recombination of this generated e^{-}-h^{+} pair before proceeding the redox reaction. Therefore, it is vital to avoid this recombination. The recombination of e^{-}-h^{+} pair in semiconductors can be avoided by the following methods.

- *Surface modification by metal ion impregnation*
- *Composite formation*
- *Metal ions (M^{n+}) doping or nonmetal (S and N) doping*
- *Use of organic molecules such as methanol for hole capture*

In aqueous media, the semiconductor follow the following mechanism in the presence of light. They produces OH and O_2^{-} radicals which are highly reactive and can easily degrade the organic pollutants.

The rate of a photocatalytic reaction depends on various factors such as temperature, pH, amount of catalyst, amount of pollutants *etc.*

Photocatalyst can be divided on the basis of light absorption into UV photocatalyst and solar photocatalyst. Those photocatalysts which absorb UV light and only encourage photocatalysis upon irradiation by UV light because it absorb only in the UV region of round about 375 nm with the band gap (~3.2 ev) in UV region. TiO_2, ZnO, Fe_2O_3, CdO, CeO_2, CdS, WO_3, SnO_2, *etc.* are UV catalysts which absorb only UV light below the visible range of light spectrum [95].

Solar photocatalyst are those which absorb light in visible range and promote photocatalysis by irradiation with light. For solar photocatalysis, a photocatalyst must promote photocatalysis by irradiation with visible light because solar spectrum consists of 46% of visible light while the UV light is only 5-7% in the solar spectrum. This least coverage of UV light in the solar spectrum, the high band gap energy (3.2 eV), and fast charge carrier recombination (within nanoseconds) of ZnO confines its extensive application in the solar light [77]. Therefore, it is an urgent demand to develop an active photocatalyst which can promote photocatalysis in visible region. For this purpose, several attempts have been made to tune the absorption range of TiO_2 and ZnO to visible region of the solar spectrum by doping with various materials. Several doped metal oxide nanomaterials, LDH and nanocomposites have been developed which work as a solar photocatalysts [92]. We have recently published several LDH and different doped nanomaterials which degraded different dyes under solar light and can be utilized for the treatment of polluted water [92]. Multi-walled carbon nanotube–TiO_2 composite catalysts can be used as catalysts in photocatalytic processes for water treatment. The introduction of increasing amounts of CNTs into the TiO_2 matrix prevents particles from agglomerating, thus increasing the surface area of the composite materials. A synergy effect on the photocatalytic degradation of phenol was found mostly for the reaction activated by near-UV to visible light irradiation. This improvement on the efficiency of the photocatalytic process appeared to be proportional to the shift of the UV–vis spectra of the CNT–TiO_2 composites for longer wavelengths, indicating a strong interphase interaction between carbon and semiconductor phases. This effect was explained in terms of CNTs acting as photosensitizer agents rather than an adsorbents or dispersing agents. Surface defects at the surfaces of carbon nanotubes provide advantages not only for the anchoring of the TiO_2 particles but also for the electron transfer process to the semiconductor. Original carbon nanotubes, containing moderate amounts of oxygen surface groups, produced the highest synergistic effect for the degradation of phenol under near-UV to visible irradiation. The efficiency of CNT–TiO_2 catalysts in the photocatalytic oxidation of mono-substituted organic compounds under visible irradiation was dependent from the ring activating/deactivating properties of the aromatic molecules. A higher kinetic synergy effect was observed for compounds presenting electron donor groups, such as phenol and aniline. For nitrobenzene and benzoic acid a synergy factor near to 1 was obtained, indicating the inexistence of any synergy effect between the CNTs and TiO_2 in the photocatalytic degradation of these pollutants [96]. A comparison of the photocatalytic activity of TiO_2 and TiO_2/CNTs composites for acetone degradation in air shows that the presence of a small amount of CNTs can enhance the photocatalytic activity of TiO_2 greatly [97]. Electrons excited by TiO_2 may easily move to the nanostructure of the CNTs due to the strong interaction between TiO_2 and CNTs. Then, CNTs raise the band gap of TiO_2, which can prevent recombination of the e^-/h^+ pairs. Moreover, the abundant hydroxyl groups adsorbed on the large surface of the composites can lead to the formation of more $^-$OH radicals, which result in an enhancement of the photocatalytic activity of TiO_2 [98].

Zero Valent Metal Nanoparticles

Different zero valent nanoparticles such as gold, silver, copper, iron, cobalt, nickel, palladium, platinum, bimetallic, *etc.* have widely been utilized in waste water treatment which is due to their size, surface area and specific chemistry. They have played important roles in different fields among them, the fast growing field is nanoparticle based catalysis. In catalysis, the particle size of nanoparticles play important role and generally the catalytic activity of zero valent nanoparticles is directly related to the particle size. The catalytic activity is only observed at the nanometric scale but not micrometric scale. The particles with smaller size have shown higher catalytic activity as compared to nanoparticles with larger particle size because the nanoparticles with small particle size have broad surface area and that is why smaller nanoparticles have excellent catalytic properties. Small particles most often have a high activity compared with relatively large particles, because the proportion of active surface metal atoms *vs.* the total number of metal atoms in nanoparticles is all the higher as the nanoparticle is smaller. These catalysts are active under mild conditions, even at ambient temperature or less [17].

However, the "naked" nanoparticles are unstable and sometime do not function as catalysts or work with low catalytic activities. The instability of nanoparticles during catalytic reaction is the major problem which reduce the catalytic properties of nanoparticles. The instability of nanoparticles caused by some external stimulus lead to aggregation and this aggregation cause change in the size, shape, surface area and activated surface electron during the catalytic process which affect the catalytic activity of nanoparticles. The second problem of nanoparticles is their economic viability because nanoparticles can be grown by the reduction of their metal precursor. During the reduction of metal precursor, the growth of nanoparticles takes place however some metal ions do not partake in the reduction process and just wasted without conversion into nanoparticles. Therefore, it is highly required to control the aggregation of nanoparticles and loss of metal ions, especially gold and other precious metals. To overcome these problems, a proper support is necessary. The supporting materials avoid the aggregation of nanoparticles and lose of metal ions and also improve their catalytic properties because functional groups of the supporting materials cause the interaction between the nanoparticle and supporting material. Thus, a real challenge for the design of an active and selective zero valent catalysts is the selection of the proper stabilizer and supporting materials. The stabilizer and supporting materials should have strong affinity toward metal ions of the concerned zero valent nanoparticles, stabilization ability and large surface area. The stabilizers and supporting materials directly affect the catalytic activity of nanoparticles and it has been found that some stabilizer gave faster kinetics than other [17].

Various kinds of stabilizers and supporting materials have been used such as:

- Organic molecules:
  - N,N-dimethylformamide (DMF)
  - Cetyltrimethylammonium bromide (CTAB)
- Carbon materials:
  - Carbon nanotubes
  - Graphene
  - Graphene oxide
  - Carbon black
  - Active carbon
  - Carbides
  - Carbon dots
- Polymer:
  - Chitosan
  - Cellulose, cellulose acetate, CMC
  - Polyethylene glycol
  - Poly vinylpyrrolidone
  - Cyclodextrin
  - Polyaniline (PANI)
  - Polythiosemicarbazide (PTSC)
  - Polyethyleneimine (PEI)
  - Polyvinyl alcohol (PVA)
  - Polysilsesquioxane (PSQ)
  - Ionic polymers

- ▪ Poly(allylamine hydrochloride) (PAH)
- ▪ Methyl-imidazolium-based ionic polymers
- Metal oxide:
  - ○ TiO_2
  - ○ CeO_2
  - ○ Al_2O_3
  - ○ Fe_3O_4/Fe_2O_3
  - ○ MgO
  - ○ ZnO
  - ○ Mixed oxides
- Dendrimer:
  - ○ Poly(amidoammine) dendrimers (PAMAM)
  - ○ Poly(propyleneimine)dendrimers (PPI)
- Silicone:
  - ○ SiO_2
  - ○ $Fe_3O_4@SiO_2$
  - ○ Silicates
- Zeolite:
- Metal-organic frameworks (MOF) and carbon organic frameworks
- Peptide and protein
- Plant extracts

Nanoparticles have widely been used as catalyst for the removal of organic pollutants such as nitrophenols, nitroaromatics, dyes *etc.* and have been deliberated as tremendous catalysts. Zero valent nanoparticles catalyzes different reactions but among them, 4-NP reduction by $NaBH_4$ is a "model catalytic reaction". Such a reaction should be well controlled yielding a single product from a single reactant at mild temperatures, and it does not proceed in the absence of the catalyst. This reaction have generally used by a large number of researchers as a test reaction to evaluate the catalytic properties of various zero valent nanoparticles. The common and most efficient way for the reduction of 4-NP is to introduce $NaBH_4$ as a reductant along with zero valent nanoparticles as a catalyst [17, 99 - 101].

Nitrophenols catalytic reduction using borohydride ions (BH_4^-) in the presence of nanoparticles has become one of the typical reactions. This reaction is thermodynamically feasible, but kinetically it is not achievable in the absence of a catalyst because of high kinetic barrier between the equally repelling negative ions of p-nitrophenoxide and BH_4^-. According to traditional theory, the electrons are transferred from BH_4^- ion to 4-NP through adsorption of reactant molecules onto the metallic nanoparticle surface during catalytic reduction of 4-NP by metallic nanoparticles. Metallic nanoparticles then transmit and pass electrons to complete the oxidation-reduction reaction and thus contribute to overcome the kinetic barrier of the reaction [101].

The catalytic potential of zero valent nanoparticles in the catalytic reduction of 4-NP in the presence of excess NaBH$_4$ at room temperature can be monitored by UV-vis spectrometry. The pure 4-NP solution displays a strong absorption peak at 317 nm under neutral or acidic conditions with a pale yellow color. But the absorption peak of 4-NP immediately red shift from 317 to 400 nm upon the addition of freshly prepared aqueous solution of NaBH$_4$, and the color of the solution changed from pale yellow to bright yellow owing to the formation of 4-nitrophenolate ions. The reaction does not proceed for a couple of days even using excess of NaBH$_4$ in the absence of the catalyst [22, 24, 102], confirming that the use of NaBH$_4$ alone does not affect the catalytic reaction. Upon addition of a small amount of zero valent nanoparticles catalyst, the absorbance band intensity at 400 nm consecutively decrease as the reaction proceed while at the same time, a new absorption peak at 300 nm appear and its intensity increases. This decrease in absorption peak at 400 nm and appearance of new peak at 300 nm demonstrate the reduction of the –NO$_2$ group of 4-NP to an –NH$_2$ group [99, 100].

Kinetics for the Reduction of Nitrophenols

Pseudo-first-order Kinetic Equation

The catalytic potential of zero valent nanoparticles in the catalytic reduction of nitrophenols with excess NaBH$_4$ at room temperature can be monitored by decrease in the strong absorption of 4-nitrophenolate ions at 400 nm in UV vis. First order reaction is generally apply to compare the catalytic activity of zero valent nanoparticles and decay rate of nitrophenols in the presence of excess NaBH$_4$. The following equation is generally applied to determine the reduction rate kinetics and the catalytic activity of nanoparticles are usually compare by the variation in the decay rates [99, 100].

$$r = dc/dt = ln(C_t/C_o) = -k_{app}t$$

where r is the reduction rate of the reactant; c is the concentration of the reactant; t is the reaction time; k is the reaction rate constant; C_t is denoted as the concentration of nitrophenol which is equal to relative intensity of absorbance at time t, and the initial concentration is regarded as C_0 which is equal to relative intensity of absorbance at time 0. The equation can be re-write in the modified form as follow:

$$r = dc/dt = ln(C_t/C_o) = ln(A_t/A_o) = -k_{app}t$$

The linear relationships between $\ln(C_t/C_0)$ and the reaction time (t) is usually found for the reduction reaction catalyzed by zero valent nanoparticles, where rate constant (K_{aap}) can be estimated from the slopes of $\ln(C_t/C_0)$ *vs.* t. The reaction rate constants are used for the comparison of catalytic activity of different nanocatalysts. After an induction time t in which no reduction takes place, the reaction starts. The apparent rate constant k_{app} is found to be proportional to the total surface area S available in the metal nanoparticles. Hence, two kinetic constants k_{app} and k_1 can be defined by:

$$-dc_t/dt = k_{app}C_t = k_1SC_t$$

Where $C_t \rightarrow$ concentration of nitrophenol at time t

$k_1 \rightarrow$ rate constant normalized to S

$S \rightarrow$ the total surface area of metal nanoparticles normalized to the unit volume of the solution.

Kinetic Control vs. Diffusion Control

In order to realize whether the reaction is kinetic control or diffusion control, *second Damköhler number (DaII)* (which is the ratio among the reaction rate and the diffusion rate for a first order reaction) can be determined. One has to consider that both catalytic reaction and diffusion occurred on zero valent nanoparticles surface. One can say the reaction to be diffusion control, if turn over number at surface is higher than diffusion of the reactants. In thermally triggered reactions, rate-determining step is measured for the reaction because diffusion in this case is very faster. The rivalry among chemical reaction and mass transport can be assessed through the *DaII* that is the ratio of the rate of the reaction to the rate of diffusion [100].

$$DaII = kC^{n-1}/\beta a$$

$k \rightarrow$ reaction rate constant

$\beta \rightarrow$ mass transport coefficient = diffusion coefficient divided by the length scale δ over which the reaction takes place

$a \rightarrow$ surface area of zero valent nanoparticles.

If *DaII* < 1 $\rightarrow$ Diffusion of the reactants to and from the nanoparticles is much faster than the chemical reaction on their surface.

If $DaII > 1 \rightarrow$ Diffusion control must be taken into account.

In case of nitrophenol reduction, $DaII$ is much smaller than unity and thus the rate-limiting step at the surface of zero valent nanoparticles is slower than diffusion of the reactants. Thus the reduction of nitrophenol is not controlled by diffusion. In contrast, hexacyanoferrate reduction is diffusion-controlled.

Langmuir–Hinshelwood Analysis

A well known classical Langmuir–Hinshelwood model has used for investigation the kinetics of nitrophenol reduction. Several groups have presented Langmuir-Hinshelwood model for the comprehensive analysis of the rate constants of this model reaction. BH_4^- first get adsorbed onto zero valent nanoparticles and give hydrides species. The nitrophenol molecules also get adsorbed onto the surface of zero valent nanoparticles. The adsorption of both BH_4^- and nitrophenol is fast and reversible. The adsorbed species then react where transfer of hydrogen and electrons occurred from zero valent nanoparticles to nitrophenol and finally the reaction product dissociates from the surface. The adsorption of both BH_4^- and nitrophenol can be modelled by a Langmuir isotherm. If we assume that the diffusion of BH_4^- and nitrophenol to the zero valent nanoparticles and also adsorption–desorption steps are fast. Then the reduction of nitrophenols adsorbed on the surface by the surface-hydrogen species becomes the rate-determining step. Rate of electron transfer strongly depend on the adsorption of nitrophenol onto the surface of zero valent nanoparticles and then desorption of aminophenol. When nitrophenol get adsorb on the surface of zero valent nanoparticles, then interfacial electron transfer takes place which convert nitrophenol into aminophenol and gets away from the surface leaving a free nanoparticles which is then available for a new catalytic cycle to proceed [100, 101]. This model shows the exact kinetics of the reaction. Especially, the rate constant is inversely related with the nitrophenol concentration *i.e.*, it decreases as the nitrophenol concentration increases and reaches to a highest value upon increasing the $NaBH_4$ concentration. This describes a LH model as a competition of the two reactants for zero valent nanoparticles surface adsorption sites. The thermodynamic adsorption constants $K_{4\text{-}NP}$ and K_{BH4}^- and the rate constant of the reaction at the zero valent nanoparticles surface can be obtained from the data. Moreover, the reaction generally starts after an induction time. The borohydride reacts with the zero valent nanoparticles surface to reversibly produce zero valent nanoparticles hydrogen bonds [100, 101], and formation of aminophenol is the rate limiting step which occurs through reaction between zero valent nanoparticles hydrogen and the adsorbed 4-NP (*i.e.* the diffusion and adsorption of the reactants as the desorption are fast compared with the reaction on the nanoparticle surface) [100, 101]. The rate constant k is directly proportional to the nanoparticle surface S in the LH model:

$dc/dt = -kC = -k_1 CS$; C = 4-NP concentration

Observing the LH model, k_1 is not dependent on nanoparticle surface, but it shows both the constants ($K_{4\text{-NP}}$ and K_{BH4}^-) illustrating adsorption of the substrate and the reaction rate at the nanoparticle surface correspondingly [24, 101].

In leaching mechanism, nanoparticle catalyzed 4-NP reduction is supposed to ensue in solution *i.e.* instead of nanoparticle surface, it occurs on metal atoms or small clusters already detached from the mother nanoparticle [100, 101] as in Pd nanoparticle catalyzed Suzuki-Miyaura reactions [100, 101]. Typically, the metallic nanocatalysis in solution involves two different mechanisms *i.e.* heterogeneous mechanism which takes place on the nanoparticle surface or by homogeneous mechanism in which the reaction proceed in solution by leached atoms or ions from the nanoparticle surface. The heterogeneous mechanism always results in causing the blue shift of nanoparticle, thus showing the production of hydrogen gas. Conversely, the homogenous mechanism always give a red shift as the atoms (or ions) leaking out of the nanocatalyst surface forms the aggregation of nanoparticle [100, 101]. Hence, it was concluded that the process of metal catalyzed 4-NP to 4-AP reduction using $NaBH_4$ progressed according to the LH mechanism [100, 101]. In the nanoparticle catalyzed reduction of 4-NP to 4-AP, the $NaBH_4$ first undergoes hydrolysis, then in next step the formation of the $B(OH)_4^-$ and active hydrogen occurs then, the active hydrogen is adsorbed at the surface after transferring it to the nanoparticle. In the final step, the H reacts with 4-NP to give the product 4-AP.

Nanomaterials in Water Filtration

Filtration is one of the common methods which is usually used for the cleaning of waste water to make water safe to drink. Filtration is the process of passing a fluid through porous materials in order to separate out matter in suspension [63, 103 - 107]. Filtration is the primary process utilized to clean water for human use. There are different kinds of filtration processes using different materials for filtration but the mostly commonly technique is membrane based filtration in which membrane is using as a filter. A membrane can selectively allowed the solute molecules through diffusion and sieving. Membranes can be semi-permeable and separate particles and molecules and over a wide particle size range and molecular weights. A membrane has specific size tiny holes with thin morphology. Larger particle are rejected by the membrane which are not fitted in the pore size. The hazardous particles are entrapped without any chemical change and so therefore, it is consider as one of the more efficient technology for water purification. A good membrane is the one, which is permeable to solvent molecules but impermeable to the solute molecule.

There are four common types of pressure-driven membrane filtration systems that have been applied for water treatment, reuse and desalination throughout the world.

Microfiltration (MF)

Microfiltration is essentially membrane processes that rely on pure straining through porosity in the membranes. Pressure required is lower than R.O. and due entirely to frictional head loss.

- Typical pore size: 1.0~0.01 microns
- Very low pressure < 30 psi
- Removes bacteria, some large viruses, clay, suspended solids
- Does not filter small viruses, protein molecules, sugar, and salts

Ultrafiltration (UF)

- Typical pore size: 0.01~0.001 microns
- Moderately low pressure < 20~100 psi
- Removes viruses, protein, starch, colloides, silica, dye, fat, and other organic molecules
- Does not filter ionic particles like lead, iron, chloride ions; nitrates, nitrites; other charged particles

Nanofiltration (NF)

- Typical pore size: 0.001~0.0001 microns
- Pressure < 50~300 psi
- Removes sugar, pesticides, herbicides, divalent anions
- Does not filter monovalent salts

Reverse Osmosis (RO)

- Typical pore size: 0.001~0.0001 microns
- Pressure < 225~1000 psi
- Removes monovalent salts

RO is very efficient for retaining dissolved inorganic and small organic molecules. NF can effectively remove hardness (*i.e.*, Ca(II)) and natural organic matter. The Reverse Osmosis membrane is semi-permeable where the pore size of the membrane decided the passage of only specific types of materials. The size of Na and Cl ions are same to water molecules, and seemed that it is also passed to the membrane however, it is infact repelled by the membrane due to its charged surface, because water molecule quickly adsorbed and block the passage and exclude the ions. Under pressure attached water will be transferred through the pores. Reverse osmosis methods use a selectively-permeable membrane to separate water from dissolved substances. Relatively high pressure is required to make water flow against normal osmotic pressure. The transportation of solute molecule from a region of high toward low concentration through semipermeable membrane is called osmosis through presser. The transportation of solute through semipermeable membrane make them impractical due to high pressure, where the big particles block the small pore size. High pressures from 225–1000 psi are some of the limitation needed RO process for water purification. UF and MF does not membranes dissolved organic contaminants ions and required a 5–60 psi. However, dissolved ions are preferentially removed from the aqueous solution by UF and MF due to the advancement in nanotechnology.

Low molecular weight molecules such as glucose, salts, lactose and other can be efficiently removed by nanofiltration (NF). Worldwide in the 21st century clean water availability is one of the serious concern. Water treatment systems typically involve a series of coupled processes, each designed to remove one or more different substances in the source water, with the particular treatment process being based on the molecular size and properties of the target contaminants. NF membranes selectively reject substances as well as enable the retention of nutrients in water. Therefore, the advantages of NF are comparable to the RO process. The adsorption of pollutants onto the membrane can be (1) physical in nature, which is a completely reversible process, or (2) chemical in nature, which is irreversible for strong chemical bonds, such as polymerization, or reversible for weak secondary chemical bonds, such as hydrogen bonds and complexation or (3) both. NF is capable of removing hardness, natural organic matter, particles and a number of other organic and inorganic substances *via* one single treatment. Water hardness is caused by calcium and magnesium ions, while strontium and barium rarely occur in substantial concentrations. NF membranes can reject bivalent ions in significantly high amounts. Cellulose acetate, polyvinyl alcohol (PVA), polyamide, sulfonated polysulfone, inorganic metal oxides *etc.* can be used to form NF membranes. Polyamide membrane composite have shown higher Ca^{2+} and Mg^{2+} rejection.165 Polyamide membrane NF could remove the major fraction of hardness from groundwater. NF membranes separated 60% of cations while the separation efficiency for anions was larger. For NF, the ion size plays a role for membranes with small pores, leading to a large selectivity. Nitrate rejection was 76%.

Removal of Microbes

Nanomaterials have also been used for the removal of microbes. A few studies are available detailing SWNTs with antimicrobial activity toward Gram-positive and

Gram-negative bacteria due to either a physical interaction or oxidative stress that compromises the cell membrane integrity [63, 106]. Carbon nanotubes may therefore be useful for inhibiting microbial attachment and biofouling formation on surfaces. However, the degree of aggregation the stabilization effects by NOM and the bioavailability of the nanotubes will have to be considered for these antimicrobial properties to be fully effective [108, 109]. Carbon nanotube clusters a unique material largely applied for the removal of bacteria from wastewater through adsorption. The decontamination of *E. coli* cells was found more effective the of AgNPs–CNTs–PAMAM tricomponent mixture as compared to PAMAM and acid (–COOH) MWCNTs. The superior activity of the tricomponent mixture was due to favoring of the debundling of MWCNTs by PAMAM and increasing the accessible surface area for bacterial interaction. Furthermore, the PAMAM-grafted MWCNTs contain several terminated amine groups which reduced the Ag^+ ions to AgNPs as Ag^+ ions enter through bacteria cell walls [110 - 113].

CONCLUSION

Nanomaterials have high degree of sensitivity for the detection of toxic pollutants, removal of toxic pollutants and amputation of pathogens. Due to their small sizes lower cost, high catalytic performance, advanced throughput and high performance in environmental treatment make them more suitable catalyst for practical applications. Moreover, metal oxide or metal nanostructured materials are employed efficiently for the removal of organic, inorganic, heavy toxic metals and microorganism. Their efficiency for the removal of pollutants can be enhanced by redox reactions and modifications in the chemical groups that can selectively remove pollutants from the environment. The engineering of nanomembranes is extremely important for the environmental remediation because of their extensive applications in potable water productions, water recovery, and removal of pollutants from the polluted water. Further improvements must be made in the application of environmental remediation to selectively remove materials, which have highly pH and concentrations resistance in polluted water, larger stability and cost optimization. Nanomaterials especially nanofibers have small pore size and large absorptivity which make them suitable for diverse materials for filtration applications. Polymer supported nanomaterials, for instance in chemical degradation and photocatalytic, pollutants adsorption, detection and sensing consequence in a greener environment. We can simply summarize that nanomaterials are auspicious materials at large and industrial scales for the removal of pollutants.

CONSENT FOR PUBLICATION

Not applicable.

CONFLICT OF INTEREST

The author (editor) declares no conflict of interest, financial or otherwise.

ACKNOWLEDGEMENTS

The authors are grateful to the Center of Excellence for Advance Materials Research (CEAMR), King Abdulaziz University, Saudi Arabia for providing research facilities.

REFERENCES

[1]   Khan, S.B.; Faisal, M.; Rahman, M.M.; Jamal, A. Exploration of CeO_2 nanoparticles as a chemi-sensor and photo-catalyst for environmental applications. *Sci. Total Environ.,* **2011**, *409*(15), 2987-2992.
[http://dx.doi.org/10.1016/j.scitotenv.2011.04.019] [PMID: 21570707]

[2]   Khan, S.B.; Faisal, M.; Rahman, M.M.; Jamal, A. Low-temperature growth of ZnO nanoparticles: photocatalyst and acetone sensor. *Talanta,* **2011**, *85*(2), 943-949.
[http://dx.doi.org/10.1016/j.talanta.2011.05.003] [PMID: 21726722]

[3]   Khan, S.A.; Khan, S.B.; Asiri, A.M. Layered double hydroxide of Cd-Al/C for the Mineralization and De-coloration of Dyes in Solar and Visible Light Exposure. *Sci. Rep.,* **2016**, *6*, 35107.
[http://dx.doi.org/10.1038/srep35107] [PMID: 27841277]

[4]   Khan, S.A.; Khan, S.B.; Asiri, A.M. Toward the design of Zn–Al and Zn–Cr LDH wrapped in activated carbon for the solar assisted de-coloration of organic dyes. *RSC Advances,* **2016**, *6*, 83196-83208.
[http://dx.doi.org/10.1039/C6RA10598J]

[5]   Arshad, T.; Khan, S.A.; Faisal, M.; Shah, Z.; Akhtar, K.; Asiri, A.M.; Ismail, A.A.; Alhogbi, B.G.; Khan, S.B. Cerium based photocatalysts for the degradation of acridine orange in visible light. *J. Mol. Liq.,* **2017**, *241*, 20-26.
[http://dx.doi.org/10.1016/j.molliq.2017.05.079]

[6]   Faisal, M.; Khan, S.B.; Rahman, M.M.; Jamal, A.; Asiri, A.M.; Abdullah, M. Smart chemical sensor and active photo-catalyst for environmental pollutants. *Chem. Eng. J.,* **2011**, *173*, 178-184.
[http://dx.doi.org/10.1016/j.cej.2011.07.067]

[7]   Khan, S.B.; Marwani, H.M.; Asiri, A.M.; Bakhsh, E.M. Exploration of calcium doped zinc oxide nanoparticles as selective adsorbent for extraction of lead ion. *Desalination Water Treat.,* **2016**, *57*, 19311-19320.
[http://dx.doi.org/10.1080/19443994.2015.1109560]

[8]   Khan, S.B.; Alamry, K.A.; Bifari, E.N.; Asiri, A.M.; Yasir, M.; Gzara, L.; Ahmad, R.Z. Assessment of antibacterial cellulose nanocomposites for water permeability and salt rejection. *J. Ind. Eng. Chem.,* **2015**, *24*, 266-275.
[http://dx.doi.org/10.1016/j.jiec.2014.09.040]

[9]   Liu, Y.; Deng, Y.; Sun, Z.; Wei, J.; Zheng, G.; Asiri, A.M.; Khan, S.B.; Rahman, M.M.; Zhao, D. Hierarchical Cu_2S microsponges constructed from nanosheets for efficient photocatalysis. *Small,* **2013**, *9*(16), 2702-2708.
[http://dx.doi.org/10.1002/smll.201300197] [PMID: 23420805]

[10]   Asiri, A.M.; Akhtar, K.; Seo, J.; Marwani, H.M.; Kim, D.; Han, H.; Khan, S.B. Development of polymer based nanocomposites as a marker of cadmium in complex matrices. *J. Nanomater.,* **2015**, *16*, 73.

[11]   Khin, M.M.; Nair, A.S.; Babu, V.J.; Murugan, R.; Ramakrishna, S. A review on nanomaterials for environmental remediation. . *Energy Environ. Sci.,* **2012**, *5*, 8075-8109.

[http://dx.doi.org/10.1039/c2ee21818f]

[12]   Khan, S.B.; Asiri, A.M.; Rahman, M.M.; Marwani, H.M.; Alamry, K.A. Evaluation of cerium doped tin oxide nanoparticles as a sensitive sensor for selective detection and extraction of cobalt. *Physica E,* **2015**, *70*, 203-209.
[http://dx.doi.org/10.1016/j.physe.2015.02.014]

[13]   Khan, S.B.; Lee, J-W.; Marwani, H.M.; Akhtar, K.; Asiri, A.M.; Seo, J.; Khan, A.A.P.; Han, H. Polybenzimidazole hybrid membranes as a selective adsorbent of mercury. *Compos., Part B Eng.,* **2014**, *56*, 392-396.
[http://dx.doi.org/10.1016/j.compositesb.2013.08.056]

[14]   Rahman, M.M.; Khan, S.B.; Alamry, K.A.; Marwani, H.M.; Asiri, A.M. Detection of trivalent-iron based on low-dimensional semiconductor metal oxide nanostructures for environmental remediation by ICP-OES technique. *Ceram. Int.,* **2014**, *40*, 8445-8453.
[http://dx.doi.org/10.1016/j.ceramint.2014.01.055]

[15]   Rahman, M.M.; Khan, S.B.; Marwani, H.M.; Asiri, A.M.; Alamry, K.A.; Rub, M.A.; Khan, A.; Khan, A.A.P.; Qusti, A.H. Low dimensional Ni-ZnO nanoparticles as marker of toxic lead ions for environmental remediation. *J. Ind. Eng. Chem.,* **2014**, *20*, 1071-1078.
[http://dx.doi.org/10.1016/j.jiec.2013.06.044]

[16]   Kamal, T.; Khan, S.B.; Asiri, A.M. Synthesis of zero-valent Cu nanoparticles in the chitosan coating layer on cellulose microfibers: evaluation of azo dyes catalytic reduction. *Cellulose,* **2016**, *23*, 1911-1923.
[http://dx.doi.org/10.1007/s10570-016-0919-9]

[17]   Khan, S.B.; Khan, S.A.; Marwani, H.M.; Bakhsh, E.M.; Anwar, Y.; Kamal, T.; Asiri, A.M.; Akhtar, K. Anti-bacterial PES-cellulose composite spheres: dual character toward extraction and catalytic reduction of nitrophenol. *RSC Advances,* **2016**, *6*, 110077-110090.
[http://dx.doi.org/10.1039/C6RA21626A]

[18]   Ullah, R.; Dutta, J. Photocatalytic degradation of organic dyes with manganese-doped ZnO nanoparticles. *J. Hazard. Mater.,* **2008**, *156*(1-3), 194-200.
[http://dx.doi.org/10.1016/j.jhazmat.2007.12.033] [PMID: 18221834]

[19]   Bhatkhande, D.S.; Pangarkar, V.G.; Beenackers, A.A. Photocatalytic degradation for environmental applications–a review. *J. Chem. Technol. Biotechnol.,* **2002**, *77*, 102-116.
[http://dx.doi.org/10.1002/jctb.532]

[20]   Dom, R.; Subasri, R.; Radha, K.; Borse, P.H. Synthesis of solar active nanocrystalline ferrite, MFe_2O_4 (M: Ca, Zn, Mg) photocatalyst by microwave irradiation. *Solid State Commun.,* **2011**, *151*, 470-473.
[http://dx.doi.org/10.1016/j.ssc.2010.12.034]

[21]   Raja, P.; Bensimon, M.; Klehm, U.; Albers, P.; Laub, D.; Kiwi-Minsker, L.; Renken, A.; Kiwi, J. Highly dispersed PTFE/Co3O4 flexible films as photocatalyst showing fast kinetic performance for the discoloration of azo-dyes under solar irradiation. *J. Photochem. Photobiol. Chem.,* **2007**, *187*, 332-338.
[http://dx.doi.org/10.1016/j.jphotochem.2006.10.033]

[22]   Wawrzyniak, B.; Morawski, A.W. Solar-light-induced photocatalytic decomposition of two azo dyes on new TiO_2 photocatalyst containing nitrogen. *Appl. Catal. B,* **2006**, *62*, 150-158.
[http://dx.doi.org/10.1016/j.apcatb.2005.07.008]

[23]   Wang, Y.; Li, J.; Peng, P.; Lu, T.; Wang, L. Preparation of $S-TiO_2$ photocatalyst and photodegradation of L-acid under visible light. *Appl. Surf. Sci.,* **2008**, *254*, 5276-5280.
[http://dx.doi.org/10.1016/j.apsusc.2008.02.050]

[24]   Mohapatra, L.; Parida, K. Zn–Cr layered double hydroxide: Visible light responsive photocatalyst for photocatalytic degradation of organic pollutants. *Separ. Purif. Tech.,* **2012**, *91*, 73-80.
[http://dx.doi.org/10.1016/j.seppur.2011.10.028]

[25]   Zhu, J.; Xie, J.; Chen, M.; Jiang, D.; Wu, D. Low temperature synthesis of anatase rare earth doped titania-silica photocatalyst and its photocatalytic activity under solar-light. *Colloids Surf. A Physicochem. Eng. Asp.*, **2010**, *355*, 178-182.
[http://dx.doi.org/10.1016/j.colsurfa.2009.12.016]

[26]   Parida, K.; Mohapatra, L. Carbonate intercalated Zn/Fe layered double hydroxide: a novel photocatalyst for the enhanced photo degradation of azo dyes. *Chem. Eng. J.*, **2012**, *179*, 131-139.
[http://dx.doi.org/10.1016/j.cej.2011.10.070]

[27]   Zhao, X.; Li, Z.; Chen, Y.; Shi, L.; Zhu, Y. Enhancement of photocatalytic degradation of polyethylene plastic with CuPc modified TiO_2 photocatalyst under solar light irradiation. *Appl. Surf. Sci.*, **2008**, *254*, 1825-1829.
[http://dx.doi.org/10.1016/j.apsusc.2007.07.154]

[28]   Im, J.S.; Bai, B.C.; In, S.J.; Lee, Y.S. Improved photodegradation properties and kinetic models of a solar-light-responsive photocatalyst when incorporated into electrospun hydrogel fibers. *J. Colloid Interface Sci.*, **2010**, *346*(1), 216-221.
[http://dx.doi.org/10.1016/j.jcis.2010.02.043] [PMID: 20227710]

[29]   Lianos, P. Production of electricity and hydrogen by photocatalytic degradation of organic wastes in a photoelectrochemical cell: the concept of the Photofuelcell: a review of a re-emerging research field. *J. Hazard. Mater.*, **2011**, *185*(2-3), 575-590.
[http://dx.doi.org/10.1016/j.jhazmat.2010.10.083] [PMID: 21111532]

[30]   Marwani, H.M.; Lodhi, M.U.; Khan, S.B.; Asiri, A.M. Cellulose-lanthanum hydroxide nanocomposite as a selective marker for detection of toxic copper. *Nanoscale Res. Lett.*, **2014**, *9*(1), 466.
[http://dx.doi.org/10.1186/1556-276X-9-466] [PMID: 25258599]

[31]   Khan, S.B.; Rahman, M.M.; Marwani, H.M.; Asiri, A.M.; Alamry, K.A. Exploration of silver oxide nanoparticles as a pointer of lanthanum for environmental applications. *J. Taiwan. Inst. Chem. Eng.*, **2014**, *45*, 2770-2776.
[http://dx.doi.org/10.1016/j.jtice.2014.07.005]

[32]   Rahman, M.M.; Khan, S.B.; Marwani, H.M.; Asiri, A.M.; Alamry, K.A.; Al-Youbi, A.O. Selective determination of gold(III) ion using CuO microsheets as a solid phase adsorbent prior by ICP-OES measurement. *Talanta*, **2013**, *104*, 75-82.
[http://dx.doi.org/10.1016/j.talanta.2012.11.031] [PMID: 23597891]

[33]   Marwani, H.M.; Bakhsh, E.M.; Al-Turaif, H.A.; Asiri, A.M.; Khan, S.B. Enantioselective separation and detection of D-phenylalanine based on newly developed chiral ionic liquid immobilized silica gel surface. *Int. J. Electrochem. Sci.*, **2014**, *9*, 7948-7964.

[34]   Khan, S.B.; Alamry, K.A.; Marwani, H.M.; Asiri, A.M.; Rahman, M.M. Synthesis and environmental applications of cellulose/ZrO2 nanohybrid as a selective adsorbent for nickel ion. *Compos., Part B Eng.*, **2013**, *50*, 253-258.
[http://dx.doi.org/10.1016/j.compositesb.2013.02.009]

[35]   Khan, S.B.; Rahman, M.M.; Asiri, A.M.; Marwani, H.M.; Bawaked, S.M.; Alamry, K.A. Co_3O_4 co-doped TiO_2 nanoparticles as a selective marker of lead in aqueous solution. *New J. Chem.*, **2013**, *37*, 2888-2893.
[http://dx.doi.org/10.1039/c3nj00298e]

[36]   Khan, S.B.; Rahman, M.M.; Marwani, H.M.; Asiri, A.M.; Alamry, K.A.; Rub, M.A. Selective adsorption and determination of iron (III): Mn_3O_4/TiO_2 composite nanosheets as marker of iron for environmental applications. *Appl. Surf. Sci.*, **2013**, *282*, 46-51.
[http://dx.doi.org/10.1016/j.apsusc.2013.03.180]

[37]   Krasovska, M.; Gerbreders, V.; Tamanis, E.; Gerbreders, S.; Bulanovs, A. The Study of Adsorption Process of Pb Ions Using Well-Aligned Arrays of ZnO Nanotubes as a Sorbent. *Latvian J. Phys. Technical Sci.*, **2017**, *54*, 41-50.

[38]   Rahman, M.M.; Khan, S.B.; Marwani, H.M.; Asiri, A.M.; Alamry, K.A.; Rub, M.A.; Khan, A.; Khan, A.A.P.; Azum, N. Facile synthesis of doped ZnO-CdO nanoblocks as solid-phase adsorbent and efficient solar photo-catalyst applications. *J. Ind. Eng. Chem.,* **2014**, *20*, 2278-2286. [http://dx.doi.org/10.1016/j.jiec.2013.09.059]

[39]   Rahman, M.M.; Khan, S.B.; Marwani, H.M.; Asiri, A.M. SnO_2–TiO_2 nanocomposites as new adsorbent for efficient removal of La (III) ions from aqueous solutions. *J. Taiwan Inst. Chem. Eng.,* **2014**, *45*, 1964-1974. [http://dx.doi.org/10.1016/j.jtice.2014.03.018]

[40]   Tseng, R-L.; Wu, F-C.; Juang, R-S. Liquid-phase adsorption of dyes and phenols using pinewood-based activated carbons. *Carbon,* **2003**, *41*, 487-495. [http://dx.doi.org/10.1016/S0008-6223(02)00367-6]

[41]   Rengaraj, S.; Moon, S-H.; Sivabalan, R.; Arabindoo, B.; Murugesan, V. Agricultural solid waste for the removal of organics: adsorption of phenol from water and wastewater by palm seed coat activated carbon. *Waste Manag.,* **2002**, *22*(5), 543-548. [http://dx.doi.org/10.1016/S0956-053X(01)00016-2] [PMID: 12092764]

[42]   Ge, F.; Li, M-M.; Ye, H.; Zhao, B-X. Effective removal of heavy metal ions $Cd2^+$, Zn^{2+}, Pb^{2+}, Cu^{2+} from aqueous solution by polymer-modified magnetic nanoparticles. *J. Hazard. Mater.,* **2012**, *211-212*, 366-372. [http://dx.doi.org/10.1016/j.jhazmat.2011.12.013] [PMID: 22209322]

[43]   Bahaj, A.; James, P.; Moeschler, F. Efficiency enhancements through the use of magnetic field gradient in orientation magnetic separation for the removal of pollutants by magnetotactic bacteria. *Sep. Sci. Technol.,* **2002**, *37*, 3661-3671. [http://dx.doi.org/10.1081/SS-120014825]

[44]   Peng, X.; Li, Y.; Luan, Z.; Di, Z.; Wang, H.; Tian, B.; Jia, Z. Adsorption of 1, 2-dichlorobenzene from water to carbon nanotubes. *Chem. Phys. Lett.,* **2003**, *376*, 154-158. [http://dx.doi.org/10.1016/S0009-2614(03)00960-6]

[45]   Machado, L.C.; Lima, F.; Paniago, R.; Ardisson, J.D.; Sapag, K.; Lago, R. Polymer coated vermiculite–iron composites: novel floatable magnetic adsorbents for water spilled contaminants. *Appl. Clay Sci.,* **2006**, *31*, 207-215. [http://dx.doi.org/10.1016/j.clay.2005.07.004]

[46]   Kanel, S.R.; Greneche, J-M.; Choi, H. Arsenic(V) removal from groundwater using nano scale zero-valent iron as a colloidal reactive barrier material. *Environ. Sci. Technol.,* **2006**, *40*(6), 2045-2050. [http://dx.doi.org/10.1021/es0520924] [PMID: 16570634]

[47]   Agarwal, S.; Al-Abed, S.R.; Dionysiou, D.D.; Graybill, E. Reactivity of substituted chlorines and ensuing dechlorination pathways of select PCB congeners with Pd/Mg bimetallics. *Environ. Sci. Technol.,* **2009**, *43*(3), 915-921. [http://dx.doi.org/10.1021/es802538d] [PMID: 19245036]

[48]   Virkutyte, J.; Varma, R.S. Green synthesis of metal nanoparticles: biodegradable polymers and enzymes in stabilization and surface functionalization. *Chem. Sci. (Camb.),* **2011**, *2*, 837-846. [http://dx.doi.org/10.1039/C0SC00338G]

[49]   Qi, L.; Xu, Z.; Chen, M. *In vitro* and *in vivo* suppression of hepatocellular carcinoma growth by chitosan nanoparticles. *Eur. J. Cancer,* **2007**, *43*(1), 184-193. [http://dx.doi.org/10.1016/j.ejca.2006.08.029] [PMID: 17049839]

[50]   Burkhard, L.P. Estimating dissolved organic carbon partition coefficients for nonionic organic chemicals. *Environ. Sci. Technol.,* **2000**, *34*, 4663-4668. [http://dx.doi.org/10.1021/es001269l]

[51]   Alvarez-Ayuso, E.; García-Sánchez, A.; Querol, X. Purification of metal electroplating waste waters using zeolites. *Water Res.,* **2003**, *37*(20), 4855-4862.

[http://dx.doi.org/10.1016/j.watres.2003.08.009] [PMID: 14604631]

[52]   Moreno, N.; Querol, X.; Ayora, C.; Pereira, C.F.; Janssen-Jurkovicová, M. Utilization of zeolites synthesized from coal fly ash for the purification of acid mine waters. *Environ. Sci. Technol.,* **2001**, *35*(17), 3526-3534.
[http://dx.doi.org/10.1021/es0002924] [PMID: 11563657]

[53]   Ahmed, S.; Chughtai, S.; Keane, M.A. The removal of cadmium and lead from aqueous solution by ion exchange with Na☐ Y zeolite. *Separ. Purif. Tech.,* **1998**, *13*, 57-64.
[http://dx.doi.org/10.1016/S1383-5866(97)00063-4]

[54]   Kim, T-Y.; Park, S-K.; Cho, S-Y.; Kim, H-B.; Kang, Y.; Kim, S-D.; Kim, S-J. Adsorption of heavy metals by brewery biomass. *Korean J. Chem. Eng.,* **2005**, *22*, 91-98.
[http://dx.doi.org/10.1007/BF02701468]

[55]   Antonietti, M.; Lohmann, S.; Eisenbach, C.D.; Schubert, U.S. Synthesis of metal-complexing latices *via* polymerization in microemulsion. *Macromol. Rapid Commun.,* **1995**, *16*, 283-289.
[http://dx.doi.org/10.1002/marc.1995.030160409]

[56]   Pena, M.E.; Korfiatis, G.P.; Patel, M.; Lippincott, L.; Meng, X. Adsorption of As(V) and As(III) by nanocrystalline titanium dioxide. *Water Res.,* **2005**, *39*(11), 2327-2337.
[http://dx.doi.org/10.1016/j.watres.2005.04.006] [PMID: 15896821]

[57]   Chen, M-Q.; Chen, Y.; Kaneko, T.; Liu, X-Y.; Cheng, Y.; Akashi, M. Pb 2+-Specific Adsorption/Desorption onto Core-Corona Type Polymeric Nanospheres Bearing Special Anionic Azo-Chromophore. *Polym. J.,* **2003**, *35*, 688.
[http://dx.doi.org/10.1295/polymj.35.688]

[58]   Bell, C.A.; Smith, S.V.; Whittaker, M.R.; Whittaker, A.K.; Gahan, L.R.; Monteiro, M.J. Surface-Functionalized Polymer Nanoparticles for Selective Sequestering of Heavy Metals. *Adv. Mater.,* **2006**, *18*, 582-586.
[http://dx.doi.org/10.1002/adma.200501712]

[59]   Tungittiplakorn, W.; Cohen, C.; Lion, L.W. Engineered polymeric nanoparticles for bioremediation of hydrophobic contaminants. *Environ. Sci. Technol.,* **2005**, *39*(5), 1354-1358.
[http://dx.doi.org/10.1021/es049031a] [PMID: 15787377]

[60]   Tungittiplakorn, W.; Lion, L.W.; Cohen, C.; Kim, J-Y. Engineered polymeric nanoparticles for soil remediation. *Environ. Sci. Technol.,* **2004**, *38*(5), 1605-1610.
[http://dx.doi.org/10.1021/es0348997] [PMID: 15046367]

[61]   Crane, R.A.; Scott, T.B. Nanoscale zero-valent iron: future prospects for an emerging water treatment technology. *J. Hazard. Mater.,* **2012**, *211-212*, 112-125.
[http://dx.doi.org/10.1016/j.jhazmat.2011.11.073] [PMID: 22305041]

[62]   Tomalia, D.A. Birth of a new macromolecular architecture: dendrimers as quantized building blocks for nanoscale synthetic polymer chemistry. *Prog. Polym. Sci.,* **2005**, *30*, 294-324.
[http://dx.doi.org/10.1016/j.progpolymsci.2005.01.007]

[63]   Diallo, M.S.; Christie, S.; Swaminathan, P.; Johnson, J.H., Jr; Goddard, W.A., III Dendrimer enhanced ultrafiltration. 1. Recovery of Cu(II) from aqueous solutions using PAMAM dendrimers with ethylene diamine core and terminal NH2 groups. *Environ. Sci. Technol.,* **2005**, *39*(5), 1366-1377.
[http://dx.doi.org/10.1021/es048961r] [PMID: 15787379]

[64]   Lianchao, L.; Baoguo, W.; Huimin, T.; Tianlu, C.; Jiping, X. A novel nanofiltration membrane prepared with PAMAM and TMC by *in situ* interfacial polymerization on PEK-C ultrafiltration membrane. *J. Membr. Sci.,* **2006**, *269*, 84-93.
[http://dx.doi.org/10.1016/j.memsci.2005.06.021]

[65]   Bosso, S.T.; Enzweiler, J. Evaluation of heavy metal removal from aqueous solution onto scolecite. *Water Res.,* **2002**, *36*(19), 4795-4800.
[http://dx.doi.org/10.1016/S0043-1354(02)00208-7] [PMID: 12448522]

[66]   Hochella, M.F., Jr There's plenty of room at the bottom: Nanoscience in geochemistry. *Geochim. Cosmochim. Acta,* **2002**, *66*, 735-743.
[http://dx.doi.org/10.1016/S0016-7037(01)00868-7]

[67]   Yue, Z.; Economy, J. Nanoparticle and nanoporous carbon adsorbents for removal of trace organic contaminants from water. *J. Nanopart. Res.,* **2005**, *7*, 477-487.
[http://dx.doi.org/10.1007/s11051-005-4719-7]

[68]   Ebbesen, T.W. Wetting, filling and decorating carbon nanotubes. *J. Phys. Chem. Solids,* **1996**, *57*, 951-955.
[http://dx.doi.org/10.1016/0022-3697(95)00381-9]

[69]   Johnson, R. D.; de Vries, M. S.; Salem, J.; Bethune, D. S. Electron paramagnetic resonance studies of lanthanum-containing. *Physics and Chemistry of Fullerenes: A Reprint Collection,* **1993**, *1*, 215.

[70]   Li, Y-H.; Ding, J.; Luan, Z.; Di, Z.; Zhu, Y.; Xu, C.; Wu, D.; Wei, B. Competitive adsorption of Pb^{2+}, Cu^{2+} and Cd^{2+} ions from aqueous solutions by multiwalled carbon nanotubes. *Carbon,* **2003**, *41*, 2787-2792.
[http://dx.doi.org/10.1016/S0008-6223(03)00392-0]

[71]   Bina, B.; Pourzamani, H.; Rashidi, A.; Amin, M. M. *Ethylbenzene removal by carbon nanotubes from aqueous solution.,* **2012**.
[http://dx.doi.org/10.1155/2012/817187]

[72]   Šafařík, I.; Nymburská, K.; Šafaříková, M. Adsorption of water-soluble organic dyes on magnetic charcoal. *J. Chem. Technol. Biotechnol.,* **1997**, *69*, 1-4.
[http://dx.doi.org/10.1002/(SICI)1097-4660(199705)69:1<1::AID-JCTB653>3.0.CO;2-H]

[73]   Bhatnagar, A.; Minocha, A. Conventional and non-conventional adsorbents for removal of pollutants from water . *RE:view,* **2006**.

[74]   Lu, C.; Chiu, H.; Liu, C. Removal of zinc (II) from aqueous solution by purified carbon nanotubes: kinetics and equilibrium studies. *Ind. Eng. Chem. Res.,* **2006**, *45*, 2850-2855.
[http://dx.doi.org/10.1021/ie051206h]

[75]   Gupta, V.K.; Agarwal, S.; Saleh, T.A. Synthesis and characterization of alumina-coated carbon nanotubes and their application for lead removal. *J. Hazard. Mater.,* **2011**, *185*(1), 17-23.
[http://dx.doi.org/10.1016/j.jhazmat.2010.08.053] [PMID: 20888691]

[76]   Chen, C.L.; Wang, X.K.; Nagatsu, M. Europium adsorption on multiwall carbon nanotube/iron oxide magnetic composite in the presence of polyacrylic acid. *Environ. Sci. Technol.,* **2009**, *43*(7), 2362-2367.
[http://dx.doi.org/10.1021/es803018a] [PMID: 19452887]

[77]   Gong, J-L.; Wang, B.; Zeng, G-M.; Yang, C-P.; Niu, C-G.; Niu, Q-Y.; Zhou, W-J.; Liang, Y. Removal of cationic dyes from aqueous solution using magnetic multi-wall carbon nanotube nanocomposite as adsorbent. *J. Hazard. Mater.,* **2009**, *164*(2-3), 1517-1522.
[http://dx.doi.org/10.1016/j.jhazmat.2008.09.072] [PMID: 18977077]

[78]   Wang, S-G.; Gong, W-X.; Liu, X-W.; Yao, Y-W.; Gao, B-Y.; Yue, Q-Y. Removal of lead (II) from aqueous solution by adsorption onto manganese oxide-coated carbon nanotubes. *Separ. Purif. Tech.,* **2007**, *58*, 17-23.
[http://dx.doi.org/10.1016/j.seppur.2007.07.006]

[79]   Khan, M.; Tahir, M.N.; Adil, S.F.; Khan, H.U.; Siddiqui, M.R.H.; Al-warthan, A.A.; Tremel, W. Graphene based metal and metal oxide nanocomposites: synthesis, properties and their applications. *J. Mater. Chem. A Mater. Energy Sustain.,* **2015**, *3*, 18753-18808.
[http://dx.doi.org/10.1039/C5TA02240A]

[80]   Jiao, T.; Liu, Y.; Wu, Y.; Zhang, Q.; Yan, X.; Gao, F.; Bauer, A.J.; Liu, J.; Zeng, T.; Li, B. Facile and scalable preparation of graphene oxide-based magnetic hybrids for fast and highly efficient removal of organic dyes. *Sci. Rep.,* **2015**, *5*, 12451.

[http://dx.doi.org/10.1038/srep12451] [PMID: 26220847]

[81]   Asiri, A.M.; Khan, S.B.; Alamry, K.A.; Marwani, H.M.; Rahman, M.M. Growth of Mn_3O_4 on cellulose matrix: nanohybrid as a solid phase adsorbent for trivalent chromium. *Appl. Surf. Sci.,* **2013**, *270*, 539-544.
[http://dx.doi.org/10.1016/j.apsusc.2013.01.083]

[82]   Kamal, T.; Anwar, Y.; Khan, S.B.; Chani, M.T.S.; Asiri, A.M. Dye adsorption and bactericidal properties of TiO_2/chitosan coating layer. *Carbohydr. Polym.,* **2016**, *148*, 153-160.
[http://dx.doi.org/10.1016/j.carbpol.2016.04.042] [PMID: 27185126]

[83]   Ahmed, S.; Kamal, T.; Khan, S.A.; Anwar, Y.; Saeed, T. M.; Muhammad Asiri, A.; Bahadar Khan, S. Assessment of anti-bacterial Ni-Al/chitosan composite spheres for adsorption assisted photo-degradation of organic pollutants. *Curr. Nanosci.,* **2016**, *12*, 569-575.
[http://dx.doi.org/10.2174/1573413712666160204000517]

[84]   Khan, S.A.; Khan, S.B.; Kamal, T.; Asiri, A.M.; Akhtar, K. A.; Akhtar, K. Recent development of chitosan nanocomposites for environmental applications. *Recent Pat. Nanotechnol.,* **2016**, *10*(3), 181-188.
[http://dx.doi.org/10.2174/1872210510666160429145339] [PMID: 27136929]

[85]   Khan, S.A.; Khan, S.B.; Kamal, T.; Yasir, M.; Asiri, A.M. Antibacterial nanocomposites based on chitosan/Co-MCM as a selective and efficient adsorbent for organic dyes. *Int. J. Biol. Macromol.,* **2016**, *91*, 744-751.
[http://dx.doi.org/10.1016/j.ijbiomac.2016.06.018] [PMID: 27287771]

[86]   Kim, H.; Hong, H-J.; Jung, J.; Kim, S-H.; Yang, J-W. Degradation of trichloroethylene (TCE) by nanoscale zero-valent iron (nZVI) immobilized in alginate bead. *J. Hazard. Mater.,* **2010**, *176*(1-3), 1038-1043.
[http://dx.doi.org/10.1016/j.jhazmat.2009.11.145] [PMID: 20042289]

[87]   Zhao, X.; Lv, L.; Pan, B.; Zhang, W.; Zhang, S.; Zhang, Q. Polymer-supported nanocomposites for environmental application: a review. *Chem. Eng. J.,* **2011**, *170*, 381-394.
[http://dx.doi.org/10.1016/j.cej.2011.02.071]

[88]   Khan, S.B.; Marwani, H.M.; Seo, J.; Bakhsh, E.M.; Akhtar, K.; Kim, D.; Asiri, A.M. Poly (propylene carbonate)/exfoliated graphite nanocomposites: selective adsorbent for the extraction and detection of gold (III). *Bull. Mater. Sci.,* **2015**, *38*, 327-333.
[http://dx.doi.org/10.1007/s12034-015-0885-0]

[89]   Munir, S.; Dionysiou, D.D.; Khan, S.B.; Shah, S.M.; Adhikari, B.; Shah, A. Development of photocatalysts for selective and efficient organic transformations. *J. Photochem. Photobiol. B,* **2015**, *148*, 209-222.
[http://dx.doi.org/10.1016/j.jphotobiol.2015.04.020] [PMID: 25974905]

[90]   Khan, S.B.; Faisal, M.; Rahman, M.M.; Akhtar, K.; Asiri, A.M.; Khan, A.; Alamry, K.A. Effect of particle size on the photocatalytic activity and sensing properties of CeO_2 nanoparticles. *Int. J. Electrochem. Sci.,* **2013**, *8*, 7284-7297.

[91]   Asif, S.A.B.; Khan, S.B.; Asiri, A.M. Efficient solar photocatalyst based on cobalt oxide/iron oxide composite nanofibers for the detoxification of organic pollutants. *Nanoscale Res. Lett.,* **2014**, *9*(1), 510.
[http://dx.doi.org/10.1186/1556-276X-9-510] [PMID: 25246877]

[92]   Asif, S.A.B.; Khan, S.B.; Asiri, A.M. Visible light functioning photocatalyst based on Al_2O_3 doped Mn_3O_4 nanomaterial for the degradation of organic toxin. *Nanoscale Res. Lett.,* **2015**, *10*(1), 355.
[http://dx.doi.org/10.1186/s11671-015-0990-4] [PMID: 26353934]

[93]   Faisal, M.; Khan, S.B.; Rahman, M.M.; Jamal, A.; Abdullah, M. Fabrication of ZnO nanoparticles based sensitive methanol sensor and efficient photocatalyst. *Appl. Surf. Sci.,* **2012**, *258*, 7515-7522.
[http://dx.doi.org/10.1016/j.apsusc.2012.04.075]

[94]   Jamal, A.; Rahman, M.M.; Khan, S.B.; Faisal, M.; Akhtar, K.; Rub, M.A.; Asiri, A.M.; Al-Youbi, A.O. Cobalt doped antimony oxide nano-particles based chemical sensor and photo-catalyst for environmental pollutants. *Appl. Surf. Sci.,* **2012**, *261*, 52-58.
        [http://dx.doi.org/10.1016/j.apsusc.2012.07.066]

[95]   Seo, J.; Jeon, G.; Jang, E.S.; Bahadar Khan, S.; Han, H. Preparation and properties of poly (propylene carbonate) and nanosized ZnO composite films for packaging applications. *J. Appl. Polym. Sci.,* **2011**, *122*, 1101-1108.
        [http://dx.doi.org/10.1002/app.34248]

[96]   Hadjar, H.; Hamdi, B.; Ania, C.O. Adsorption of p-cresol on novel diatomite/carbon composites. *J. Hazard. Mater.,* **2011**, *188*(1-3), 304-310.
        [http://dx.doi.org/10.1016/j.jhazmat.2011.01.108] [PMID: 21339051]

[97]   Zhu, Z.; Zhou, Y.; Yu, H.; Nomura, T.; Fugetsu, B. Photodegradation of humic substances on MWCNT/nanotubular-TiO$_2$ composites. *Chem. Lett.,* **2006**, *35*, 890-891.
        [http://dx.doi.org/10.1246/cl.2006.890]

[98]   Leary, R.; Westwood, A. Carbonaceous nanomaterials for the enhancement of TiO$_2$ photocatalysis. *Carbon,* **2011**, *49*, 741-772.
        [http://dx.doi.org/10.1016/j.carbon.2010.10.010]

[99]   Kamal, T.; Ahmad, I.; Khan, S.B.; Asiri, A.M. Synthesis and catalytic properties of silver nanoparticles supported on porous cellulose acetate sheets and wet-spun fibers. *Carbohydr. Polym.,* **2017**, *157*, 294-302.
        [http://dx.doi.org/10.1016/j.carbpol.2016.09.078] [PMID: 27987930]

[100]   Zhao, P.; Feng, X.; Huang, D.; Yang, G.; Astruc, D. Basic concepts and recent advances in nitrophenol reduction by gold-and other transition metal nanoparticles. *Coord. Chem. Rev.,* **2015**, *287*, 114-136.
        [http://dx.doi.org/10.1016/j.ccr.2015.01.002]

[101]   Hervés, P.; Pérez-Lorenzo, M.; Liz-Marzán, L.M.; Dzubiella, J.; Lu, Y.; Ballauff, M. Catalysis by metallic nanoparticles in aqueous solution: model reactions. *Chem. Soc. Rev.,* **2012**, *41*(17), 5577-5587.
        [http://dx.doi.org/10.1039/c2cs35029g] [PMID: 22648281]

[102]   Liu, H.; Wang, X.; Wang, L.; Lei, F.; Wang, X.; Ai, H. Corrigendum to" Effect of fluoride-ion implantation on the biocompatibility of titanium for dental applications". *Appl. Surf. Sci.,* **2009**, *255*, 3912-3912. [Applied Surface Science 254 (20)(2008) 6305-6312].
        [http://dx.doi.org/10.1016/j.apsusc.2008.08.004]

[103]   Van der Bruggen, B.; Schaep, J.; Wilms, D.; Vandecasteele, C. Influence of molecular size, polarity and charge on the retention of organic molecules by nanofiltration. *J. Membr. Sci.,* **1999**, *156*, 29-41.
        [http://dx.doi.org/10.1016/S0376-7388(98)00326-3]

[104]   Košutić, K.; Furač, L.; Sipos, L.; Kunst, B. Removal of arsenic and pesticides from drinking water by nanofiltration membranes. *Separ. Purif. Tech.,* **2005**, *42*, 137-144.
        [http://dx.doi.org/10.1016/j.seppur.2004.07.003]

[105]   Orecki, A.; Tomaszewska, M.; Karakulski, K.; Morawski, A. Surface water treatment by the nanofiltration method. *Desalination,* **2004**, *162*, 47-54.
        [http://dx.doi.org/10.1016/S0011-9164(04)00026-8]

[106]   Schaep, J.; Van der Bruggen, B.; Uytterhoeven, S.; Croux, R.; Vandecasteele, C.; Wilms, D.; Van Houtte, E.; Vanlerberghe, F. Removal of hardness from groundwater by nanofiltration. *Desalination,* **1998**, *119*, 295-301.
        [http://dx.doi.org/10.1016/S0011-9164(98)00172-6]

[107]   Van der Bruggen, B.; Everaert, K.; Wilms, D.; Vandecasteele, C. Application of nanofiltration for removal of pesticides, nitrate and hardness from ground water: rejection properties and economic evaluation. *J. Membr. Sci.,* **2001**, *193*, 239-248.

[http://dx.doi.org/10.1016/S0376-7388(01)00517-8]

[108]   Fersi, C.; Gzara, L.; Dhahbi, M. Treatment of textile effluents by membrane technologies. *Desalination,* **2005**, *185*, 399-409.
[http://dx.doi.org/10.1016/j.desal.2005.03.087]

[109]   Kang, S.; Pinault, M.; Pfefferle, L.D.; Elimelech, M. Single-walled carbon nanotubes exhibit strong antimicrobial activity. *Langmuir,* **2007**, *23*(17), 8670-8673.
[http://dx.doi.org/10.1021/la701067r] [PMID: 17658863]

[110]   Narayan, R.J.; Berry, C.; Brigmon, R. Structural and biological properties of carbon nanotube composite films. *Mater. Sci. Eng. B,* **2005**, *123*, 123-129.
[http://dx.doi.org/10.1016/j.mseb.2005.07.007]

[111]   Wick, P.; Manser, P.; Limbach, L.K.; Dettlaff-Weglikowska, U.; Krumeich, F.; Roth, S.; Stark, W.J.; Bruinink, A. The degree and kind of agglomeration affect carbon nanotube cytotoxicity. *Toxicol. Lett.,* **2007**, *168*(2), 121-131.
[http://dx.doi.org/10.1016/j.toxlet.2006.08.019] [PMID: 17169512]

[112]   Hyung, H.; Fortner, J.D.; Hughes, J.B.; Kim, J-H. Natural organic matter stabilizes carbon nanotubes in the aqueous phase. *Environ. Sci. Technol.,* **2007**, *41*(1), 179-184.
[http://dx.doi.org/10.1021/es061817g] [PMID: 17265945]

[113]   Yuan, J.; Liu, X.; Akbulut, O.; Hu, J.; Suib, S.L.; Kong, J.; Stellacci, F. Superwetting nanowire membranes for selective absorption. *Nat. Nanotechnol.,* **2008**, *3*(6), 332-336.
[http://dx.doi.org/10.1038/nnano.2008.136] [PMID: 18654542]

CHAPTER 2

Nanotechnology for Safe and Sustainable Environment: Realm of Wonders

Saima Sohni[1,2,*], **Hafiz Nidaullah**[1], **Kashif Gul**[2], **Imtiaz Ahmad**[2] and **A.K. Mohd Omar**[1]

[1] *School of Industrial Technology, University Sains Malaysia, 11800, Minden, Penang, Malaysia*

[2] *Institute of Chemical Sciences, University of Peshawar, Peshawar, Pakistan*

Abstract: Recent years have witnessed the devastating impact of climate change on mankind and environment. Explosive growth in human population, indiscriminate burning of fossil fuels and rapid expansion of industrial sector have significantly contributed in worsening the current status of environment and the situation is highly anticipated to aggravate on extended time scale. To avert any further extreme environmental deterioration, adaptation of eco-friendly and cost effective alternatives is indispensable in order to address the global challenges in energy, industry and environmental sectors. These concerns have led to numerous scientific endeavours from multidisciplinary areas aimed at modifying existing technologies and designing new approaches with distinct economic and environmental benefits. Nanotechnology is genuinely one such revolutionary field of science with unlimited potential applications. This is evident from some of the greatest breakthroughs in the area of environmental protection, remediation and pollution prevention. With this idea, the present chapter is set to explore the fundamental aspects and promising applications of nanotechnology with emphasis on its role in environmental sustainability. In short, nanotechnology has significantly contributed in benefiting society and shaping the nature of modern life, hence it would be appropriate to consider it as the "Realm of Wonders".

Keywords: Agriculture, Environmental monitoring, Environmental protection, Membrane technology, Nanotechnology, Nano-enhanced industry technology, Nano-enhanced energy technologies, Photocatalysis, Remediation.

INTRODUCTION

Environmental issues have taken precedence in gaining a great deal of attention from political and scientific community in the 21st century, also named as the "Century of Environment" [1]. Extensive research in the area of nanotechnology

* **Corresponding author Saima Sohni:** School of Industrial Technology, University Sains Malaysia, 11800 Pulau Penang, Malaysia; Tel: +6011-3307-5305;
Institute of Chemical Sciences, University of Peshawar, Pakistan; Email: saima.sohni@gmail.com

has resulted in the establishment of green, reliable, efficient and economical approaches with far reaching impact in the domain of energy, industry and environment. When Neil Armstrong stepped onto the surface of the moon, he referred to it as a small step and a giant leap in the history of mankind. Likewise, the emergence of nano represents another giant leap [2]. The prefix nano has come from nanos, a Greek word meaning "dwarf" and in terms of nanotechnology, it refers to the things in ballpark of one-billionth of a meter in dimension. In the past, Albert Einstein as a graduate student performed certain experiments on the diffusion of sugar in aqueous solution and proposed that the dimension of a single molecule of sugar is about 1 nm. Regarding the idea of exploiting materials at nano-scale, the Nobel Prize Laureate Richard Feynman, an American theoretical physicist stated that "the principles of physics as far as I can see, do not speak against the possibility of manoeuvring things atom by atom" and later, in future, it will be possible to "write the entire 24 volumes of the Encyclopaedia Britannica on the head of a pin" (1959). However, it was Norio Taniguchi who originally coined the term nanotechnology in 1974 in his paper entitled "On the Basic Concept of Nano-Technology" [3]. Although Feynman and Norio Taniguchi have been considered as the pioneers of this field, but a large number of individuals and communities in the timeline of history have long wondered about the nature of materials on small length scales and investigated nano-scale materials, *e.g.* Roman glassblowers used small gold particles, which appeared ruby red once the dimension of metal particles was more or less smaller than 30 nm. In 1857, Michael Faraday observed some unusual colors displayed by the solutions of gold salts under study and attributed his observation to the presence of highly dispersed metal particles of gold. It was Gustav Mie who in 1908 highlighted that if the particles were made small enough, their properties would deviate from that of the bulk material. However, it was in 1980s era where major breakthrough took place by two very important inventions, *i.e.* scanning tunnelling microscope followed by atomic force microscope which revolutionized the field of nanotechnology, thus making it possible to manipulate a single molecule [4]. In brief, nanotechnology deals with the research and technology development at atomic, molecular or macromolecular levels in the range of 1 to 100 nm, providing basic understanding of the phenomena and materials, and to fabricate and use devices, structures and systems that possess unusual properties and functions owing to their smaller dimensions. National Nanotechnology Initiative (NNI), a US government agency, has pointed out four generations of nanotechnology;

1st generation: Passive nanostructures (~2000) such as dispersed and contact nanostructures, *e.g.* colloids, aerosol as well as products incorporating nanostructures, for instance coatings, nanoparticle (NPs) reinforced composite materials.

2nd generation: Active nanostructures (~2005) with bioactive health effects, *e.g.* bioactive devices, targeted drugs and physico-chemically active structures (such as 3D transistors, adaptive structures).

3rd generation: Systems of nanosystems (~2010), *e.g.* 3D networking, novel hierarchical structures.

4th generation: Molecular nanosystems (~2015-2020), *e.g.* molecular devices [5].

It is indeed true that extraordinary advances in chemistry have made it possible to design and manipulate chemical properties at molecular level. Furthermore, this progress has led to the development of valuable materials and ground-breaking technologies. Noteworthy, our understanding of molecular-level designs has been triggered by multidisciplinary input from across the fields of physical sciences as well as engineering. At this moment, it would be appropriate to define the term nanoscience and its relationship to the field of nanotechnology. Nanoscience refers to the study of materials with dimensions on the length-scale of 100 nm or smaller. Fundamentally, compared to bulk material, nanostructured counterpart exhibits unique physical, chemical and biological properties that solely depend on its structure and dimensions. Fabrication of the materials with ultra-small sizes exhibiting unusual properties have resulted in the monumental growth of this area [6]. Nanoscience has been defined by the Royal Society and Royal Academy of Engineering as the study of phenomena and manipulation of materials at atomic, molecular and macromolecular scales, while nanotechnology refers to the design, characterization, production and utilization of structures, devices and systems through controlling size and shape in nano-domain range [7]. In other words, nanotechnology refers to the application of science in controlling materials on their molecular levels. Implausible progress in nanotechnology has opened up new gateways towards novel fundamental and applied front lines in the fields of material science and engineering, for instance bio-nanotechnology, nano-biotechnology, surface-enhanced Raman scattering (SERS) as well as applied microbiology [8].

Perhaps the most notable feature related to materials is that in nano-scale range they behave differently compared to their micro-sized counterparts. This novel behaviour has been attributed to their high specific surface area and hence, higher reactivity [9]. By rule, there is going to be certain modification in the property of a substance when its size reaches the dimension related to the property of that material. In the nano-scale range, physical, magnetic, electrical, thermal, kinetic and other aspects related to different materials undergo dramatic changes because of the physical dimensions of the matter. As the size approaches nano dimensions, different interactions between atoms making up matter would change accordingly.

These interactions are averaged out of existence in the bulk materials. It is for this reason nano gold behaves differently than bulk gold. Scientists exploring the field of nanoscience and nanotechnology are in active search for understanding this alteration in different properties across the vague area between quantum and classical (bulk) domains [2, 10]. The phenomena occurring on nano-length scale have been intriguing for chemist, physicists, biologists, computer scientists as well as electrical and mechanical engineers [11]. In addition to surface area, quantum effects also confer unique properties on nanomaterials. Compared to bulk material, a given quantity of nanomaterial is more accessible in a chemical reaction. Further, higher surface energies together with corners, edges and surface defects add to the intrinsic reactivity of the nanomaterials. Further, as mentioned earlier, the quantum effects responsible for imparting unique properties to the nanomaterials predominate as the size of a material approaches nano-scale dimensions. This would in turn affect optical, magnetic and electrical behaviour of these materials [12]. The size particularity gives nanomaterials these novel attributes enabling them to adopt new comportments because of the laws of quantum physics existing at nano-scale range [13]. It is noteworthy to mention that this alteration in the properties of matter as the size reduces to nano-scale range has led to the exploitation of nanomaterials in a wide spectrum of consumer products, for example; food, medicines, cosmetics, suntan lotions, paints and other important materials [14].

A key emerging trend in science and technology sector has been to consider nanotechnology in conjunction with other technologies. At one symposium held in recent past under the Organisation for Economic Co-operation and Development (OECD)/NNI on the assessment of the economic impact of nanotechnology, it was highlighted that *via* integration of different key enabling technologies, it would be most likely to realise their anticipated potential [15]. Nanotechnology has been currently ranked as one of the most attractive and challenging fields of research and development. Precisely, it can be referred to as a multidisciplinary science involving research from every scientific and engineering disciplines providing phenomenal prospects for innovation *via* integration of knowledge in diverse fields such as materials, electronics, photonics, biology and medicine with technology-driven and application-driven approaches [16].

Generally, two different approaches are used to describe nanotechnology; top-down and bottom-up. In top-down approach, nano-scale architecture is designed from the smallest structures *via* machining, templating and lithographic techniques (photonics applications in nanoelectronics and nanoengineering). On the other hand, bottom-up, or molecular nanotechnology, deals with fabricating different organic and inorganic materials into definite structures, atom by atom or molecule by molecule, usually by self-assembly/self-organization (application in several biological processes) [17]. Nano-sized materials have been fabricated with controlled size and morphology in order to fully exploit them [18]. Classification has been done on the basis of their origin such as natural and anthropogenic (intentionally produced and engineered), chemical composition (organic, inorganic or carbon-based nanostructures, metal oxide nanoparticles, elemental metallic NPs and organic polymers), formation (biogenic, geogenic, anthropogenic, atmospheric), size, shape (fibres, rings, tubes, spheres, and planes, *etc.* as depicted in Fig. (**1**), state (single, fused, aggregated or agglomerated), phase composition properties (single phase, multiphase solids and multiphase systems) and applications (biological, environmental, industrial) [19]. Nanostructures have at least one dimension lesser than 100 nm. They are modulated over nano-scale at different dimensions and can be categorized into zero dimensional (0D), one dimensional (1D), two dimensional (2D) and three dimensional (3D) nanostructures. Briefly, 0D nanostructure, such as quantum dots (QDs), nanospheres and NPs, with all its dimensions being on the nano-range, 1D nanostructures correspond to the structures that have two dimensions (nanowires, nanotubes and nanobelts) in nano-scale range, 2D nanostructures possess only one dimension on nano-scale which in most of the cases is perpendicular to their layer plane (nanoflakes, nanoplatelets and superlattices). Moreover, 3D nanostructures are constructed by an ensemble of nanostructures connected through single-crystalline junctions with complete geometrical dimensions in nano or micro range (nanowires or nanorods). Ranging from 0D to 3D structures, different categories of nanomaterials have been investigated and gradually introduced into the industry and everyday life [5, 20]. However, 2D nanomaterials, especially the inorganic ultrathin nanosheets possessing unique structure with single or few-atomic layers have been widely investigated for their excellent physical properties due to quantum confinement of electrons [21].

Generally, NPs have been categorized based on the number of components into single nanomaterials (consisting of one component only) and multiple nanomaterials. The later type is based on two or more components that have been further classified into composite and core–shell NPs. The core–shell NPs can be generally defined as the type of NPs composed of a central core surrounded by the shell [23]. Using NNI platform, scientists, engineers and academician are together striving to build a future in which the potential to understand and control the matter at nano-scale has led to a revolution in the field of technology and industry. To solve important national or global problems, nanotechnology R&D community is set to achieve the following goals developed by NNI agencies working with National Nanotechnology Coordination Office and Office of Science and Technology Policy (OSTP) by 2025;

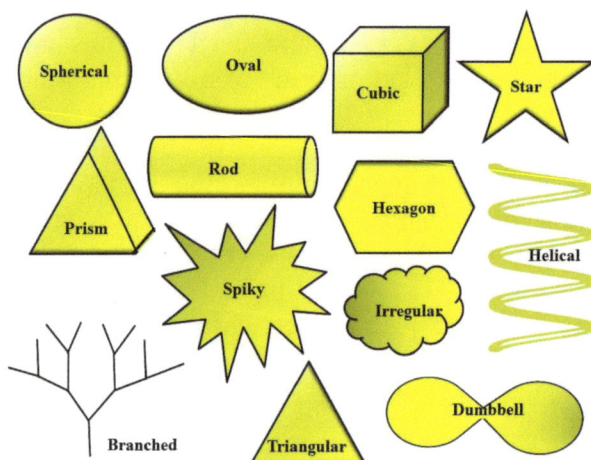

Fig. (1). Different shapes of nanomaterials, adapted from Sources; Gatoo *et al.*, Tisch, U. and H. Haick, Liu, Xu *et al.* [7, 22].

i. Create miniaturized devices capable of sensing, computing, and communicating without wires or maintenance at least for a decade.

ii. Construct 100 times faster computer chips that are less power consuming.

iii. Manufacture atomically-precise materials stronger than aluminium (50 times) at half the weight but with the same price.

iv. Developing cost effective (4 times cheaper) strategies for turning sea water into drinkable water.

v. Determine environmental, health, and safety aspects related to the use of nanomaterials [24].

We really hope that our readers would enjoy the spectacular contributions of nanotechnology with regard to safe and sustainable environment. This chapter goes on to describe some selected examples of nanotechnology targeted at improving the overall quality of environment. It is indeed a cutting edge technology that has led to a rapid growth of nano-enhanced industrial processes with far reaching impact on the chemical transformations such as environmentally benign catalysis *etc.* Importantly, nano-enhanced energy technologies that are aimed at exploiting renewable energy resources by incorporating nanostructures in different product devices that have displayed improved performance characteristics at low costs together with addressing environmental and health related concerns. Especially, the concept of nanoenergy as a newly progressing area for powering micro-nano systems intended at harvesting energy from environment for self-sufficient operation is enthralling. On agricultural side, the use of nano-engineered materials has proven to be vastly effective with little or no adverse effects on environment. Lastly, a wide range of nanosized materials and their hybrid architectures that have brought about significant improvements in environmental treatment, remediation and monitoring have also been discussed.

APPLICATIONS OF ENVIRONMENTAL NANOTECHNOLOGY

At present, conserving and improving the quality of environment (air, water and soil) are among the most pressing global challenges. Ever-increasing release of pollutants from wide variety of sources such as chemical and oil spills, fertilizer and pesticide runoff, abandoned industries and mining sites and gaseous and particulate matter released as automobiles exhaust, among others, has seriously deteriorated environmental quality together with its direct implications on human health. Presently, the detection of toxic chemicals, removal of contaminants from polluted environment, prevention of new pollution sources along with associated financial constraints have been featured among the burning environmental issues worldwide [25]. NNI established a long-term plan by 1999 which was revised in 2004 for 2006–2010. It has highlighted several grand challenges of environmental improvement including;

1. Better understanding of molecular processes that occurs in the environment.
2. Development of innovative "green" technologies that could curtail pollution including manufacturing and transport of waste products.
3. Improved environmental clean-up *via* new efficient ways for eliminating air and water borne contaminants, continuous monitoring and mitigation of pollution in large regions [26].

Nanotechnology despite being diminutive in size has proven to be a promising approach which offers tremendous potential for effective and economical remediation of the environment [25].

The term green nanotechnology can be interpreted in two ways. On one hand, it is possible to transform a number of general applications and processes to environmentally friendlier routes using nanomaterials *via* adopting resource/energy saving methods and removal of toxic substances. Alternately, this term can be directly related to the application of these materials in the domain of environmental technology [27]. In order to approximate different products and processes to the natural practices, as well as rendering them as safe and environmentally friendly as possible, four main areas have been identified with varying degree of impact on the production process, energy and environmental technologies including;

i. **Biomimetics**: Biomimetics involve local materials as well as energy sources and renewable resources through molecular self-organization as a manufacturing paradigm and physiological manufacturing conditions, *i.e.* aqueous synthesis.

ii. **Resource Efficiency**: Resource efficiency can be ensured by preventing the occurrence of side reactions, use of enzyme mediated reactions, precision manufacturing, miniaturization/dematerialization, to skip cleaning steps and avoiding the use of rare materials. Additionally, energy efficiency in terms of improved production efficiency by employing low process temperature, lightweight construction and recyclability by preventing losses using limited range of materials, waste collection, lesser use of additives and processing aids, and to minimize diffuse emissions as well as contaminations are some of the ways to achieve resource efficiency.

iii. **Minimum Risk–design Safety**: For this purpose, toxic substances and nanofunctionalities which present serious threats to health, safety and environment are avoided. Further, prevention and reduction of potential exposure to the contaminants by avoiding mobility and bioavailability is also important in this respect.

iv. **Energy and Environmental Technologies**: The area of energy and environmental technologies points to emissions reduction, environmental monitoring, improving environment by remediation (*in* and *ex situ*) as well as a shift towards renewable materials and energy sources [28] as illustrated in Fig. (**2**).

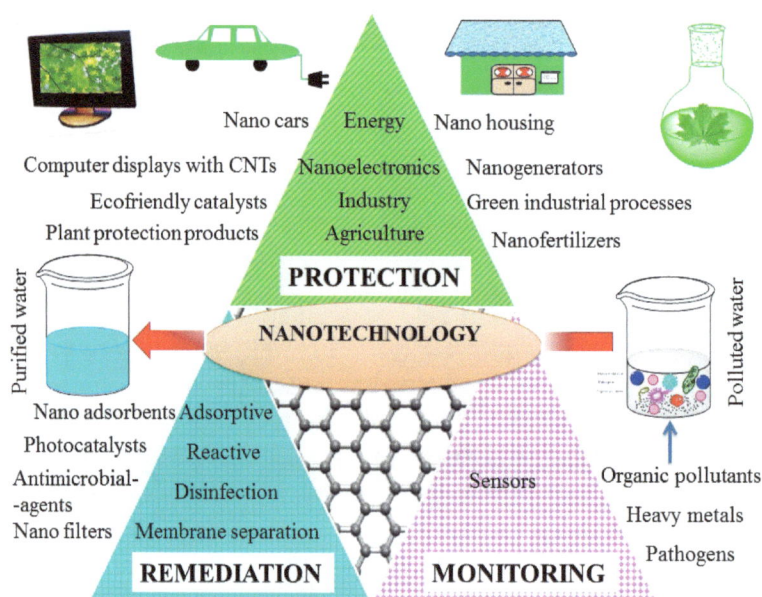

Fig. (2). Applications of environmental nanotechnology.

Most of the applications of nanotechnology for biological and non-biological materials in environment fall into three classes as listed below [5, 25, 29];

1. Environmental protection (eco-friendly and/or sustainable products)
2. Environmental remediation using engineered nanomaterials (nanoremediation)
3. Environmental monitoring (nanotechnology enabled sensors)

A wide array of nanoengineered materials have been fabricated and extensively utilized for environmental remediation including reactive metals, semiconductor NPs, dendrimers, carbon nanotubes (CNTs) and nano-scale biopolymers [30]. Below, we will discuss some of the important applications as well as prospects of nanomaterials in the area of environmental protection, remediation and monitoring.

ENVIRONMENTAL PROTECTION (ECO-FRIENDLY AND/OR SUSTAINABLE PRODUCTS)

The definitive potential of both green chemistry and nanoscience can be realized by following an integrated approach referred to as green nanoscience. For example, new approaches aimed at manufacturing nanoelectronics are likely to address the technological difficulties faced by electronics industry and reduce environmental and human health related problems associated with their products and processes. The integration of these two emerging fields is considered far more promising in reducing the negative impact of industrial development on the society and environment [6].

Nano-Enhanced Green Technologies

Owing to the growing concerns regarding energy, health, raw materials and environment, different industries have been motivated to adopt green processes and technologies. This Section emphasizes the role and prospects of nanoengineered materials in industrial processes.

Nanomaterials Based Extraction Method

Recent trends in extraction methods have been dedicated towards developing strategies that could minimize solvent consumption along with process intensification and economical production of valuable extracts. In line with this aim, the idea of green extraction has been put forward by scientific community. Green extraction is related to the discovery and design of extraction processes that are based on environmentally friendly solvents and renewable natural products. These methods consume less energy and ensure a safe and high quality of extracted materials [31]. For example, in vegetable oil extraction industries, the fundamental limitations of traditional extraction practices, *i.e.* mechanical pressing and solvent extraction, are financial, environmental as well as safety aspects [32]. The release of volatile organic compounds (VOCs) in the conventional process of solvent extraction from vegetable oil extraction industries is particularly of grave concern as these can be a potential precursor for more pollutants such as ozone and other photochemical oxidants. VOCs are considered equally harmful for human health as well as industrial crops. Besides being greenhouse gases, some are carcinogenic and toxic in nature [33].Vegetable oil extraction plants represent potential chief source of hexane which is ranked as a hazardous air pollutant by the US Environmental Protection Agency (EPA). According to recent estimates, 0.7 kg of hexane/ton of seed is released into the environment (EPA, 2015). Exposure to hexane at 125 ppm for 3 months would cause peripheral nerve damage, muscle wasting, and atrophy (Agilent Technology, MSDS, Agilent Technology, 2008) [34]. The application of microemulsions during extraction process increases the solubility of otherwise immiscible water and oil systems. Microemulsion systems are aqueous-based environmentally benign alternatives to the organic solvents for dry cleaning, degreasing and hard surface cleaning applications. It is noteworthy to mention that microemulsion can be regarded as "small" vessels or nanophases with domain sizes in nano-scale range, hence can be used for carrying out reactions [35]. Microemulsion based extraction of vegetable oil offers numerous advantages. Firstly, since organic solvent is not required for extraction, this aspect reduces the release of toxic substance to the environment during extraction. Secondly, the microemulsion based extraction can be carried out at room temperature, thus it consumes less energy. Being cost effective and green, it is likely that the aqueous microemulsion-based system would be considered as an attractive substitute in the domain of green chemistry or clean technology for oil extraction industries [36]. A number of vegetable oils (palm oil, flaxseed oil, sunflower oil, soya bean oil, castor oil, medium chain triglycerides *etc.*) have been investigated for generating submicron/nano emulsions [37]. Nguyen *et al.* used biosurfactant mixtures in the extraction of vegetable oil for prospective biofuel applications [38].

Environmentally Friendly Catalysis

Sustainable chemistry puts emphases on the design of chemical processes that have the potential to minimize associated environmental issues along with reduction as well as prevention of pollution at its source [39]. Optimization of existing chemical processes together with the development of new environmentally benign processes are related to the improvement of catalyst performance [40]. Catalysis is thus central to the sustainable process management [41]. In addition, both depletion in petroleum reserves and climate change concerns have stirred considerable efforts in the domain of energy. For this reason, efficient catalytic processes have been introduced for the conversion of non-petroleum carbon feedstocks such as coal, natural gas and biomass into clean liquid fuels for fulfilling ever-growing energy demands [42]. Solid heterogeneous catalyst is considered to be a convenient approach for the recovery and recycling of catalysts from reaction environment, thus ensuring far better, cost effective and environmentally friendly manufacturing [43]. Nanocatalysts are considered to be the excellent candidates compared to conventional catalysts because of their higher surface area. Among prominent features are; highest catalytic activity, selectivity and stability depending upon shape, size, composition as well as the nature of nanocatalyst used [44]. In addition to their role as catalyst, NPs can also be used as support materials in heterogeneous catalysis, ascribed to their larger surface-to-volume ratios [44c, 45]. Among nanocatalysts, different categories such as magnetic nanocatalysts, nano-mixed metal oxides, core-shell nanocatalysts, nano-supported catalysts, graphene-based nanocatalysts have been proposed for different applications.

Magnetic nanocatalysts offer great advantages in terms of economy, exceptional activity, high selectivity, enhanced stability and efficient recovery as well as recyclability [46]. The application of magnetic NPs as catalyst or support using wide variety of solid matrices permits the merger of well-known methods for catalyst heterogenization and magnetic separation. To deal with the aggregation problem of NPs and achieve the grafting of catalyst species on prepared magnetic NPs, modification and functionalization with coating/encapsulating materials such as silica, polymers, carbon, graphene and CNTs has been attempted by many researchers [47]. Besides, other nanosized materials such as nanocrystalline zeolites, porous nanomaterials with dimension smaller than 100 nm possessing unique external and internal surface reactivity have found numerous useful applications. The nanocrystalline zeolites have been used in catalysis and separations. The surface properties of nanocrystalline zeolites are tailored *via* functionalization of surface silanol groups for diverse uses [48]. One of the most attractive features of using nanosized zeolites to catalyse different industrial reactions is that the process is environmentally friendly. Firstly, energy consumption is minimized as oxidation is driven by visible light. Secondly, wasteful secondary photoreactions are eliminated using visible light, hence triggering only the low energy reaction pathways, resulting in higher yields of the desired products. In addition, nanosized catalysts offer enhanced selectivity for the desired products as compared to conventional catalysts, thus making chemical manufacturing a highly efficient process [49].

Amid pressing global challenges of climate change and depleting fossil fuel reserves, a great deal of interest has been developed in exploiting lignocellulosic biomass as green and sustainable resource of energy. Sustainability demands the exploitation of every component of biomass residues and with the advantages of abundance and no food *vs.* fuel competition, biomass conversion represents a highly affordable way to produce clean fuel and other valuable products [50]. Recently, David W. Wakerley and his team at the University of Cambridge successfully performed photoreforming of lignocellulosic biomass for generating H_2 by solar light driven catalysis using semiconducting cadmium sulphide QDs for potential application in the synthesis of renewable liquid fuel or in fuel cells [51].

Considering nano-TiO_2, its use in producing photodegradable polymers is known to reduce the release of toxic byproducts during polymer incineration. Kim *et al.* demonstrated the suppression of dioxin emission for polyvinylchloride (PVC) incineration when TiO_2 nanocomposites were used. Compared to pure PVC (without TiO_2), the GC results on exhaust gases from incineration showed reduced emissions of dioxin and precursors as the content of TiO_2 in PVC/TiO_2 nanocomposite was increased [52]. A review of some previous work is provided below in which nanoengineered materials and their hybrids have been employed as green and highly efficient substitute to the conventional catalysts. Zhang and team synthesized Pd@Pt core@shell octahedral NPs and investigated its structural stability and shell thickness dependent catalytic performance for p-chloronitro-benzene hydrogenation with H_2. The enhanced catalytic performance of Pd@Ptoctahedral NPs for selective p-chloronitrobenzene hydrogenation certainly originated from the core–shell interaction [53]. Moushoula *et al.* have synthesized five different types of calcium oxide-based catalysts supported gold NPs (AuNPs) including commercial CaO, egg shell, mussel shell, calcite and dolomite for biodiesel synthesis. It was found that a high quality biodiesel product was synthesized by the supported catalysts as compared to the conventional CaO based catalysts [54]. In another investigation, $Cs_{2.5}H_{0.5}PW_{12}O_{40}$ NPs catalyst has been reported as new, recyclable, reusable and efficient solid acid catalyst for cost-effective and green synthesis of (S)-(-)-propranolol in high yields and good selectivity by Gharib *et al.* [55]. Similarly, highly efficient and environmentally benign chemical conversion system for the p-nitrophenol hydrogenation on magnetically recyclable nanocatalyst (Pd-Fe_3O_4) synthesized through a simple and low cost procedure has been suggested by Zhang *et al.* [56]. A simple and useful approach of gel-deposition-precipitation (G-D-P) was studied by Chen *et al.* to prepare core–shell-like silica@nickel species as catalysts for dehydrogenation of 1, 2-cyclohexanediol to catechol. The nanocatalysts thus prepared by proposed technique displayed better activity and stability, ascribed to the higher dispersion of metallic nickel stabilized by nickel phyllosilicates compared to the catalyst fabricated by following deposition–precipitation (D–P) approach [57].

The nanoassembly of approximately monodisperse NPs as uniform building blocks to fabricate zirconia (ZrO_2) nanostructures was proposed by Das and El-Safty with mesoscopic ordering using a template as a fastening agent. Interestingly, the as-prepared nanomaterials showed exceptional catalytic activity in the transformation of long-chain fatty acids to their methyl esters with maximum biodiesel yield of 100 % [58]. A new library of mesoporous composite materials containing Ga_2O_3 NPs were synthesised by Lueangchaichaweng and his team. The mesoporous composites of NPs showed higher catalytic activity than their unstructured counterpart. The catalytic activity and recovery of as-prepared catalysts was investigated by epoxidation of different alkenes with hydrogen peroxide and recyclability studies, respectively [59]. In another study, $Fe_3O_4@SiO_2$-imid-$H_3PMo_{12}O_{40}$ NPs were fabricated as heterogenic catalyst to synthesize 3, 4-dihydropyrimidinones under solvent-free conditions by Javidi *et al.* [60]. Zhang and co-workers have assembled different noble metals (Pd, Pt, Au and Ag) based nanocatalysts and graphene with the chemically inert nylon rope by one-step hydrothermal method to fabricate a novel 3D noble metal/graphene/nylon rope as a highly efficient catalyst for continuous-flow organic reactions [61]. Another green, rapid, convenient and eco-friendly approach for the synthesis of 2,4,5-trisubstituted imidazoles using graphene oxide (GO)-chitosan based bio-nanocomposite as an efficient nanocatalyst has been described by Maleki and Paydara. This protocol offered many advantages in terms of short reaction time, greater yield, convenient separation of the catalyst as well as the use of solvent-free conditions [62].

Eco-Friendly Alternatives in Agriculture

The world population is projected to increase by more than one billion people in the next 15 years, reaching 8.5 billion in 2030, increasing further to 9.7 billion by 2050 and 11.2 billion in 2100 [63]. Over the coming three decades, agricultural sector is expected to face an unprecedented confluence of pressures with a 30% rise in the global population, intensified competition for increasingly limited land, water and energy resources as well as the climate change. It is estimated that the agricultural sector would need to produce 60% additional food globally and 100% more in the developing countries to meet the demand at current levels of consumption by 2050. Food and Agriculture Organization (FAO) of UN has proposed five interconnected and complementary principles for enhancing agricultural productivity and ensuring sustainability.

1. Improving the efficiency in exploiting resources is critical to sustainable agriculture.
2. Sustainability necessitates direct action to conserve, protect and improve natural resources.
3. Agriculture that is unsuccessful in protecting and improving rural livelihoods

and social well-being is unsustainable.
4. Enhanced resilience of people, communities and ecosystems is also important.
5. Sustainable food and agriculture demand responsible and efficient governance
   machineries.

It is notable to mention that profound changes in the agricultural system would be required to enhance agricultural productivity and sustainability [64]. There has been an upsurge in the research based on using nanotechnology in agriculture over the past decade [65]. Presently, nanotechnology has transformed the entire state of agricultural and food industry by offering attractive applications including role as fertilizer and pesticides, use of new tools to treat various plant diseases, nano-based kits for rapid detection of pathogens and improving the uptake of nutrients by plants are some of the examples. Nano biosensors and other smart delivery systems have the potential to improve agricultural productivity by targeting different crop pathogens. Aside from these systems, nano-catalysts can have prospective role in increasing the efficacy of commonly used pesticides and insecticides, and likewise reducing their dosage required by different crops thereby preventing the excessive use of pesticides and their negative implications on our environment [66]. Being smart delivery systems, these materials have rightly being called the Magic bullets by Noble laureate in Physiology, Paul Ehrlich [67].

Nanofertilizers

Certainly, a modern approach of applying nanostructured materials as fertilizer carriers or controlled-release vectors for building "smart fertilizer" has proven to be a new facility in order to enhance the nutrient use efficiency with corresponding reduction in the cost of environmental protection [68]. Nitrogen is considered as a key ingredient in fertilizer formulations. However, nitrogen fertilizers are quite expensive on account of the energy needed during their synthesis as well as the ample quantity required during field applications. Further, the particle size of around 70% of the nitrogen applied using conventional fertilizers exceeds 100 nm resulting in its loss to the soil because of leaching as the nitrogen utilization efficiency (NUE) of plants is typically lower [69]. Contrary to this, nanofertilizers have caught the attention of many as a promising approach. Because of high surface area to volume ratios, these nanomaterials are considered more effective compared to polymer-coated conventional fertilizers. These nanostructured formulations reduce the loss of fertilizer nutrients into the soil by leaching and/or leaking [70]. In order to encapsulate NPs within fertilizers, three approaches can be followed; i) nutrient can be encapsulated inside nanoporous materials, ii) coated with the thin film of polymer or iii) delivered as particle or emulsions of nano-scale dimensions [71].

NPs-mediated Gene Transfer for Insect Pest Management

The strategies practiced traditionally in agricultural sector such as integrated pest management are considered to be insufficient nowadays. Moreover, the use of different pesticides like dichlorodiphenyltrichloroethane (DDT) has caused a decline in soil fertility with a negative impact on animals and human beings. In comparison, the emergence of new insect resistant varieties by NPs-mediated gene transfer would possibly provide green and eco-friendly substitute to insect pest management with potentially no harm to the environment [17].

Nano Plant Protection Products

Nanopesticides or nano plant protection products represent a new technological progress in the field of agriculture that offers a number of attractive features in terms of augmented efficacy and durability along with the reduction in dosage of active ingredients required. A number of formulation types including nanocapsules (with polymers), emulsions (nanoemulsions) and NPs supported products such as metals, metal oxides and nanoclays have been used [72].

Antimicrobial activity of different metal NPs, in particular, copper and silver NPs have been investigated against different plant pathogens [73]. Owing to the large scale use in agriculture, pesticides are one of the key agrochemicals that have caused the contamination of surface and ground water sources [74]. As compared to conventional formulations, nanopesticides are less harmful to the environment. The high proportion of nanoformulations for insecticidal purposes can be explained by the fact that the active ingredients (AIs) of many conventional insecticides are scarcely water soluble and thus need a delivery system for field applications. Additionally, AIs are less detrimental to non-target organisms and may also reduce the development of resistances. Many of these alternative AIs being unstable require protection against premature deterioration which is possible by nanoformulations [65].

Nanomaterials Based Biosensors

Application of various kinds of biosensors based on nanostructured materials and their composites represents another exciting aspect of nanotechnology in agricultural sector for remote sensing devices in precision farming. It has greatly helped in reducing agricultural wastage, hence keeping environmental pollution to a minimum [17].

The use of green and efficient alternatives in terms of nanomaterials and technology has the potential to bring in a revolution in the field of agriculture by sustainable production with minimal negative impact on environment.

Green Synthesis of Nanomaterials

The key aspects considered from green chemistry perspective encompass the choice of solvent, environmentally friendly reducing agent and non-toxic materials for stabilizing NPs during their synthesis [75]. In recent years, the development of efficient and green alternatives has emerged as a major focus of scientific attention in order to quest for an eco-friendly approach to prepare well-characterized NPs. To achieve this target, one of the most useful ways for the production of metal NPs is based on organisms and plants. This seems to be a suitable approach for the large-scale biosynthesis of NPs due to several reasons [8]. The key benefits associated with bio-based synthesis of NPs include; easy availability of plant extracts as biogenic agents and synthesis of NPs at room temperature and pressure [76]. In addition, it has been found that the rate of synthesis of NPs from plant source is faster when compared to the processes involving microorganisms. Moreover, these NPs also demonstrated remarkable stability as well as diversity in its shape and size compared to those produced using other organisms [8, 75].

Green synthesis of silver NPs has been reported from different plant sources including *Jatropha curcas* [77], *Capsicum annuum* [78], *Hibiscus Sabdariffa* [79], *Ficus benghalensis* [80], *Artocarpus heterophyllus Lam* [81], *Eucalyptus chapmaniana* [82], *Terminalia chebula* [83], *Iresine herbstii* [84], *Nyctanthes arbortristis* [85], *Thevetia peruviana* [86], *Rosmarinus officinalis* [87], *Quercus brantii* [88], *Ficus benghalensis, Azadirachta indica* [89], *Euphorbia prostrate* [90], *Ziziphora tenuior* [91], *Skimmia laureola* [92] and *Limonia acidissima* [93]. Besides silver NPs, many green schemes have been proposed for the synthesis of other metal or metal oxide NPs including Pd [94], Pd/CuO [95], Pd/Fe$_3$O$_4$ [96], CdO [97], Cu [73 b, 98], CuO [99], Au [100], ZnO [101], CeO$_2$ [79b], TiO$_2$ [90], Al$_2$O$_3$ [102] and Au–Ag bimetallic NPs [76]. In recent years, *in situ* green route has been proposed for the synthesis of copper NPs assisted by plant extracts [98b, 103]. An excellent review on metal NPs synthesis *via* plant based sources has been presented by Iravani [8]. Aside from using plants, Dhanasekar *et al.* synthesized gold NPs using fungus *Alternaria sp* [104]. Another exciting example of green synthesis of nanomaterial is the fabrication of metallic nanorods and nanowires. For example, synthesis of gold, silver nanorods and silver nanowires has been carried out at room temperature assisted by low cost surfactant in aqueous media. QDs have found promising applications in the field of medical imaging, sensing, electronics and solar cells. Synthesizing QDs employing less toxic chemicals to replace cadmium selenide nanocrystals due to their toxicity is a good example. A fast, green and economical method for the production of water-soluble CNT using microwaves has been proposed by S. Mitra involving functionalization of CNT to tailor them for diversified applications [105]. In a recent study, green synthesis of CNT/Polyaniline nanocomposites has also been proposed by Nguyen and Shim [106]. Green synthesis of metal oxide nanomaterials, for example, crystalline ceramics at low temperature using molten salt synthesis method has been reported by Wong *et al.* [107]. In another investigation, green aqueous based method has been proposed by Vigneshwaran *et al.* for fabricating highly stable copper NPs in the presence of polyvinyl pyrrolidone (PVP) as a stabilizer with no inert gas protection. The as-prepared NPs were narrowly distributed with average size of less than 5 nm. In this method, ascorbic acid (natural vitamin C) has been utilized (i) as an excellent oxygen scavenger and (ii) as reducing agent. It reduces the metallic ion precursor into NPs while also preventing the undesirable conversion of copper nano clusters into oxides [108]. In one study, green synthesis of silver NPs-CNTs based reduced graphene oxide (rGO) composite has been carried out by one step hydrothermal method [108b, 109]. Zhang *et al.* reported one pot green synthesis of silver NPs-graphene nanocomposites for sensing applications in the detection of SERS, H_2O_2 and glucose [110]. In another work, green synthesis of silver nanoribbons employing waste X-ray films has been proposed by Shankar *et al.* [111].

Nanoelectronics and Nano-Enhanced Energy Technologies

Fuel cell is a highly efficient and useful electrochemical energy converter device. Unlike internal combustion-engine that transforms chemical energy of the fuel *via* combustion into mechanical energy, this device changes the chemical energy of fuel directly into electrical energy. It is noteworthy to mention that the useful energy of power or fuel is much better implemented in fuel cell, resulting in a higher efficiency. Fuel cells are therefore considered as a promising technology for electricity and heat generation as well as powering automobiles [112]. With regard to fuel cell technology, four major areas of application have been identified; (i) stationary applications (domestic energy supply, thermal power stations); (ii) mobile application (vehicles); (iii) portable applications (consumer electronics, leisure sector); (iv) special applications (special vehicles, emergency power or uninterruptible power supply) [113]. Nanomaterials and nanodevices have been playing a crucial role in all forms of energy harvesting, conversion, storage and utilization. Nanostructures have gained remarkable popularity in electronic industry due to their sizes, shapes and unique properties compared to larger dimension counterparts [12]. The utilization of nanoenhanced products in renewable energy based technologies offers an efficient and green approach of fulfilling energy deficit compared to the conventional fossil fuels and nuclear energy as energy resources. Currently, greater interest has been witnessed in the field of renewable energy resources that is estimated to fulfill 50% of the world's primary energy demand by 2040. Moreover, renewable energy has the potential to play influential role in cutting down gas emissions into the environment up to 70% by 2050. The application of nanoproducts in the area of renewable energy in addition to being a source of 2.5 trillion revenue by 2015 can lead to higher efficiency of lighting and heating, improved electrical storage capacity as well as would contribute in reducing pollution level [114]. The decisive role played by nanotechnology in reducing pollution is obvious from the information provided by NNI, stating that nanotechnology-based home lighting is likely to reduce the energy consumption by an estimated 10% in the US, thus saving $100 billion per annum and cutting down carbon emissions by 200 million tons annually [25]. With the incredible expediency offered by nanotechnology in divergent areas, drive technologies are not exempted. Mobility in all its forms, concepts for megacities on the individual mobility to vehicle concepts is a matter of concern for the future. In relation to the criteria of resource efficiency or CO_2 reduction as applied to the automotive, aviation and energy technology sector, certain objectives have been pointed out; (i) joining technologies for multi substantive process chain; (ii) manufacturing of additives for customized components; (iii) development of new materials for energy storage; (iv) materials for energy and energy conversion [115]. On the way towards alternative drive technologies in operational use, battery buses have been tested that offer a number of advantages such as energy efficiency, low noise as well as being locally emission-free and this development can be anticipated in other transportation modes as well [116]. Vessels can be predominantly environmentally friendly means of transport if they not based on diesel engines, the latter being a source of exhaust, fine dust and noise emissions. It requires considerable technical advancement to reduce these emissions but 100% emission-free scenario will be possible if alternative fuels based on renewable energy or electricity are utilized. The future thus lays with ferries, passenger and cargo ships driven by batteries or fuel cells. In one special application to submarines, use of fuel cells has eliminated toxic emissions from their service. However in conventional ships, electric drives are rare so far although feasibility studies and initial practical projects have shown promising results [117]. Conjunction of innovative technologies are prerequisite on the avenue of transition from a society that is fulfilling its energy demands by burning fossil fuels towards a green society relying on clean renewable alternatives. Below, we will discuss some of the innovatory uses of nanotechnology with selected examples that have stirred great enthusiasm in the electronics and energy sector.

Development of New Nanostructure-based Solar Cells (Nanophotovoltaics)

The effective utilization of non-conventional energy resources including solar, wind, tidal and hydropower is indeed of great significance to reduce dependency on fossil fuels and minimize environmental deterioration. Non-conventional resources being clean and inexhaustible are regarded as the green alternatives of energy. Harnessing solar energy for fulfilling energy demands has attracted considerable attention from the scientific community. This process however requires an infrastructure that is capable of solar energy collection and conversion [118]. Currently, the development in solar cells technology sector is facing two major challenges; conversion efficiency and cost. Research targeted at using nanowires and other nanostructures to fabricate potentially cheaper and more efficient solar cells called third generation solar cells (quantum-based and dye-sensitized solar cells) compared to conventional planar silicon solar cells is in progress. Further, bringing new highly efficient solar cells with greater efficiency/cost ratio into use would possibly satisfy the rising global energy demands [119]. The inclusion of nano-scale components in photovoltaic cells offers some distinct advantages in terms of its ability to control the energy band gap providing flexibility and inter-changeability. Secondly, nanomaterials enhance the effective optical path and prevent electron hole recombination. Innovation in this sector has sparked the advent of fourth-generation photovoltaics [120]. Due to the associated toxicity of cadmium based solar cells, a growing trend towards the development of organic materials based solar cells has been observed. The distinctive features of these systems include flexibility and eco-friendliness. Moreover, they are light weight and disposable as well [121]. In this respect, bulk heterojunction photovoltaic cells have been anticipated to offer great promise as fourth generation electronics [122]. Combining both inorganic and organic components, Pathaka *et al.* have recently developed a novel fourth generation inorganic organic hybrid bulk heterojunction photovoltaics based on $P_3HT.PCBM.AIS$ (P_3HT-poly[3-hexylthiophene-2,5-diyl], PCBM-(phenyl-C_{61}-butyric acid methyl ester), AIS-$AgInSe_2$ nanopowder) [118]. On the other hand, Brooks *et al.* have developed a novel hybrid solar cell comprising a percolated network of single-crystalline Bi_2S_3 nanowires and P_3HT [123].

Amid global energy crisis and climate change concerns, researchers are striving to explore new green hybrid nanostructured systems with desired flexibility and performance characteristics for efficient and economical exploitation of the solar energy.

Nanotechnology-based Hydrogen Fuel Cells

The use of hydrogen based fuel cell technology offers a great deal of potential to address energy issues for sustainable future ahead. Unlike coal or nuclear power plants which produce continuous on demand oriented amounts of energy, the renewable sources are unable to provide a constant supply of energy. Owing to the strong fluctuation observed in electricity generation *via* wind and solar energy resources and the associated temporal disparity of power generation and power consumption, additional energy storage systems would be needed in future. Utilizing this excess and renewable electricity, hydrogen thus generated *via* electrolysis can be stored until its consumption. Inevitably, hydrogen based fuel cells would undoubtedly contribute towards sustainable energy supply and efficient electro-mobility in the coming future [124]. Hydrogen is ideal energy storage medium for the following reasons; abundance, high energy yield (122 kJ/g) which is higher than the corresponding value for gasoline (40 kJ/g), eco-friendly nature as well as the possibility of its storage and distribution. In general, hydrogen generation *via* solar water splitting can be categorized into three: (i) thermochemical; (ii) photobiological and (iii) photocatalytic water splitting [125]. In a typical fuel cell, the chemical energy of hydrogen is converted into electricity and water with the release of heat. This reaction is the reverse of electrolysis, where H_2O is decomposed into its constituents (hydrogen and oxygen) by the imposition of electric current [126].

Fuel cell: $2H_2$ (g) + O_2 (g) → $2H_2O$ (l) ………………….. Reaction i

Electrolysis: $2H_2O$ (l) → $2H_2$ (g) + O_2 (g)…………………… Reaction ii

Hydrogen generated through the electrolysis of water is employed as fuel in a fuel cell for producing electricity during the time of low power production or peak demand, or in the fuel cell vehicles as a clean alternative with efficiency greater than 70% (Fig. **3**). The bottleneck here is the consumption of a substantial amount of electricity, directly affecting its cost and potential commercialization. Consequently, the future of hydrogen generation is intended towards its production from the renewable sources including solar, thermal, wind energy, thermochemical cycles or biomass gasification, hence preventing electrical, heat and mechanical losses. Moreover, water splitting in the presence of nanophotocatalyst is one of the most cutting edge approaches for the direct generation of hydrogen from a primary renewable energy source on economic and technical basis [120]. The first solar-driven water splitting in a photoelectrochemical (PEC) cell was pioneered by Fujishima and Honda, using TiO_2 anode and Pt cathode for oxygen and hydrogen production. PEC water splitting is certainly the most attractive approach compared to other hydrogen production methods [127].

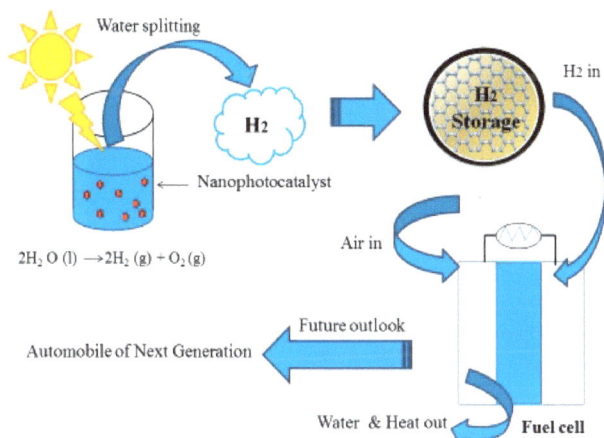

Fig. (3). Nanotechnology in hydrogen production and storage for prospects in the transport sector, adopted from Source Mao *et al.* [127a].

Over the past decade, numerous attempts have been directed towards the development of nanomaterials to achieve applicable renewable energy conversion and utilization, which is highly anticipated to contribute in the ultimate transition towards renewable energy (solar and hydrogen) based economy [127a]. Although TiO_2 has been widely used as photocatalysts for hydrogen evolution, but because of the wide bandgap of 3.0–3.2 eV, only 2–3% of the solar light could be utilized. For developing visible light driven photocatalysts for hydrogen generation, nanomaterials have shown great promise [128]. Maeda and Kazunari signified the role of modifying photocatalysts with suitable co-catalysts to improve the overall efficiency of water splitting. It is believed that these co-catalysts provide reaction sites and lower the activation energy required for gas evolution [129]. For this purpose, some noble metals (Pt, Ru, Au, *etc.*) and metal oxides (NiOx, Rh/Cr_2O_3 *etc.*) have been reported to work well as water reduction cocatalysts by entrapping electrons from the semiconductors; whereas many other metal oxides (IrO_2, RuO_2, Rh_2O_3, Co_3O_4 and Mn_3O_4) act as efficient oxidation co-catalysts by entrapping holes [127a]. In addition, co-catalyst using nanotechnology core/shell-structured NPs (such as a noble metal or metal oxide core and chromia, Cr_2O_3 shell or Ni core/NiO shell NPs) have also been used for photocatalytic water splitting [129]. Recently, nanosized silicon (10 nm diameter) has been reported to generate hydrogen at a faster rate (1000 times) than bulk counterpart after undergoing a reaction with water which is 100 times faster than the earlier reported Si structures, and considered to be 6 times faster than competing metal formulations. These results suggested that nanosilicon has the potential to provide on-demand hydrogen production without consuming heat, light or electrical energy [130].

Being highly efficient energy converters for electricity and heat using hydrogen as a clean energy resource, fuel cell provides ideal conditions for safe, efficient and eco-friendly energy available at competitive costs [131].

For the successful introduction of hydrogen as a fuel in transport sector, two basic conditions have been identified: further expansion of the retail network and an attractive price range of fuel cell automobiles. Scientists at the Institute of Vehicle Concepts German Aerospace Centre (DLR) are working with two scenarios; in the conservative scenario, a significant market ramp-up of fuel cell vehicles is carried out only after 2030, while the ambitious scenario in 2030 where already 140,000 fuel cell based cars, 6,600 truck, drive up to 900 city buses and 50 locomotives on the roads and rails of Baden-Württemberg [132].

New Methods for Hydrogen Storage

Scientists have been keenly interested in looking for greener alternatives to drive power plants, vehicles and other equipment in view of environmental challenges in the form of global warming and climate change triggered by excessive reliance on fossil fuels. In this scenario, hydrogen being cleanest burning fuel is going to be the fuel of tomorrow [133]. In an effort to routinely use hydrogen as a fuel, major emphasis has been placed on improving its manufacturing and storage. Indeed, hydrogen storage has been considered the greatest obstacle in way of its commercial utilization. Interestingly, owing to the ease and consistency with which hydrogen is accepted and released in large quantities by physisorption and chemisorption, nanostructured carbon such as carbon nanomaterials and in particular CNTs have been greatly investigated [134].

Numerous researchers have reported their work on hydrogen storage using pure as well as modified carbon nanostructures [135]. According to some reported studies, nanostructured composites made up of CNTs and metal oxides have proven to be a good hydrogen storage material. Recently, Silambarasana *et al.* reported a new single walled CNTs (SWCNTs)-tin oxide (SnO_2) nanocomposite films for hydrogen storage [136]. In addition, nanoconfinement has the potential to alter different physicochemical properties of metal hydrides, thus finding promising applications in hydrogen storage and rechargeable batteries. The reversibility of hydrogen release is eased by nanoconfining the materials in a carbon or metal–organic framework, specifically for reactions concerning multiple solid phases, *i.e.* the decomposition of $LiBH_4$, $NaBH_4$, and $NaAlH_4$ [137].

Paper Electronics

Contemporary science and technology have contributed in transforming modern society by introducing soft portable electronic devices, for example wearable devices, rollup displays, smart cards, electronic paper, chip and basic components in numerous devices. A great deal of research has been done on using paper or paper-like materials as an excellent alternative substrates for basic electronics components because of their exceptional technological attributes and commercial perspectives for various existing substrates [138]. Amongst the most intriguing electrical properties, volume and surface resistivity are important; however, the dielectric constant, dielectric loss factor, charging potential and decay rate, and dielectric breakdown strength are sometimes considered essential as well. These properties vary with relative humidity, electric field, measurement frequency and morphology, temperature and composition of the paper substrate [139]. Nanotechnology could change how the world views and uses even the simplest objects such as paper. Nowadays, paper electronics is gaining popularity as one of the fast-developing sectors of nanotechnology for applications in electronic displays and sensors. Paper is a promising material to fabricate flexible electronics for a number of reasons such as its light weight nature, portability, low production cost and availability of standard production methods. Owing to the growing demand for green technology, paper is the material of choice because it can be recycled and renewed, thus presents minimal threats to the environment. In addition, it does not need high process temperatures as required in crystalline silicon technologies. Nanotechnology researchers and engineers are currently investigating different methods to incorporate nanomaterials onto and into paper electronics using techniques such as standard screen or ink-jet printing. Using these established methods, it would be possible to integrate circuits or microcontrollers directly onto paper. These devices with its low power energy consumption would allow densely packed integrated circuits for a wide range of applications such as computer memory chips, digital logic and microprocessors, to (linear) analogue circuits, hence promoting the next-generation microelectronics revolution in information and communication technologies [140]. Three types of paper electronics have been introduced. The paper itself can have electronic or electrical properties. Secondly, electronics can be placed onto paper. Thirdly, electronics and electrics can be firmly covered up in the paper or operated on both sides of the paper [141].

Recently, a work on bacterial cellulose nanopaper (BC), a multifunctional material known for a number of valuable properties such as high surface area, flexibility, high mechanical strength, hydrophilicity, high porosity, sustainability, biocompatibility, biodegradability, optical transparency, thermal properties and broad chemical-modification capabilities has been reported. In this study, several nanopaper-based optical sensing platforms have been designed and tuned, incorporating nanomaterials (Ag, Au, CdSe@ZnS quantum dots, $NaYF_4$: $Yb^{3+}@Er^{3+}$ & SiO_2) to display plasmonic or photoluminescent characteristics that can be exploited for applications in sensor technologies [142].

Zhong *et al.* have recently demonstrated a transparent paper-based, self-powered and human-interactive flexible system for potentially useful applications in anti-theft and anti-fake systems. The underlying mechanism for this system is based on an electrostatic induction with no additional power system. This original work joins the transparent nanopaper with a self-powered as well as human-interactive electronic system and in this way leads towards the emergence of smart transparent paper electronics [143]. Advancement in paper technology as substrates in the host of electronics is anticipated to permit intelligent packages as well as applications, and considerable work is ongoing to make paper electronic device applications feasible despite numerous challenges.

Nanoproducts in Display Technology

A range of eco-friendly nanomaterials have been fabricated that have replaced other commonly used toxic and expensive materials. Newer liquid crystalline displays are compact, do not involve toxic materials (such as lead) and are energy efficient compared to the traditional cathode ray tube (CRT) based computer monitors. The use of CNTs in computer displays have reduced the negative impact of conventional displays on environment by replacing toxic heavy metals [25]. Moreover, incorporating newer nanotechnology in computer screens has resulted in decreasing material and energy demands, while offering enhanced performance that complies with the consumer needs. Field emission displays (FEDs) based on CNT (0.5 g/monitor) represent further advancement in display technology from conventional CRT based monitors [25, 144]. FED with its huge commercial market has been ranked as one of the most promising nanotechnology based product utilizing CNT as a key ingredient [145]. Another important materials that has shown great promise for next-generation displays are the particles in the range of few to several tens of nanometers termed as quasi zero-dimensional mesoscopic system, QDs, quantized or Q-particles [146]. These particles possess several desirable features such as color tunability, narrow emission and extraordinary luminescence efficiency. Novel optoelectronic properties of QDs have been attributed to the size quantization effect. QDs with controllable energy bandgaps, high quantum efficiency (QE), exquisite color purity, extended lifetime and multifunctionalities can be exploited in many potential applications, especially for display devices: Hybrid LEDs using QDs as active electroluminescent materials based on direct electron–hole injection have been thoroughly studied and the color-converting LEDs which employ inorganic InGaN LEDs with QDs as color converters have been vastly studied [147].

Electronic paper represents advanced, affordable and durable display technologies. Unlike conventional flat panel displays which use a power-consuming backlight to illuminate its pixels, electronic paper reflects light like ordinary paper and it is possible to hold the text and images on screen for an indefinite period without drawing electricity, although the image can change later on [148]. Electrophoretic image display (EPID) is one prime example of electronic paper category offering a number of desirable features such as low cost, good stability, flexibility and reliability as a replacement to conventional paper. Investigations by nanotechnology researchers have revealed that the organic electrophoretic ink NPs ensure improved electronic ink fabrication technology, resulting in e-paper with high brightness, good contrast ratio and low production price [148, 149].

Supercapacitors

There has been a growing need to develop cheaper and readily available energy sources with minimal or no harmful impact on the environment. Nanotechnology has been successful in overcoming some of the technological obstacles while exploiting various substitutes to non-renewable energies. Electrochemical capacitors (ECs), also referred to as supercapacitors or ultracapacitors are the devices used to store electrical energy. The use of nanostructured materials as NPs or nanofilms in ECs has made it possible to combine both high power and high energy performance parameters [120].

Replacement of Tinlead Solders with Newly Developed Nanomaterials

For several decades, tin-lead (Sn-Pb) solders have been widely used in electronic industry as interconnect material. However, environmental problems and health issues related to Pb being toxic have resulted in its ban by electronics manufacturing sector in many countries. These concerns called for urgent need to utilize lead-free solders as suitable alternatives to Sn-Pb based solders [150]. Several compositions incorporating NPs have been proposed as a substitute for the traditional lead-based solders to achieve desired level of performance. Among lead free solders, Sn/Ag/Cu (SAC) system has found widespread use. Herein, to enhance the performance characteristics, NPs paste formulations containing an organic matrix and a plurality of metal NPs dispersed in the organic matrix have been suggested. The NPs paste formulation maintained a fluid state and was found to be dispensable through a micron-sized aperture [151]. In one study, Nai *et al.* synthesized novel lead-free composite solders using different weight percentages of multiwall CNT into 95.8Sn-3.5Ag-0.7Cu solder. Melting temperatures of the composite solders remained unchanged whereas the mechanical properties in terms of microhardness and tensile properties were improved after the addition of CNT [152]. The nanostructured materials technology of polyhedral oligomeric silsesquioxanes (POSS) with suitable organic groups was utilized by Lee *et al.* that could promote bonding between nano-reinforcements and the metallic matrix.

Their study validated the idea of using surface active, inert nanostructured chemical rigid cages for pinning the grain boundary of solder alloys. This led to the improvement of lead-free electronic solders in terms of mechanical performance at high temperatures and service reliability [153]. The low melting point tin/silver alloy NPs with various sizes were synthesized by Jiang *et al.* for low-temperature lead-free interconnect applications [154]. Shu *et al.* synthesized low melting temperature tin/indium (Sn/In) nanosolders with varying compositions based on surfactant-assisted chemical reduction method in aqueous medium under ambient conditions [155]. In another study, Sn-Zn based solder were prepared as a possible substitute for Pb solder by Wadud *et al.* that were found to show improved mechanical properties [156]. The Advanced Technology Centre of the Lockheed Martin Corporation has reported the utility and potential of nanocopper as a replacement for SnPb and Pb-free solders that can be processed around 200 °C [157]. Recently, co-deposition of functionalized CNT along with lead free solder using electroplating process has been reported [158] and further research in this area is on-going.

Concept of Nanoenergy and Nano Generator (NG)

Since the mid of 20th century, an emerging trend in the field of electronics has been directed towards miniaturization and portability so that each individual can use a dozens to hundreds of electronic systems. But, powering these sensor networks exclusively using batteries would possibly result in environmental and health hazards. Therefore, alternate power sources are needed to assure independent, sustainable, maintenance-free and continuous operation of such miniature electronic systems utilized in ultrasensitive chemical and biomolecular sensors, nanorobotics, micro-electromechanical systems, remote and mobile environmental sensors, homeland security and even portable/wearable personal electronics. In this regard, new technologies that can harvest energy from the environment as sustainable self-sufficient micro/nano-power sources can be a potential solution. Today, this newly emerged field has been called nanoenergy which deals with the applications of nanomaterials and nanotechnology in generating electricity for operating micro/nano-systems. The concept of a nanogenerator (NG) first came in 2006 which was based on ZnO nanowire (NW) [159]. Various approaches and mechanisms have been suggested for harvesting different types of energies such as solar, thermal, mechanical, chemical, biological and magnetic energy. To utilize the available energy at any time and location, hybrid cells (HC) were developed which can transform the energy into electricity [160]. A nanosensor system is made-up of components with the capability of sensing, controlling, communicating and responding. Among the important functions of this system include; sensing, data processing and transmitting components as well as power harvesting. Using nanofabrication technologies, it is possible to design a nanosensor system which is estimated to be of small size with low power consumption, thus making it possible to exploit the energy harvested from environment to power such system for wireless and self-sustainable operation. Noteworthy, the harvested energy should be regulated and stored to maintain the functioning of nanosystems [161]. In powering micro/nano-systems, the importance of energy storage cannot be side-lined. In this regard, Wang's group initially devised a self-charging power cell (SCPC) as an extraordinary hybrid of two different processes; energy conversion and energy storage, ensuring direct conversion and storage of mechanical energy as chemical energy [162]. Inevitably, the energy sources in nature (wind and sound) and human body (heart beats, blood flow and muscle movement) can be an infinite source of clean energy for powering NG [163]. Hansen *et al.* invented a bio NG made up of a piezoelectric nanofiber nanogenerator and a flexible enzymatic biofuel cell individually arranged on a plastic substrate [162, 163]. While considering other renewable alternatives, wind energy is one of the chief environmentally friendly resources of energy [164]. NGs technology has played a significant role in efficiently harvesting wind energy. Recently, triboelectric nanogenerators (TENGs) with outstanding features such as simple design, exceptional reliability, great output power, excellent efficiency and low price have been devised [165]. Similarly, Yang *et al.*, reported a TENG design performing dual roles; a sustainable power source by means of harvesting wind energy and as a self-powered wind vector sensor system for detecting wind speed and direction [166]. In another study, a rotary structured triboelectric nanogenerator (R-TENG) has been developed by Xie *et al.* for scavenging weak wind energy from the surroundings [167]. Fan *et al.* designed a new high-output, flexible and transparent nanogenerator from transparent polymeric materials that can be used as a self-powered pressure sensor for detecting a water droplet (8 mg, ~3.6 Pa in contact pressure) and a falling feather (20 mg, ~0.4 Pa in contact pressure) with a low-end detection limit of ~13 mPa [168]. The concept of NG is extremely vital as self-sufficient alternate power source for providing sustainable operation of small electronics and therefore holds great potential.

Nano Cars

In modern-day world, nanotechnology has extended the frontiers of technological breakthroughs in automotive sector that are promisingly sustainable, safe, comfortable, eco-friendly and economically viable. Nanotechnology offers many attractive features, thus setting up the stage for a new era in the automobile industry as schematized in Fig. (**4**) [169].

Among the key features include; CO_2-free engines, safe driving, noiseless cars, self-cleaning body and windscreens [170]. Nanotechnology has been implemented in the automotive industry in the form of coatings, fabrics, structural materials, fluids, lubricants, tires, smart glass/windows and video display systems [171]. Green cars are complex products incorporating green nanotechnology in several different ways; as a constituent in the tires, chassis and windscreen. An automobile is a complex combination of numerous components that can be transformed into a greener vehicle in different ways. Green nano-electronic manufacturing has enabled the components of a car and its production to reduce energy wastage together with the monitoring and reduction of emissions by using sensors. Importantly, battery material, electrodes and overall battery features have been modified by incorporating nanotech to give us the automobile of next generation called "Nano enhanced green car" [172].

Fig. (4). Nanotechnology in automobiles, adapted from Source Hessen Nanotech [169].

Importantly, vehicles designed by employing nanomaterials can be light weight, hence more fuel efficient. Nano-enhanced cars would not only be safe (coated windshield improving visibility) and cost effective (in terms of fuel consumption) but also require little maintenance. In addition, they would also help in preventing air pollution, reduce energy use and minimize wastage [173]. Further, use of NPs enables tires last longer, ensure better grip, reduce resistance and as a result save the fuel. Some of the most important nanomaterials including nanosilica, organoclay and CNTs used as fillers are now replacing various traditional fillers like carbon black and silica in tyres [172]. The incorporation of nanomaterials in tires has the potential to save millions of gallons of gas per annum wasted by under inflated tires and at the same time would lower the rate of road accidents [174]. Nickel metal hydride or lithium ion technology is currently utilized in the batteries of electric automobiles. Still, no battery technology available is in fact satisfactory enough in making electric vehicles a viable like-for-like replacement for conventional internal combustion. Another disadvantage is related to the range and recharging durations, which make electric vehicles unsuitable for long journeys. Numerous attempts have been made to solve these problems by incorporating nanostructured materials to improve battery's performance. Various studies have suggested that modified electrolytes using NPs and nanocomposite materials significantly improved specific features of lithium ion batteries as well as other new battery technologies. Also, polymer nanocomposites, nanostructured metals, nano-enhanced sensors and power electronics can contribute in improving battery performance (weight reduction, energy efficiency, improved control and communications). A wide range of nanostructured materials have been used as new candidate materials in the fabrication of electrodes for batteries such as nanotubes, nanowire, nanopillars, NPs and mesopores [175]. In addition, the use of nanofluids as fuel additives (nanofuel) can help to increase fuel economy and bring down the production of harmful gases [176]. The research in developing nanostructured materials exploiting their size-dependent phenomena for prospective use in improving the performance of automobile is in progress. Future research would be targeted towards reducing the cost of nanomaterials for batteries and ensuring that they could stand up to the criteria of commercial applications [175].

Nano Housing

Funding in nanotechnology sector has prompted an unprecedented increase in the nanotechnology R & D across manufacturing sector and construction industry is no exception. Numerous nanoproducts such as high performance structural material/functional materials have been designed and commercialized for potential use in construction and built environment [177]. Nowadays, green buildings have created a lot of buzz in the global arena. It has been speculated that bringing nanotechnology into the buildings could transform our living style. This idea is not only going to reduce waste and toxicity, but also minimize the consumption of energy and raw material in the building industry. As a result, we would get cleaner and healthier buildings. In addition to these advantages, nanotechnology in green building would open up new doors of economic benefits for both building and nanomaterials industry. For nanotechnology based companies, green building indeed represents one of the potentially largest markets for new products and processes [178].

To sum up, Grebler and Nentwich have highlighted some of the fascinating attributes of nanotechnology and its prospects that would transform every aspect of science and technology.

1. Nanotechnology can be used to tune different materials for instance, plastics or metals with CNTs, making airplanes and vehicles lighter than before with lower energy consumption.
2. Adding nano-scale carbon black to modern automobile tires results in the reinforcement of material and minimal rolling resistance savings fuel up to 10%.
3. "Easy-to-clean" nanomaterials based coatings onto glass or other surfaces can save both energy and water required in cleaning process.
4. Nanotribological wear protection products such as fuel or motor oil additives can reduce the fuel consumption of vehicles together with extended life of engine.
5. NPs such as flow agents allow plastics to melt and cast at lower temperature.
6. Nanoporous insulating materials utilized in the construction industry would be of great help in minimizing the energy demands for heating and cooling purposes.
7. Using nanoceramic based corrosion coatings for metals, for example in vehicles can be environmentally benign substitute for toxic chromium (VI) layers.
8. Nano-scale TiO_2 can replace hazardous bromine in flame retardants [114, 179].

ENVIRONMENTAL REMEDIATION USING ENGINEERED NANOMATERIALS (NANOREMEDIATION)

The quality of environment can be restored by variety of ways. The field of study that deals with the clean-up methods in order to eliminate and/or degrade environmental pollutants in contaminated soils, surface waters, groundwater as well as sediments has been referred to as environmental remediation [180]. An upsurge in population, urbanization, migration and industrialization has continued to increase the demand for freshwater resources. According to the recent figures provided by United Nations Educational, Scientific and Cultural Organization (UNESCO), a wave of increase in global water demand estimated to be 55% has been anticipated by 2050, mainly due to the ever increasing needs from manufacturing industries, thermal electricity generation and domestic usage (Fig. **5**) [181].

Over the past decade, considerable efforts have been devoted towards fabricating and designing nanosized materials. However, a shift in the trend from "What shapes can be made?" to "What can these shapes be made to do?" has been witnessed in recent years. One of the sectors that has shown tremendous potential for exploiting functional nanomaterials is the area of environmental remediation. At present, substantial progress has been made in fabricating functional nanomaterials that has created exciting new possibilities for environmental clean-up [182]. Typically, the classification of conventional remediation techniques is equally applicable to the nanotechnology for environmental remediation, *i.e.* adsorptive *vs*. reactive and *in situ vs. ex situ*. Decontamination using adsorptive technologies involves removal of the pollutants by sequestration, whereas reactive technologies primarily involve the complete degradation of contaminants into some harmless products. *In situ* technologies are based on the treatment of contaminants on site, whereas *ex situ* is based on removing the polluted sample to a more convenient locality followed by its treatment [183]. The utilization of novel engineered materials such as NPs possessing extraordinary capabilities has

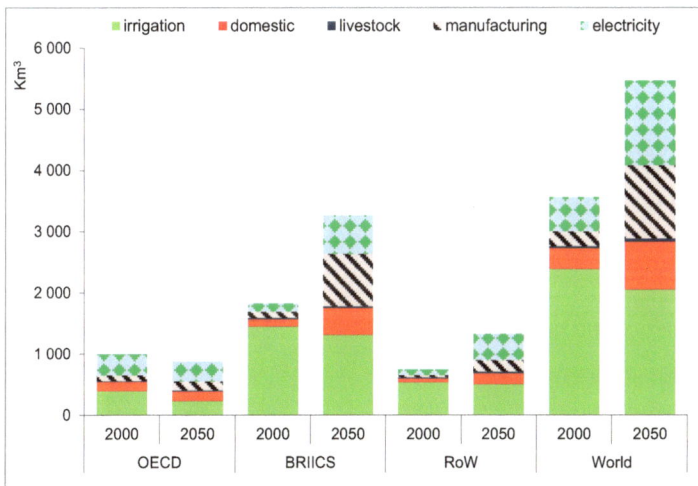

Fig. (5). Global water demand: Baseline scenario, 2000 and 2050.
Note: BRIICS (Brazil, Russia, India, Indonesia, China, South Africa); OECD (Organisation for Economic Co-operation and Development); ROW (rest of the world). This graph only measures 'blue water' (freshwater in aquifers, rivers, lakes, that can be withdrawn to serve people, for example as water for irrigation, manufacturing, human consumption, livestock, generation of electricity) demand and does not consider rainfed agriculture.
Source: OECD (2012a, Fig. 5.4, p. 217, output from IMAGE). OECD Environmental Outlook to 2050 © OECD.

gained tremendous popularity as attractive and cost-effective remediation substitute for the conventional methods [180]. It is worth stating that nanotechnology would enable scientists to establish next-generation water supply systems. Among important aspects of nanomaterials include unique electrochemical, magnetic as well as optical properties, high surface area, catalytic properties, photosensitivity, antimicrobial activity (disinfection and bio-fouling control) tuneable pore size and surface chemistry. These versatile traits provide solid ground for their application in the domain of environmental remediation for diverse uses, for instance sensors, adsorbents, solar disinfection/ decontamination and high performance membranes [184]. Nowadays, the research on nanotechnology-enabled water treatment has been categorized into four main categories; 1) adsorptive removal of contaminants; 2) breakdown of pollutants by reactive nanotechnologies (such as catalytic processes) 3) disinfection and microbial control and 4) membrane nanofiltration and desalination [185]. Amongst the unique properties of nanostructures, their low cost, dispersibility, facile surface modification and ease of fabricating supported catalyst for *in situ* as well as *ex situ* applications make these materials highly suitable for diverse applications in the area of environmental decontamination [30].

Adsorptive Removal of Pollutants

Adsorption involves the transfer of a constituent from liquid phase to solid phase where adsorbent refers to the solid, liquid or gas phase on which the molecules (pollutant) have been concentrated. The substance which has been removed from liquid phase (wastewater) is known as adsorbate (pollutant) [186]. What makes these nanomaterials highly useful and promising as adsorbents is their exceptionally high specific surface area as well as unique physical and chemical properties, thus enabling highly efficient adsorptive removal of wide variety of pollutants from wastewater.

Metal Based Nano-adsorbents

Metal oxides such as iron oxide, titanium dioxide and alumina are highly efficient and economical adsorbent materials for the removal of heavy metals and radionuclides. The sorption mechanism involves complexation between dissolved metals and oxygen in metal oxides. Interestingly, their nanosized counter-parts show remarkable removal efficiency and faster kinetics attributed to the greater specific surface area, shorter intra-particle diffusion distance and higher number of surface reaction sites (*i.e.* edges, corners, vacancies) [187]. In addition to high adsorption capacity, some iron oxide NPs, for instance nano-magnetite and nano-maghemite also exhibit superparamagnetism [188]. Because of magnetic properties, magnetic nano-adsorbents have been applied to the nano-scale magnetic separations as they can be separated and recovered from aqueous solution with the great ease. In addition, high adsorptive efficiencies of magnetite NPs make them suitable candidates for the adsorption of heavy metals and radionuclides from contaminated water. Moreover, the super-paramagnetic properties permit easy and quick separation under low magnetic field strength for the recovery and reuse of magnetic nanomaterials [189]. In such applications, magnetic NPs can be employed solely or in the form of core material in a core-shell NPs where shell performs the desired function while magnetic core allows magnetic separation as shown in Fig. (**6**) [187].

Metal based nanomaterials have been explored for the removal of variety of pollutants such as heavy metals and organic pollutants. Furthermore, magnetic NPs have also been applied to many other fields as well, particularly in analytical chemistry, for the preconcentration and removal of several harmful and risky pollutants from environmental water, soil, and biological samples [191]. Over the past several years, magnetite NPs have attracted tremendous attention by the researchers for their unusual size- and morphology-dependent physical as well as chemical characteristics, biocompatibility and remarkable magnetic properties. Nevertheless, small particle size and ease of oxidation of Fe_3O_4 NPs limit their applicability to the continuous flow systems. To deal with these problems, Fe_3O_4 NPs@ cross-linked polymeric matrix based on either natural or synthetic polymers have been fabricated for catalytic degradation of wide ranging organic pollutants in wastewater treatment [192].

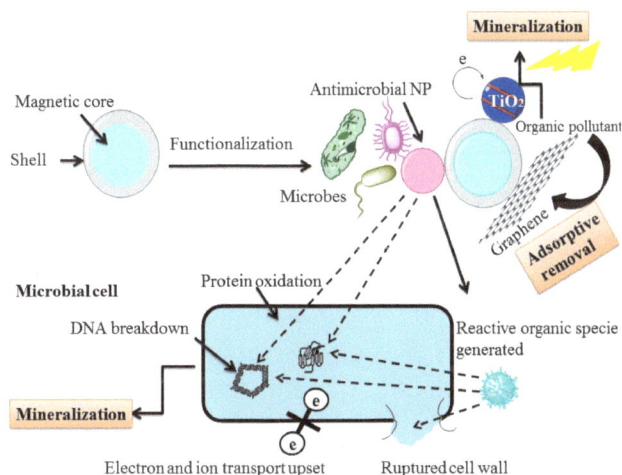

Fig. (6). Multifunctional core-shell NPs for water decontamination and disinfection: where core and silica shell would allow magnetic separation and functionalization, respectively, adopted from Sources Li *et al.*, Qu *et al.*, Brame *et al.* [185, 187, 190].

Carbon Nanostructures as Adsorbents

Wide variety of carbon nanostructures have been effectively utilized as adsorbent for trapping or separation of organic compounds due to their extremely large surface area and excellent adsorption capacity [193]. For instance, CNTs, discovered by Wiles and Abrahamson in 1978 [194] and Iijima in 1991 [195], have received special attention due to their exceptional prospects in wastewater remediation and treatment of chemical and biological pollutants. CNTs have been successfully applied as efficient adsorbent material for the removal of a large number of contaminants such as heavy metals including Cr^{3+} Pb^{2+} and Zn^{2+}, metalloids such as arsenic compounds, organic compounds such as polycyclic aromatic hydrocarbons (PAHs), atrazine, pharmaceuticals, personal care products, endocrine disrupting chemicals [196], bacteria, viruses and cyanobacteria toxins.

Also, biological contaminants, especially pathogens can be removed ascribed to the exceptional physical, cytotoxic and surface functionalizing properties of CNTs [197]. A number of attractive features such as well-defined cylindrical hollow structure, enormous surface area, high aspect ratios, hydrophobic wall and easily functionalized surfaces have enabled CNTs to be a promising candidate for the elimination of different toxic chemicals from wastewater [198]. In one study, Wang *et al.* reported the use of magnetic CNTs (MCNTs) for rapid physical separation of the oil-water emulsion [199].

Graphene, initially synthesized in 1855 [200], is considered to be a highly studied material soon after its electronic properties were revealed in 2004 [201]. It possesses unique electrical, electrochemical, optical and mechanical properties [202]. Graphene and its derivatives such as GO and rGO have been extensively investigated for pollution abasement [203]. It is considered to be the mother element of some carbon allotropic forms and as a fundamental building block for graphitic materials of all other dimensionalities which can be transformed into fullerenes, CNT, or 3D graphite *via* wrapping, rolling, or stacking, respectively [204, 205]. GO is a very important member of graphene family due to its excellent mechanical and physicochemical properties. Moreover, it can be reduced to obtain rGO by a simple chemical route. Noteworthy, the conversion of GO to rGO shows a fractional restoration of mechanical and electronic properties when compared to pristine graphene. Both forms of graphene based nanomaterials can be synthesized with great ease *via* chemical exfoliation of graphite, without the aid of complex apparatus or metallic catalysts. Graphene material thus obtained is free of catalyst residues without further purification steps being required [202, 206]. The readers are referred to an article by Ayrat M. Dimiev and James M. Tour to understand the mechanism of GO formation [207]. Graphene exhibits comparable or superior adsorption capacities than CNTs [208]. Compared to CNT, single-layered graphene materials have two basal planes available for the adsorptive removal of contaminant [202, 209]. On the other hand, in case of CNTs, the inner walls are inaccessible to the adsorbate. Secondly, GO and rGO can be synthesized through a simple chemical route based on exfoliation of graphite [202]. Prospective role of graphene and its derivatives discussed in the scientific communities was notably wide before physicists Andre K. Geim and Konstantin S. Novoselov from the University of Manchester (UK) were honoured with the 2010 Nobel Prize for their innovative work regarding graphene. The popularity of this material has skyrocketed in the recent times [210]. Tremendous work dealing with the application of graphene-based materials as adsorbents has been undertaken for the removal of metallic pollutants and dyes from wastewater [202, 211].

Moreover, the combination of magnetic NPs and graphene into a nanocomposites has become a hot topic among researchers in recent past for their new and/or enhanced incorporated functionalities that cannot be accomplished by either component solely and therefore has yielded promising results in diverse applications such as removal of contaminants from wastewater [212, 213] as highlighted in Fig. (**7**). It is believed that the resulting hybrid of magnetic NPs and graphene/graphene derivatives would present improved functionalities and performances for many applications [192]. However, a careful optimization of the synthesis step is essential for the optimal performance of graphene-based composite materials [214]. These materials have found extensive applications in the decontamination of wastewater and solid phase extraction of pollutants for environmental monitoring. Silica/graphite oxide based composite material has also been used for the removal of heavy metal ions from aqueous solutions [215]. In one study, Luoa *et al.* immobilized 2D planar graphene sheets by simple adsorption onto silica-coated magnetic micro-spheres ($Fe_3O_4@SiO_2$) to fabricate $Fe_3O_4@SiO_2$/graphene which were applied for extracting sulfonamide antibiotics from environmental water samples [216].

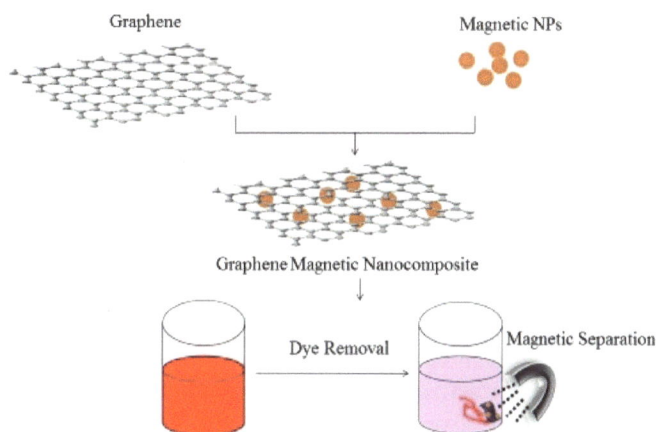

Fig. (7). Decontamination of textile effluents using graphene magnetic nanocomposite material.

Nanocrystalline Zeolites

Zeolites are inorganic crystalline porous materials with extremely ordered structure comprising of silicon (Si), aluminium (Al) and oxygen (O). The outstanding physicochemical characteristics of these materials have formed the basis for their widespread use in catalysis, separation and ion-exchange. These include high mechanical and chemical resistance in addition to their high surface area. As compared to micron-scale zeolites, the nano-scale zeolites show more external surface areas, lesser diffusion path lengths and a greater aversion to the formation of coke [217].

Nano-scale Tuneable Biopolymers

Genetic and protein engineering have emerged as a promising technology for the fabrication of nano-scale materials that can be controlled precisely at the molecular level. Recombinant DNA techniques have made it possible to build artificial protein polymers with essentially new molecular organization. The unique aspect of these nano-scale biopolymers is that they are precisely pre-programmed within a synthetic gene template and it is possible to control their size, composition and function at the molecular level and thus design protein-based nano-biomaterials with both metal-binding and tuneable properties that can be applied for the clean-up of heavy metals laden wastewater [218].

Dendrimer and Dendrimer-conjugated Magnetic NPs

Dendrimer represents a novel class of highly branched nanostructured materials in the form of globular macromolecule with three-dimensional configuration including hyper branched polymers, dendrigraft polymers and dendrons. Fundamentally, dendrimers consist of three covalently bonded components; core, interior branch cells and terminal branch cells [219]. The dimensions of dendrimers range from 2 to 20 nm with cones, spheres and disc-like morphologies. Dendrimers offer extraordinary adsorption capacities towards heavy metals [220]. Dendrimers have also been utilized as templates for synthesizing NPs including metallic (Au, Ag, Cu, Pd, and Pt), bimetallic (AuAg, AuPd, PdPt, PtAu, and PdRh) and semiconducting (CdS) based NPs [221]. Poly (amidoamine) (PAMAM) dendrimers are highly branched macromolecular compounds representing a new class of nanomaterials that can function as water-soluble chelators. They are made up of three main structural components; a core, interior repeating units and terminal functional groups [222]. Environmental applications of PAMAM dendrimers were initially described by Diallo in 1999 for the elimination of copper ions from wastewater [223]. The utilization of PAMAM dendrimers as environmentally benign alternatives for remediation has been reported for the adsorptive removal of heavy metals from water and soil [220, 223, 224]. They can be synthesized from rather inexpensive and easily available materials and have been considered essentially non-toxic. In particular, hydroxyl and amine-terminated PAMAM dendrimers could possibly be regarded as the most suitable chelating agents for metal complexation applications. In a recent study, PAMAM dendrimers with ethylenediamine cores immobilized on TiO_2 were evaluated as novel metal chelation material for the chelation and elimination of Cu (II), Ni (II) and Cr (III) from synthetic solutions as model pollutants [225]. In another study done by Chou and Lien, efficient and recyclable dendrimers-conjugated magnetic NPs were fabricated, combining the advantages of both components for application in removing Zn (II) from aqueous media [226].

Self-assembled Monolayer on Mesoporous Supports (SAMMS)

Progress in the field of nanostructured ceramics predominantly in the area of mesoporous silica has provided a powerful and versatile foundation for the fabrication of extraordinary surface area materials such as heterogeneous catalysts, environmental sorbents, sensors and molecular recognition materials [227]. By lining the pore surfaces of mesoporous silica with self-assembled monolayers of organosilanes terminated with chemically selective ligands give rise to a promising new class of heavy metal sorbents known as self-assembled monolayers on mesoporous supports (SAMMS) [182, 227, 228]. Silane-based self-assembled monolayers provide a simple and straightforward way of chemically transforming mesoporous ceramic oxide surfaces. These nano-hybrid structures are indeed considered to be highly effective adsorbent materials and their interfacial chemistry can be fine-tuned for selective sequestration of a specific contaminant [227]. As a result of attaching a monolayer of molecules to mesoporous ceramic support material, SAMMS with highly ordered nanostructure is obtained owing to three molecular-self-assembly stages. Herein, the first stage involves aggregation of surfactant molecules to make a micelle template. The second step (Fig. **8**) comprises the aggregation of silicate-coated micelles into mesostructured body, followed by the third phase involving self-assembly of silane molecules as an ordered monolayer structure across the pore interface [229].

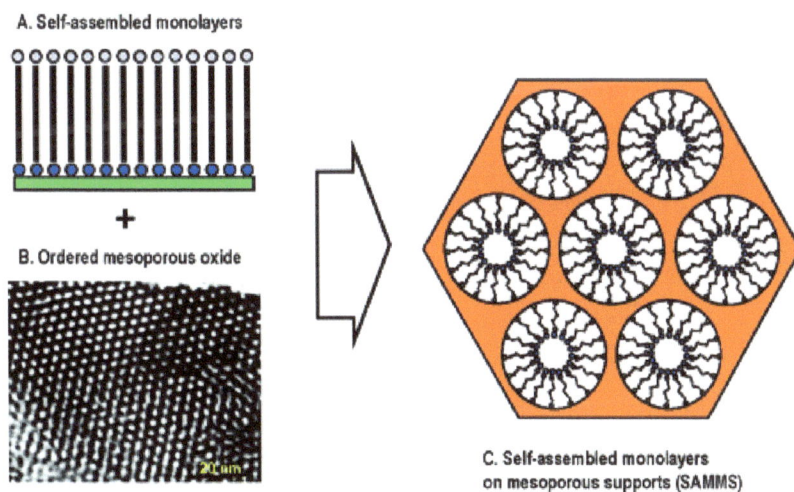

A. Self-assembled monolayers

+

B. Ordered mesoporous oxide

C. Self-assembled monolayers on mesoporous supports (SAMMS)

Fig. (8).   Showing a schematic illustration of the SAMMS material that is a combination of the self-assembled functional monolayers (Fig. **1A**) and mesoporous oxides (Fig. **1B**) showing a transmission electron microscopy micrograph of the mesoporous silica copyright American Chemical Society Fryxell *et al.* [182].

As shown in Fig. (**8**), ceramic structure bears a striking resemblance to a hexagonal honeycomb [182]. Notably, high surface area associated with mesoporous ceramic and high population density of binding groups build a great loading capacity in the SAMMS material. The rigid ceramic backbone gives robust physical stability and prevents pore closure due to solvent swelling. The dense monolayer coating ensures higher capacity. The close proximity of the binding sites permits multi-ligand chelation of the target specie in the monolayer interface, increasing the binding affinity and chemical stability of the metal laden adduct. Furthermore, open pores structure promotes facile diffusion into the mesoporous matrix, thereby allowing faster adsorption kinetics. All these attributes make SAMMS potentially very useful as sorbent with unprecedented binding affinities, excellent chemical selectivity and exceptionally fast sorption kinetics [228c, 230]. The monolayers formed within the porous surfaces in fact adsorb or bind target molecules. It is noteworthy to mention that as highly sorbent materials, SAMMS have the potential to bind different types of molecules due to their capability to alter exposed functional group of the monolayer [219]. For environmental remediation, the mesoporous silica functionalized with mercaptopropyl group, aminopropyl group, amine group and aminoethyl-aminopropyl group have been investigated to remove different heavy metals [231]. SAMMS have been applied as a useful and promising sorbents for the selective removal of heavy metals from ground-water, oils and chemical-warfare agents, greenhouse gases from air and radionuclides from nuclear waste [232]. A number of different SAMMS classes have been introduced and the first one which has widely been investigated is the thiol-SAMMS. These were designed primarily for impounding mercury including other metallic cations as well such as silver, cadmium, lead and thallium. In addition to the removal of mercury from wastewater, thiol-SAMMS have been applied for the clean-up of mercury, cadmium and lead contaminated soil as well [233]. Furthermore, a number of different SAMMS classes have been designed for the removal of Anions [234] Cesium [228c], Radioiodine [235], Lanthanide [227], Actinide [182], Arsenic [236] and Chromium [237].

The key feature of SAMMS includes fine tuning of interfacial chemistry of these materials to selectively sequester specific target specie [238]. Recently, SAMMS have been applied for the removal of Co, Sr and Cs from aqueous solutions by Park *et al.* The single-and bi-solute competitive sorption of Co, Sr and Cs onto SAMMS were explored and experimental data was fitted to several models including Freundlich, Langmuir, Dubinin-Radushkevich (D-R) competitive Langmuir, modified extended Langmuir and P-Factor models [239].

DEGRADATION OF POLLUTANTS BY REACTIVE NANOTECHNOLOGIES

Photocatalysis

Although, a range of different physical and biological methods have been employed for environmental clean-up but using the combination of clean renewable solar light and highly efficient photocatalysts for the degradation and removal of different pollutants have attracted considerable attention in the recent years [240]. Perhaps the most prominent technology in the domain of environmental remediation has been the heterogeneous photocatalysis for purifying air and water [241]. It offers many advantages over other clean-up methods; the use of environmentally friendly oxidant (O_2) and highly efficient oxidation of even low concentration level of organics at room temperature and pressure [242]. Moreover, it is possible to completely mineralize organic pollutants into H_2O, CO_2 and other non-toxic inorganic compounds without causing secondary pollution [243]. Nano-crystalline photocatalysts are ultra-small semiconductor particles with the dimensions in the nano-range. Over the years, photochemistry of nanomaterials has been considered as one of the fastest growing research areas in physical chemistry due to their unique photo-physical and photocatalytic properties [244]. Taking up the role of catalysts, magnetic nanostructures have often been considered as very competitive for environmental remediation due to a number of attractive features such as easy availability, environmentally friendly nature, economy and ease of retrieving them from the reaction medium after use. Additionally, surface modifications with reactive non-toxic compounds transform them into highly efficient catalysts capable of removing recalcitrant organic contaminants from various media [245]. In general, catalytic NPs including nanosized semiconductor materials such as nano-TiO_2, ZnO, CdS and WO_3, zero-valence metal (Fe^0, Cu^0 and Zn^0, Sn^0), bimetallic NPs (Fe/Pd, Fe/Ni, Fe/Al, Zn/Pd), polymer based nanocomposites, mixed oxides and single-enzyme NPs (SENs) have been commonly employed [229, 240a, 246]. Below, we will be discussing various types of nanostructured materials and their heteroarchitectures that have been used as photocatalyst for the degradation of contaminants present in polluted environment.

Semiconductor Nanomaterials

Photocatalysis based on semiconductors or their heteroarchitectures has found widespread use for the treatment of contaminated water because of complete degradation of wide variety of organic pollutants under ultraviolet (UV) light. Nevertheless, the sun is a rich source of photons in which UV and visible light accounts for 5% and 45%, respectively. Consequently, to use solar energy to its greatest advantage, the development of highly effective visible-light driven photocatalyst has become an actively growing area of research in photocatalysis for environmental applications [247]. TiO_2 has been widely used for preparing different types of photocatalysts because it is less toxic, chemically stable, cost effective and abundant [248]. In principle, after the absorption of UV photons by TiO_2, an electron/hole pair is generated, which either migrates to the surface and form reactive oxygen species (ROS) or undergoes electron hole recombination. The efficiency of photocatalysis of nano-TiO_2 can be increased by various strategies; morphological modifications, such as increasing surface area and porosity as well as variety of chemical modifications by incorporating additional components in the TiO_2 structure [249]. These strategies involve particle size and shape optimization, reducing electron hole recombination by noble metal doping, maximizing reactive facets and surface functionalization to enhance adsorption ability of pure TiO_2. Interestingly, TiO_2 nanotubes were found to be more effective than TiO_2 NPs in degrading different types of organics [250]. At the present time, research related to TiO_2 based photocatalysis is mainly focused on the shifting of excitation spectrum of TiO_2 from UV to the visible region. A general approach involves the doping of metal impurities, dye sensitizers, narrow band-gap semiconductors, or anions into nano-TiO_2, thus forming hybrid nanostructures or nanocomposites [187, 251]. Graphene based hybrid nanocomposites of TiO_2 have also been synthesized showing enhanced activity as photocatalysts [252].

Mixed Oxide Nanocomposites

Fundamentally, mixed oxides based catalysts have proven to be superior in performance compared to the pure catalyst. In one such study, mixed oxide catalysts based on TiO_2/SiO_2 were found to be far more efficient than pure TiO_2 [240a]. Recently, mixed oxide nanocomposites, *i.e.* $TiO_2/SiO_2/Co$ with hydroxyl propyl cellulose showing excellent photocatalytic activity have been prepared [253]. The photocatalytic degradation of organic compounds of olive mill wastewater was investigated by Ruzmanova *et al.* using core–shell $Fe_3O_4/SiO_2/TiO_2$ NPs as catalysts [254]. It has been implied that the interaction between TiO_2 and SiO_2 in mixed oxide creates new catalytic active sites. Moreover, TiO_2–SiO_2 mixed oxides display much higher thermal stability, good adsorption capability and improved redox properties [240a]. In recent years, it has been revealed that nano ZnO is superior as compared to TiO_2 attributed to its lower cost, higher quantum yield, environmental benign nature and biodegradability [255]. However, a great deal of effort has been devoted towards enhancing the photocatalytic efficiency of ZnO by reducing the rate of photogenerated electron-hole pair recombination. Research has confirmed that ZnO-based heteroarchitectures, such as metal (Ag, Au, Pt, *etc.*)/ZnO and semiconductor (TiO_2, SnO_2, CdS, NiO, *etc.*)/ZnO hybrid materials could evidently suppress this undesirable phenomena by mutual transfer of photogenerated electrons or holes in the composite material [256].

Ferrites as Catalysts and/or Support

Over the years, magnetic NPs have found widespread use as photocatalysts in environmental remediation [257]. Magnetic separating technology has gained tremendous popularity because it offers a very easy route for removing and recycling magnetic particles *via* external magnetic field. Presently, ferrite (MFe_2O_4) NPs are the focus of attention because of their inherent strong magnetic properties, thereby allowing repeated magnetic separation along with substantial cost saving. Moreover, it is possible to enhance the photocatalytic performance of ferrites by coupling it with the variety of solid matrixes such as semiconductor materials. In this regard, graphene nanomaterials have been used as a support for these catalysts [258]. In particular, the coupling between magnetic NPs and semiconductor photocatalysts, such as ZnO and TiO_2 especially as core–shell structures has also gained significant attention for their outstanding performance [259].

Carbon Nanomaterial and Related Heteroarchitectures

Multi-walled carbon nanotubes (MWNTs) have shown great potential as catalyst supports or catalyst carriers due to their great surface area, extraordinary chemical stability, special aperture structure and good electron conductivity. Xiong *et al.* used MWNTs supported nickel ferrite as highly efficient photocatalysts attributed to the synergistic effect between $NiFe_2O_4$ and CNTs for the degradation of phenols under UV light [260]. In addition, a range of graphene-based hybrid nanomaterials have been extensively used for photo-degradation of pollutants from wastewater including $CoFe_2O_4$@graphene [261], rGO–MFe_2O_4 (M = Mn, Zn, Co, and Ni) hybrids [262], $Bi_5Nb_3O_{15}$@graphene [263], $ZnFe_2O_4$/ZnO-graphene [264], $NiFe_2O_4$-graphene [261], $ZnFe_2O_4$–graphene [265], $CuFe_2O_4$–graphene [247], $MnFe_2O_4$- graphene [266] and ZnS NPs-GO composites [267].

Polymer-based Nanocomposites (PNCs)

In principle, PNCs incorporate the unique features inherent in NPs and polymers. These composite materials have received considerable attention from both academia as well as industry. Interestingly, the PNC retains the innate features of NPs, while the polymer matrix provides stability, processability and some enhancement due to the interaction between NPs and matrix. NPs have remarkably large surface area to volume ratios and high interfacial reactivity which give distinctive physical and chemical properties to the resulting hybrid material. In addition, separation and recycling aspects considered as the bottleneck in applying NPs for remediation can also be solved [268]. Notably, chitosan possesses high chemical reactivity and favorable particle- and film-forming property, thus it has been used as a host material for anchoring NPs such as CdS QDs /chitosan based composites as visible light driven photocatalyst. The resultant hybrid consists of CdS QDs and chitosan in which both the components complement each other and hence, it shows great promise in the photocatalytic treatment of contaminated water [268b, 269].

Photocatalytic Self-assembled Nanostructures

Self-assembly processes present the most powerful way of fabricating complex materials with unique structural and compositional sophistication [270]. The self-assembly technique has also been applied in preparing catalysts as thin films on a support material. The application of such films in a liquid phase is useful for various reasons: (i) coating a film of catalyst onto an inert support permits more even distribution of light and reduces the hydrodynamic pressure drop problems in case of unsupported particulate catalysts. (ii) ease of separation after the reaction is completed. (iii) immobilized semiconductor nanocrystals increase the effectiveness of photooxidation [271].

Besides, a number of methods exist for the preparation of these films but colloid chemical methods being less energy consuming are highly preferred, therefore more economical as precursor materials in solved or suspended state in aquatic medium [272]. An easy and appropriate approach of preparing ultra-thin films with precise thickness adjustment is the layer-by-layer self-assembly approaches. Szabo *et al.* reported a new process for synthesizing $Zn(OH)_2$ and ZnO NPs [271]. These nanocrystals were self-assembled with a layer of silicate to fabricate multilayer films. The photocatalytic activity of fabricated nanocatalysts films were investigated by monitoring the photodegradation of model organic pollutants namely Kerosene and β-naphthol. Xiao *et al.* prepared stable aqueous dispersion of polymer-modified graphene nanosheets (GNs) *via in situ* reduction of exfoliated graphite oxide in the presence of cationic poly (allylamine hydrochloride) (PAH). The resultant water-soluble PAH-modified GNs (GNs-PAH) and tailor-made negatively charged CdS QDs were used as nano-building blocks for sequential layer-by-layer (LbL) self-assembly of well-defined GNs–CdS QDs hybrid films. It was found that as-prepared GNs–CdS QDs multi-layered films showed improved photoelectrochemical and photocatalytic activity under visible light irradiation as compared to the native CdS QDs and GNs films [273].

A new kind of TiO_2-graphene based multifunctional nanocomposite hydrogel

(TGH) were prepared by a facile one-pot hydrothermal method by Zhang *et al.* for environmental and energy applications; photocatalyst, reusable adsorbents and supercapacitor. It was found that the resultant three-dimensional (3D) network of TGH showed synergistic effects of the assembled graphene nanosheets and TiO_2 NPs, hence displayed unusual physical and chemical properties such as high adsorption capacities, improved photocatalytic activities, and enhanced electrochemical capacitive performance as compared to the pristine graphene hydrogel and TiO_2 NPs [274]. In another study, self-assembled CdS–nanoporous TiO_2 nanotube array (CdS–NP-TNTA) hybrid nanostructures were prepared by Xiao *et al.* The as-prepared CdS–NP-TNTA hybrid nanostructures exhibited promising visible-light photoactivity and improved photostability [275]. Following self-assembly approach, CdS nanowires–rGO nanocomposites (CdS NWs–RGO NCs) were successfully synthesized combining one-dimensional (1-D) with two-dimensional (2-D) structures followed by a hydrothermal reduction process by Liu *et al.* The photocatalytic activity of the as-prepared nanocomposite was evaluated by selective reduction of nitrophenols in water using visible light irradiation. Compared to CdS nanowires (CdS NWs), CdS NWs–RGO NCs exhibited significantly enhanced photoactivity [276].

Photocatalysis is considered as a simple, low-cost and environmental friendly approach for degrading wide variety of organic contaminants. Nevertheless, researcher are facing several technical obstacles on large scale application of photocatalysis such as i) catalyst optimization to improve quantum efficiency or to efficiently use visible light; ii) effective photocatalytic reactor design and catalyst recovery/immobilization procedures [187].

Nano-scale Zero-valent Iron Particles as New Generation Technologies for the Transformation and Decontamination of Environmental Pollutants

The use of nano-scale zero-valent iron (nZVI) for remedying contaminated groundwater and soil is a promising example of engineered nanomaterials-assisted environmental clean-up. In a typical process, oxidized iron upon exposure to the air transforms into rust quite easily that can convert trichloroethylene (TCE), carbon tetrachloride and dioxins into simpler and less toxic carbon compounds [277]. This approach has gained major consideration as sustainable application of nanotechnology for *in situ* remediation in the past decade due to several reasons. Firstly, nZVI can offer a faster clean-up route compared to conventional methods because of the higher degradation rates of pollutants. Secondly, nZVI can be used for treating different categories of environmental contaminants including polycyclic aromatic hydrocarbons (PAHs), pesticides, heavy metals and many other toxic chemicals. Thirdly, these NPs may be applied for *in situ* use in areas that are difficult to access. Lastly, it might be potentially more cost-effective compared to other methods [180]. Mechanism of action of nZVI can be understood by considering the transformation of pollutant such as tetrachloroethene (C_2Cl_4), a commonly used solvent. In a typical redox reaction, C_2Cl_4 can readily accept electrons from nZVI and get reduced to ethene [278] as shown below;

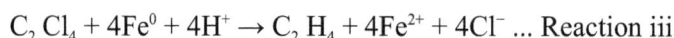

$$C_2Cl_4 + 4Fe^0 + 4H^+ \rightarrow C_2H_4 + 4Fe^{2+} + 4Cl^- \text{ ... Reaction iii}$$

High surface area together with enhanced reactivity make nZVI an excellent candidate for the degradation of contaminants present in soil and water. Besides, other important aspects related to nZVI include; low value of standard reduction potential, favorable quantum size properties and potential rise in transport efficiency through groundwater's underground matrix [279].

Three approaches have been followed for applying nZVI's in groundwater remediation; (i) conventional 'permeable reactive barrier' made with millimeter-sized construction-grade granular iron. (ii) 'reactive treatment zone' formed as a result of sequential injection of nanosized iron to form overlapping zones of particles adsorbed onto the grains of native aquifer material and (iii) treatment of dense nonaqueous phase liquid (DNAPL) contamination by injecting mobile NPs [183].

Enzymatic Biodegradation using Nano-scale Biocatalysts

Enzymatic reactions have been considered environmentally as well as user friendly [280]. Therefore, enzymatic biodegradation is considered as an attractive technology for the clean-up of contaminated environment [281]. The dimensions of enzyme molecules are of the order of nanometer, thus making it likely for enzymes to take advantage from the high surface area to volume ratios of nanomaterials. The use of NPs as enzyme supports was initially reported in late 1980s. It is noteworthy that high surface area of nanomaterials has been the principal driving force for developing nano-scale biocatalysts. Materials of varying composition, shape and structure with modified surfaces have been employed as a support for enzymes, as shown in Fig. (**9**), in particular, enzymes incorporated/encapsulated into nanostructures [282].

SENs

Enzymes offer vast capabilities due to their high specificity and targeted effectiveness. Unfortunately, unstable nature and relatively shorter life span associated with enzymes have hindered their widespread use. Nanotechnology has provided a new way to stabilize enzymes using enzyme containing NPs called

SENs or armoured enzyme NPs which have been regarded as chemically stable and environmentally persistent [284].

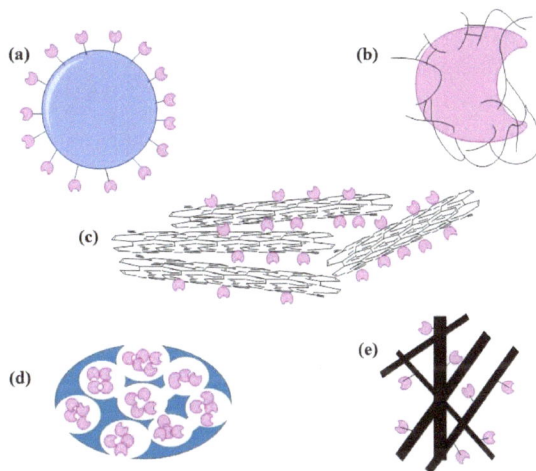

Fig. (9). Basic assemblies of nanosized biocatalysts, adapted from Sources; Wang [282] and Kim *et al.* [283] (**a**) NPs with surface-attached enzymes. (**b**) SENs (**c**) CNTs–enzyme hybrid. (**d**) Nanoporous matrix with entrapped enzymes. (**e**) Nanofibers carrying enzymes.

Each enzyme molecule is enclosed in a porous composite organic/inorganic network of less than a few nanometer dimensions. The porosity of network surrounding the enzyme allows the substrates to have an easy access to the active sites of enzyme [283]. Favourably, the nanostructured silicate cage linked to the surface of enzyme protects its catalytic function for months compared to unprotected enzymes which are active for few hours only [285]. SENs can be highly useful to treat recalcitrant compounds as they can withstand more harsh conditions such as high contaminant concentration, high salinity, extreme pH and temperature [286]. Kim *et al.* assembled the first SENs in 2003 using chymotrypsin as a model enzyme. The collection of enzymes to choose from would allow the potential remediation of a wide range of organic contaminants such as dyes, polyaromatics, phenols, chlorinated compounds, organophosphorus pesticides or nerve agents and explosives [284b]. In another investigation undertaken by Yan and team, single enzyme encapsulated into nanogels were prepared with uniformed size and controllable shell thickness by surface acryloylation of protein followed by *in situ* polymerization in aqueous media. Compared to its free equivalent, the encapsulated protein exhibited comparable biocatalytic behaviour and high stability at elevated temperature in the presence of organic solvent [287]. The basic idea of SENs has been extended to fabricate magnetic SENs encapsulated within a composite inorganic/organic polymer network *via* surface modification and *in situ* aqueous polymerization of separate enzyme molecule. Further, the evaluation of catalytic behaviour was conducted using glucose oxidase (GOD) [288]. The SENs can be deposited as films or immobilized onto the solid support material. Owing to their small size, they can penetrate and immobilize within nanostructured or nanoporous media, thereby creating hierarchical architectures. It is possibly to link them with other NPs or molecules as a part of multifunctional nano-assemblies [289].

Vault NPs Packaged with Multiple Enzymes

Although encapsulation has been regarded as a standard method to improve enzymatic stability but it can also lead to increased substrate diffusion resistance, lowered catalytic rate and higher apparent half-saturation constants. Recently, vault NPs packaged with enzymes were synthesized as suitable agents for efficiently boosting enzymatic stability and pollutants degradation. Interestingly, assembled vault NPs can simultaneously enclose multiple enzymes due to their hollow core structure. It was found that vault-packaged with enzymes can exhibit higher phenol biodegradation (3 times) in 24 h than did the corresponding unpackaged one without compromising on catalytic activity [281].

DISINFECTION AND MICROBIAL CONTROL

According to World Health Organization (WHO), 663 million people do not have access to clean water resources [290]. Water, sanitation and hygiene have a significant impact on human health. Water-related diseases include; those due to micro-organisms and chemicals in drinking water, diseases such as schistosomiasis which have part of their lifecycle in water, diseases such as malaria with water-related vectors and others such as legionellosis taken by aerosols containing micro-organisms. Inadequate drinking-water, sanitation and hygiene have estimated to cause 842 000 diarrheal disease deaths per annum (WHO 2014). Globally, it is estimated that 4% of the disease burden could be prevented by improving water supply, sanitation, and hygiene [291]. Absence or malfunctioning of centralized water treatment system in developing countries has given rise to the instant need for point of use (POU) treatment technologies such as solar disinfection (SODIS), filtration and boiling. Unfortunately, aforementioned technologies suffer from recontamination hitches. Although, chlorination brings about residual disinfection but unfortunately it ends-up in the formation of hazardous disinfection byproducts (DBPs) [292].

On the other hand, there is enormous potential for nanotechnology in providing alternative solution to obtain germ free drinking water. The availability of clean water has become increasingly challenging on account of explosive rise in population, ever growing demand for clean water and existence of additional pollutants. One of the attractive strategy offered has been the "antimicrobial nanotechnology" [293]. Li *et al.* have stated that several nanomaterials demonstrate strong antimicrobial activity *via* diverse mechanisms as illustrated in Fig. (**6**), such as (i) photocatalytic generation of reactive oxygen species that can damage the cell components and viruses (TiO_2, ZnO and fullerol), (ii) compromising bacterial cell envelope (peptides, carboxyfullerene, chitosan CNTs, ZnO and Ag NPs), (iii) interruption of energy transduction (Ag and aqueous fullerene NPs) and (iv) inhibiting enzyme activity and DNA synthesis (chitosan) [293]. But predominant antimicrobial mechanisms (Fig. **6**) involve either direct interaction of NPs with the microbial cells, *e.g.* upsetting transmembrane electron transfer, disrupting/penetrating the cell envelope, or oxidizing cell components, or the production of secondary products, *e.g.* ROS or dissolved heavy metal ions that can kill microorganisms.

The properties exhibited by an ideal disinfectant are: (i) broad spectrum antimicrobial properties at ambient temperature within short time; (ii) no harmful by-products produced during and after their use; (iii) nontoxicity of the disinfectant; (iv) cheap and easily applicable; (v) easy to store, high solubility in water and must not be corrosive towards equipment or surface; as well as (vi) the possibility of safe disposal. For intended use as nano-disinfectant, nanomaterials should comply with the above mentioned properties along with: (i) physical (no aggregation or settling problems) and chemical stability in aqueous media; (ii) if photo-excitation is mandatory, NPs must be active under solar light irradiation [190].

Antibacterial NPs discussed here can be categorized into three types; (i) naturally occurring antibacterial substances, (ii) metals and metal oxides and (iii) novel engineered nanomaterials [185]. Several nanostructured materials such as nano-TiO_2, nano-Ag, nano-ZnO, nano-Ce_2O_4, CNTs and fullerenes show antimicrobial properties without involving strong oxidation, therefore have lower tendency to yield toxic DBPs [185, 294]. Antimicrobial nanomaterials have been explored to meet three important challenges in water/wastewater systems, *i.e.* disinfection, membrane bio-fouling control and bio-film control on other surfaces. In this regard, novel CNT based filters have been employed for the removal of both bacteria and virus owing to their antimicrobial properties, fibrous shape and high conductivity (removing bacteria by size exclusion and viruses *via* depth filtration). Afterwards, retained bacteria are inactivated by CNTs within hours. Nano-Ag has also been incorporated into ceramic microfilters to act as a barrier for the pathogens. This strategy is highly beneficial especially in the remote areas of developing countries [187]. Nano-Ag has been a priority in the POU water treatment technologies for its strong and wide-spectrum antimicrobial activity and low toxicity towards humans [184]. Researchers are in tenacious quest for novel multifunctional nanomaterials in order to upgrade the existing disinfection infrastructure into highly efficient system capable of rapidly removing both organic and inorganic contaminants from wastewater [190].

MEMBRANE NANOFILTERATION AND DESALINATION

The process of removing solids by allowing wastewater to pass through a medium that blocks the particulate materials has been referred to as filtration. In this process, porous medium can be in the form of a physical barrier and/or chemical/biological process capable of removing macroscopic particles. Adversely, these systems have failed to efficiently filter out microscopic particles and microorganism. However, ongoing investigations led to the emergence of innovative filtration technologies in the form of microfiltration, ultrafiltration and nanofiltration that have played crucial role in solving aforesaid issues [295]. Membrane is referred to as a semi-permeable and selective barrier in between two phases (retentive and permeate) which only permits selected chemical species to diffuse across it [296]. Membrane technology is considered as a vital component of an integrated water treatment and reuse paradigm. This system has been widely applicable because of its capability to remove wide range of contaminants and permits the use of non-conventional water resources such as brackish water, seawater and wastewater. Moreover, membrane technology lends itself to automation, requiring little land use, less chemical consumption and the modular configuration being well adaptive at various system scales [184].

Nanotechnology researchers are on non-stop expedition to fabricate new membranes for water treatment, desalination and water reclamation based on nanomaterials including NPs made-up of alumina, nZVI and gold, among others [297]. The performance of membrane system is a function of materials from which these membranes have been constructed. Advantageously, the incorporation of functional nanomaterial into membrane improves its permeability, fouling resistance, mechanical and thermal stability. Moreover, it also renders some new functions to the membranes such as degradation of pollutants and self-cleaning. Four membrane nanotechnologies have shown great promise in water treatment and reclamation namely, nanofiber membranes, nanocomposite membranes, thin film nanocomposite (TFN) membranes and biologically inspired membranes [187]. Nanofibers and nanobiocides provide opportunity to further enhance the quality of water filtration membranes. To deal with membrane fouling problem which affects water quality, the inhibition of bacteria causing it can be accomplished by using surface-modified nanofibers. In this regard, both silver NPs laden polyvinyl alcohol (PVA) and polyacrylonitrile (PAN) nanofibers have shown excellent antimicrobial activity, with PVA nanofibers reducing 91-99% of the bacteria in contaminated water. On the other hand, PAN nanofibers demonstrated 100% efficiency in killing bacteria [293].

Desalination has long been considered as a feasible option for providing clean drinking water to the people settled in many deserted, coastal and remote areas. However, it has been unfavourably addressed as the most energy-intensive water treatment technology, offering limitations in terms of high cost and energy consumption. Importantly, membrane-based desalination has been extensively promoted as main-stream desalination technology but the associated impact on environment and resource depletion in terms of pre-treatment and recovery of membrane-based desalination are the matter of serious concern. Application of nanotechnology in the fabrication of nano-enabled membranes has been predicted to counteract the limitations associated with current materials and processes, aiding to optimize the performance of methods used in desalination, hence leading to an overall improvement in separation and productivity. Also, the application of nano-enabled membranes in desalination plants can significantly reduce the capital and operating expenses [298]. Researchers from the University of California Los Angeles (UCLA) have designed a nanomembrane based on cross-linked matrix of polymers and engineered NPs which can be used for sea water desalinization and wastewater treatment. Compared to conventional reverse osmosis membrane, the newly designed nanomembranes are less prone to clogging, thus increasing the lifetime of membranes along with distinct economic benefits as well [277]. Interestingly, the progress made in functional nanomaterials and their integration with conventional technologies have opened up the gateways towards remarkable prospects in designing nanotechnology-enabled multifunctional processes, *i.e.* water disinfection, decontamination and separation as one unit. Multifunctional system is capable of improving the overall performance of a system by creating synergy, preventing redundancy, simplifying the operation along with lowering the system footprint as well as cost. As different nanomaterials capable of performing diverse functions can easily be merged together on a very small carriers (nanofibers), thus it would be possible to design multifunctional systems employing these novel heteroarchitectures [184]. Multifunctional photocatalytic membranes are superior when compared to freely suspended NPs due to the ease with which they can be separated from purified water [299]. Much work has been carried out in developing reactive membranes, usually nano-TiO_2 based polymeric and ceramic membranes, for the simultaneous separation and breakdown of contaminants with lesser membrane fouling. In addition, nanomaterials such as nano-Ag, nano-TiO_2, nano-alumina and unaligned CNTs have been incorporated into the membranes *via* surface self-assembly or addition to the membrane casting solution to control the fouling problem associated with membrane as well as other surfaces in water treatment, storage and distribution systems [300]. Graphene-based nanomaterials represent another promising example of membrane materials for advanced desalination technologies [301].

ENVIRONMENTAL MONITORING (NANOTECHNOLOGY ENABLED SENSORS)

IUPAC defines environmental monitoring as *"the continuous or repeated measurement of agents in the (working) environment to evaluate environmental exposure and possible damage by comparison with appropriate reference values based on knowledge of the probable relationship between ambient exposure and resultant health effects"* [302]. Time has proven that the revelation made by Rachel Carson in her renowned book "Silent spring" was just a prelude to a stormy future ahead in which toxic substances would influence humankind to an unprecedented level [303]. Environmental monitoring involves the measuring aspect of environment in a repetitive way so to understand its structure and functioning. Over the course of history, humankind has exploited and modified their surrounding environment for food and shelter, becoming more and more aware of the significance to manage and preserve the quantity and quality of natural resources to sustain their livelihood. It is possible to manage our environment wisely if we understand it and use the acquired knowledge in multiple ways. The purpose of environmental monitoring is to save ecosystem and protect different organisms from extreme exposure to toxic chemicals that are dumped into the environment [304]. In recent times, this aspect has become even more critical with rapid rise in human population and expeditious pace of human activities, thus adding ever increasing strains onto the environment. (Fig. **10**) highlights some knowledge based regulations along with some benefits of environmental monitoring [305].

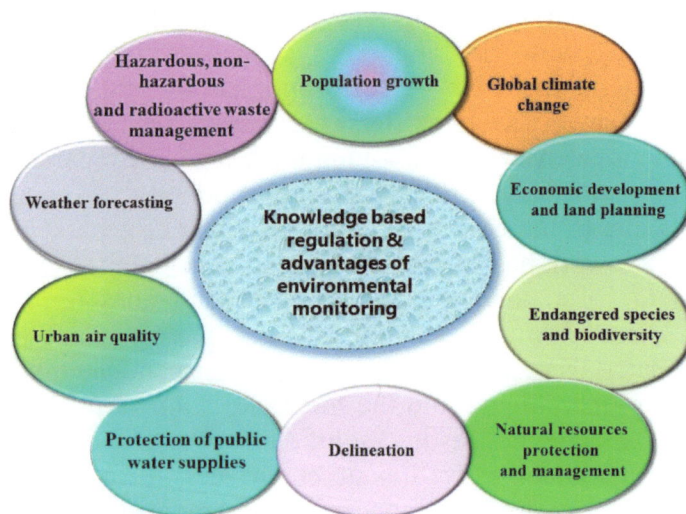

Fig. (10). Knowledge based regulations and benefits of environmental monitoring (adopted from Source; Environmental Monitoring and Characterization [305]).

Various techniques have been used for the analysis of water, air, soil and biological fluids such as chromatographic methods (gas chromatography, high performance liquid chromatography and ion chromatography), spectroscopic techniques (UV-Visible, atomic absorption, atomic emission and Fourier transform infrared spectroscopy), luminescence spectroscopy (fluorescence and phosphorescence), electrochemical methods (potentiometry, polarography *etc.*) and radioanalytical techniques (radiochemical analysis and isotopic dilution method). The choice of method is predominantly dictated by interferences, time, number of sample and accuracy desired [306]. Progress made in materials sciences and nanotechnology, electromechanical and microfluidic systems, protein engineering and biomimetics design has given unmatched boost to sensing technology from bench to market. Though conventional methodologies offer high reliability and very low limits of detection (LODs) but suffer from some serious limitations, such as high cost, longer analysis time and need for highly trained personnel [307]. Besides, sample collection and transport to the laboratory (offline analysis), sample pre-treatment steps and pre-concentration of target compounds in the collected samples are prerequisite that are indeed highly labour intensive [308]. Consequently, there is an ever growing demand for robust, fast and cheaper alternative technologies for precise *in situ* real-time monitoring of analyte molecules. Sensor represents a highly suitable substitute for conventional analytical methods by offering opportunities to meet above-stated challenges in a convenient way (Fig. **11A**). Nanotechnology and materials science has brought about significant modifications in sensor technology. It has opened up new opportunities for improving current status of sensor technology and thus has thoroughly been investigated [307]. Sensor is a device whose purpose is to monitor and learn about the system and its environment [309]. Likewise, a chemical sensor is a device that quantitatively or semi-quantitatively translates information about the presence of chemical species in the form of an analytically useful signal. Fundamentally, a sensor consists of two components; a receptor and a transducer (Fig. **11A**). Receptor is made up of organic or inorganic material that undergoes a specific interaction with one analyte or group of analytes. The second key element of a sensor, known as transducer, changes chemical information into a quantifiable signal [310].

Nanomaterials are ideal for designing sensor arrays, as highlighted in Fig. (**11B**), because they are chemically versatile, and can easily be fabricated and integrated with currently available sensing platforms [312]. In this connection, the merger of nanomaterials using various techniques and tools of surface functionalization can give rise to a highly selective, sensitive, cost-effective and disposable sensors for monitoring different contaminants [308]. As described in previous Section, nanomaterials have excellent electrical, optical, thermal as well as catalytic properties and strong mechanical strength, thus making it possible to construct nanomaterials-based sensors or devices for monitoring pollution in water, air and soil. It is possible to detect and measure potentially toxic elements (PTEs), toxic gases, pesticides and hazardous industrial chemicals with high sensitivity, selectivity and simplicity using a range of engineered nanomaterials including CNTs, gold NPs, silicon nanowires and QDs [308, 313]. To date, numerous semiconductor nanosized metal-oxides including TiO_2, ZnO, WO_3, SnO_2, In_2O_3, TeO_2, CuO, CdO, Fe_2O_3 and MoO_3 have been explored for gas sensing applications on account of their high sensitivity towards many target gases, simple fabrication, low price and high compatibility with other parts and processes. In particular, 1-D nanostructural configurations have been considered highly suitable in gas sensing applications due to their high surface-to-volume ratio as well as better chemical and thermal stability under various operating conditions [314].

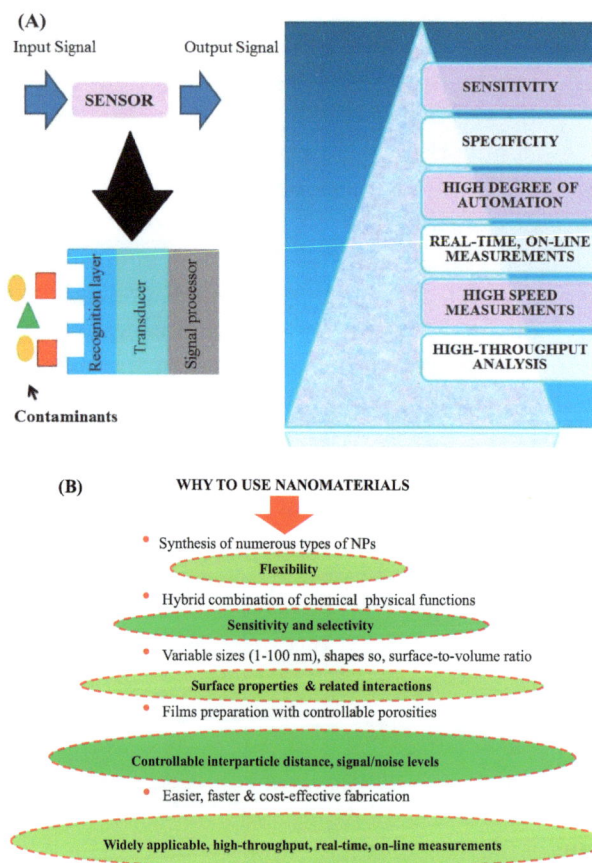

Fig. (11). (**A**) Schematic representation of a typical chemical sensor and its salient features (**B**) Advantages of incorporating nanomaterials in sensor technology.
Adopted from Sources; Scognamiglio *et al.* and Segev-Bar and H. Haick [307, 311].

Over a decade, graphene consisting of atom-thick two-dimensional conjugated structures and its derivatives have emerged as attractive materials in sensor technology because of their excellent electrical, mechanical, thermal and optical properties [315]. Besides, other properties such as high surface-to-volume ratio and superior adsorption capacity make graphene based materials highly attractive as sensing elements [316]. Considering biosensor system design, the response time is of utmost value. In particular, the response times are faster in nano-fibers and nanotubes. For instance, CNT and carbon nano-fibers (CNF) have attracted attention due to their low cost mass production, better mechanical strength, facile surface modification and more edge sites on the outer wall as compared to the metal electrodes [317]. Nanomaterial-based biosensors offer tremendous versatility in its applications, such as the analysis of organophosphorus (OP) pesticides and nerve agents in environmental and biological systems, provided these sensors are minimally susceptible to matrix effects and show selectivity towards the compound of interest [318]. Graphene based biosensors and devices offer good sensitivity and selectivity towards the detection of range of species including H_2O_2, small biomolecules, heavy metal ions and poisonous gaseous molecules present in the environment [319]. Besides, detrimental health effects of nitrites have been well documented. For this reason, environmental samples are collected and monitored to detect the presence of nitrites in environment. Recently, a nitrite sensor based on Fe_2O_3 NPs grafted onto rGO nanosheets (Fe_2O_3/rGO) has been fabricated. The as-prepared Fe_2O_3/rGO composite sensor has demonstrated higher sensitivity for the detection of nitrites as well as excellent anti-interference ability against different electroactive species and metal ions [320]. Gas sensors have gained tremendous popularity due to their selective, sensitive and efficient detection ability for gases (at ppt level), chemical vapours (at ~1 ppb) and explosives (at ~5 ppm). They have been commercialized for environmental and personalized health monitoring. The field of organic-inorganic hybrid nanocomposites, comprising of nano-scaled organic and inorganic counterparts is considered as an active area of research in advanced functional materials science [321]. Flexible sensors technology is highly anticipated to spur growth of totally new, smart sensing applications in healthcare, consumer electronics, robotics, prosthetics, geriatric care, sports, fitness, environmental monitoring, safety equipment, home-land security as well as space flight. NPs based flexible sensors are extremely promising for numerous applications and research in this area is on-going [311, 322].

Mercury is considered to be highly toxic because it accumulates in human body *via* food chains and drinking water, causing irreversible damages to the brain and central nervous system as well as might result in other chronic diseases. Recently, an extremely sensitive, highly selective and durable nanoporous gold/aptamer based surface enhanced resonance Raman scattering (SERRS) sensor has been designed with unprecedented detection sensitivity of 1 pM (0.2 ppt) for Hg^{2+} ions, a highly sensitive Hg^{2+} optical sensor known so far. The exceptionally high selectivity demonstrated by this sensor is obvious from the fact that Hg^{2+} ions can be identified in the presence of 12 other metal ions in dilute aqueous solutions, river water as well as underground water [313c]. Sensing probes based on core-shell structured nanocomposite materials have also shown some promising results. In one such study, a sensor based on nanocomposites material consisting of Fe_3O_4 NPs as core showing superparamagnetic behaviour confined by amorphous silica as shell modified with pyrene has been prepared for the detection and removal of Hg^{2+} ions in water [323]. Besides chemical pollution, microorganism contamination has also been considered precarious in wastewater-treatment system, food and pharmaceutical industries where fast detection is critical to prevent microbial outbreaks [324, 325]. The detection and analysis of pathogen are crucial for medicine, food safety, agriculture, public health and biosecurity. Illnesses caused by food borne diseases represent a persistent threat to public health as well as society. Over the past several years, a great deal of research has been carried out to develop biological sensors for the detection of microorganisms, permitting fast and "real-time" identification [326]. Recently, a nanoporous membrane based impedimetric immunosensor for the label-free detection of bacterial pathogens in whole milk has been proposed. It was found that hyaluronic acid-functionalized nanoporous membrane-based impedimetric sensor is capable of detecting pathogenic bacteria in whole milk. Notably, the samples were monitored without any pre-treatment which can be considered as a significant move towards the evaluation of safety of environmental and food samples as well as other medical diagnostics [327]. A new class of emerging nanomaterials called as the conducting polymers have been showing great potential as biological and chemical transducer platforms for food and water sensors (FWS). This is because of their ability to be decorated with biological recognition elements, direct electric signal transduction as well as biocompatible nature of these polymers. Conducting polymer based NPs and nanowires have been functionalized as molecularly imprinted polymers (MIP), phage functionalized for the detection of *Salmonella* and *Escherichia coli* O157:H7 as well as antibody functionalized for bacterial and viral immunosensing [328].

In present scenario, with regard to general safety, health care and environmental concerns, the role of nano-enhanced sensors as user friendly technology for fast, sensitive, selective and efficient detection of broad spectrum of contaminants present in the environment are especially promising.

CONCLUDING REMARKS

Present work is intended to give a comprehensive overview of the fundamental aspects and applications of nanotechnology in the field of environment, precisely called environmental nanotechnology. Over the course of human history, the advent of new technologies has altogether transformed the world we live in today. Currently, modern society is facing numerous pressing challenges in the form of climate change, energy crisis, pollution and health issues. Nanotechnology has emerged as an attractive and rapidly growing field of science and technology endeavoured to provide materials, machines and devices with exceptional properties for diverse applications. The application of nanoengineered materials for environmental protection, remediation and monitoring has decisively contributed in ensuring environmental sustainability. Based on incredible trends observed in the field of nanotechnology, it is most likely that the coming years would bring in huge technological breakthroughs reshaping our lives and leading us towards a greener future ahead. Inevitably, the references gathered in this chapter signify the positive contributions of nanotechnology towards safe and sustainable environment.

CONSENT FOR PUBLICATION

Not applicable.

CONFLICT OF INTEREST

The author (editor) declares no conflict of interest, financial or otherwise.

ACKNOWLEDGEMENTS

None Declare

REFERENCES

[1]   Gul, K.; Sohni, S.; Waqar, M.; Ahmad, F.; Norulaini, N.A.N.; A K, M.O. Functionalization of magnetic chitosan with graphene oxide for removal of cationic and anionic dyes from aqueous solution. *Carbohydr. Polym.*, **2016**, *152*, 520-531.
[http://dx.doi.org/10.1016/j.carbpol.2016.06.045] [PMID: 27516300]

[2]   Ratner, M.A.; Ratner, D. *Nanotechnology: A Gentle Introduction to the Next Big Idea*; Prentice Hall, **2003**.

[3]   (a). Schaming, D.; Remita, H. Nanotechnology: from the ancient time to nowadays. *Found. Chem.*, **2015**, *17*(3), 187-205.
[http://dx.doi.org/10.1007/s10698-015-9235-y]
(b). Rogers, B.; Adams, J.; Pennathur, S. *Nanotechnology: understanding small systems*; Crc Press, **2014**.

[4]   Mulvaney, P. Nanoscience *vs.* nanotechnology--defining the field. *ACS Nano*, **2015**, *9*(3), 2215-2217.
[http://dx.doi.org/10.1021/acsnano.5b01418] [PMID: 25802086]

[5]   Fulekar, M. *Nanotechnology: importance and applications*; IK International Pvt Ltd, **2010**.

[6]   McKenzie, L.C.; Hutchison, J.E. *Green nanoscience: An integrated approach to greener products, processes, and applications*; Chim. Oggi-Chem. Today, **2004**, p. 25.

[7]   Gatoo, M. A.; Naseem, S.; Arfat, M. Y.; Mahmood Dar, A.; Qasim, K.; Zubair, S. *Physicochemical properties of nanomaterials: implication in associated toxic manifestations.*, **2014**.
[http://dx.doi.org/10.1155/2014/498420]

[8]   Iravani, S. Green synthesis of metal nanoparticles using plants. *Green Chem.*, **2011**, *13*(10), 2638-2650.
[http://dx.doi.org/10.1039/c1gc15386b]

[9]   Suppi, S.; Kasemets, K.; Ivask, A.; Künnis-Beres, K.; Sihtmäe, M.; Kurvet, I.; Aruoja, V.; Kahru, A. A novel method for comparison of biocidal properties of nanomaterials to bacteria, yeasts and algae. *J. Hazard. Mater.*, **2015**, *286*, 75-84.
[http://dx.doi.org/10.1016/j.jhazmat.2014.12.027] [PMID: 25559861]

[10]   Hochella, M.F. There's plenty of room at the bottom: Nanoscience in geochemistry. *Geochim. Cosmochim. Acta*, **2002**, *66*(5), 735-743.
[http://dx.doi.org/10.1016/S0016-7037(01)00868-7]

[11]   Cohen, M.L. Nanotubes, nanoscience, and nanotechnology. *Mater. Sci. Eng. C*, **2001**, *15*(1), 1-11.
[http://dx.doi.org/10.1016/S0928-4931(01)00221-1]

[12]   Kent, J.A. *Handbook of industrial chemistry and biotechnology*; Springer Science & Business Media, **2013**.

[13]   Lanone, S.; Boczkowski, J. Biomedical applications and potential health risks of nanomaterials: molecular mechanisms. *Curr. Mol. Med.*, **2006**, *6*(6), 651-663.
[http://dx.doi.org/10.2174/156652406778195026] [PMID: 17022735]

[14]   Stone, V.; Nowack, B.; Baun, A.; van den Brink, N.; Kammer, Fv.; Dusinska, M.; Handy, R.; Hankin, S.; Hassellöv, M.; Joner, E.; Fernandes, T.F. Nanomaterials for environmental studies: classification, reference material issues, and strategies for physico-chemical characterisation. *Sci. Total Environ.*, **2010**, *408*(7), 1745-1754.
[http://dx.doi.org/10.1016/j.scitotenv.2009.10.035] [PMID: 19903569]

[15]   Science, O.E.C.D., Ed. Nanotechnology for Green Innovation.*Technology and Industry Policy Papers*; OECD Publishing: Paris, **2013**.

[16]   Merkoçi, A.; Kutter, J.P. Analytical miniaturization and nanotechnologies. *Lab Chip*, **2012**, *12*(11), 1915-1916.
[http://dx.doi.org/10.1039/c2lc90040h] [PMID: 22543818]

[17]   Rai, M.; Ingle, A. Role of nanotechnology in agriculture with special reference to management of insect pests. *Appl. Microbiol. Biotechnol.*, **2012**, *94*(2), 287-293.
[http://dx.doi.org/10.1007/s00253-012-3969-4] [PMID: 22388570]

[18]   Balköse, D.; Horak, D.; Šoltés, L. *Key Engineering Materials: Current State-of-the-Art on Novel Materials*; Apple Academic Press, **2014**.

[19]   (a). Nowack, B.; Bucheli, T.D. Occurrence, behavior and effects of nanoparticles in the environment. *Environ. Pollut.*, **2007**, *150*(1), 5-22.
[http://dx.doi.org/10.1016/j.envpol.2007.06.006] [PMID: 17658673]
(b). López-Serrano, A.; Olivas, R.M.; Landaluze, J.S.; Cámara, C. Nanoparticles: a global vision. Characterization, separation, and quantification methods. Potential environmental and health impact. . *Anal. Methods*, **2014**, *6*(1), 38-56.
[http://dx.doi.org/10.1039/C3AY40517F]
(c). Sutariya, V.B.; Pathak, Y. *Biointeractions of nanomaterials*; CRC Press, **2014**.
[http://dx.doi.org/10.1201/b17191]

[20]   Wu, R.; Zhou, K.; Yue, C.Y.; Wei, J.; Pan, Y. Recent progress in synthesis, properties and potential applications of SiC nanomaterials. . *Prog. Mater. Sci.*, **2015**, *72*, 1-60.
[http://dx.doi.org/10.1016/j.pmatsci.2015.01.003]

[21]   Guo, Y.; Xu, K.; Wu, C.; Zhao, J.; Xie, Y. Surface chemical-modification for engineering the intrinsic

physical properties of inorganic two-dimensional nanomaterials. *Chem. Soc. Rev.,* **2015**, *44*(3), 637-646.
[http://dx.doi.org/10.1039/C4CS00302K] [PMID: 25406669]

[22]   (a). Tisch, U.; Haick, H. Nanomaterials for cross-reactive sensor arrays. *MRS Bull.,* **2010**, *35*(10), 797-803.
[http://dx.doi.org/10.1557/mrs2010.509]
(b). Liu, R-S.; Cheng, L-C.; Huang, J-H.; Chen, H.M.; Lai, T-C.; Hsiao, M.; Chen, C-H.; Yang, K-Y.; Tsai, D.P.; Her, L-J. Highly efficient urchin-like bimetallic nanoparticles for photothermal cancer therapy. *SPIE Newsroom,* **2013**.
[http://dx.doi.org/10.1117/2.1201301.004676]
(c). Xu, T.T.; Fisher, F.T.; Brinson, L.C.; Ruoff, R.S. Bone-shaped nanomaterials for nanocomposite applications. *Nano Lett.,* **2003**, *3*(8), 1135-1139.
[http://dx.doi.org/10.1021/nl0343396]

[23]   Thatai, S.; Khurana, P.; Boken, J.; Prasad, S.; Kumar, D. Nanoparticles and core–shell nanocomposite based new generation water remediation materials and analytical techniques: A review. *Microchem. J.,* **2014**, *116*, 62-76.
[http://dx.doi.org/10.1016/j.microc.2014.04.001]

[24]   Lloyd Whitman, T.K. A Call for Nanotechnology-Inspired Grand Challenges., **2016**.

[25]   Masciangioli, T.; Zhang, W-X. Environmental technologies at the nanoscale. *Environ. Sci. Technol.,* **2003**, *37*(5), 102A-108A.
[http://dx.doi.org/10.1021/es0323998] [PMID: 12666906]

[26]   Roco, M.C. Environmentally responsible development of nanotechnology. *Environ. Sci. Technol.,* **2005**, *39*(5), 106A-112A.
[http://dx.doi.org/10.1021/es053199u] [PMID: 15787356]

[27]   (a). Schwarz, A.E. Green dreams of reason. Green nanotechnology between visions of excess and control. *NanoEthics,* **2009**, *3*(2), 109-118.
[http://dx.doi.org/10.1007/s11569-009-0061-3]
(b). Karn, B. The road to green nanotechnology. *J. Ind. Ecol.,* **2008**, *12*(3), 263-266.
[http://dx.doi.org/10.1111/j.1530-9290.2008.00045.x]

[28]   Fuchs, D. **2015**.

[29]   *Devices, N. S. T.-F. T., Nanotechnology and the Environment Applications and Implications*; American Chemical Society, **2005**.

[30]   Pradeep, T. *A textbook of nanoscience and nanotechnology*; Tata McGraw-Hill Education, **2012**.

[31]   Chemat, F.; Vian, M.A.; Cravotto, G. Green extraction of natural products: concept and principles. *Int. J. Mol. Sci.,* **2012**, *13*(7), 8615-8627.
[http://dx.doi.org/10.3390/ijms13078615] [PMID: 22942724]

[32]   Johnson, L.A.; Lusas, E. Comparison of alternative solvents for oils extraction. *J. Am. Oil Chem. Soc.,* **1983**, *60*(2), 229-242.
[http://dx.doi.org/10.1007/BF02543490]

[33]   Rosenthal, A.; Pyle, D.; Niranjan, K. Aqueous and enzymatic processes for edible oil extraction. *Enzyme Microb. Technol.,* **1996**, *19*(6), 402-420.
[http://dx.doi.org/10.1016/S0141-0229(96)80004-F]

[34]   Material Safety Data Sheet for n-Hexane *Available at: www.chem.agilent.com/ Library/msds/200-0007.pdf.,* 2016 [Accessed on: Jan 2013];

[35]   Acosta, E.J.; Nguyen, T.; Witthayapanyanon, A.; Harwell, J.H.; Sabatini, D.A. Linker-based bio-compatible microemulsions. *Environ. Sci. Technol.,* **2005**, *39*(5), 1275-1282.
[http://dx.doi.org/10.1021/es049010g] [PMID: 15787367]

[36]   Naksuk, A.; Sabatini, D.A.; Tongcumpou, C. Microemulsion-based palm kernel oil extraction using

mixed surfactant solutions. *Ind. Crops Prod.,* **2009**, *30*(2), 194-198.
[http://dx.doi.org/10.1016/j.indcrop.2009.03.008]

[37]   Parthasarathy, S.; Siah Ying, T.; Manickam, S. Generation and optimization of palm oil-based oil-i-
       -water (O/W) submicron-emulsions and encapsulation of curcumin using a liquid whistle
       hydrodynamic cavitation reactor (LWHCR). *Ind. Eng. Chem. Res.,* **2013**, *52*(34), 11829-11837.
       [http://dx.doi.org/10.1021/ie4008858]

[38]   Nguyen, T.; Do, L.; Sabatini, D.A. Biodiesel production *via* peanut oil extraction using diesel-based
       reverse-micellar microemulsions. *Fuel,* **2010**, *89*(9), 2285-2291.
       [http://dx.doi.org/10.1016/j.fuel.2010.03.021]

[39]   (a). Galian, R.; Pérez-Prieto, J. Catalytic processes activated by light. *Energy Environ. Sci.,* **2010**,
       3(10), 1488-1498.
       [http://dx.doi.org/10.1039/c0ee00003e]
       (b). Corma, A. Materials chemistry: Catalysts made thinner. *Nature,* **2009**, *461*(7261), 182-183.
       [http://dx.doi.org/10.1038/461182a] [PMID: 19741695]

[40]   De Vos, D.E.; Dams, M.; Sels, B.F.; Jacobs, P.A. Ordered mesoporous and microporous molecular
       sieves functionalized with transition metal complexes as catalysts for selective organic
       transformations. *Chem. Rev.,* **2002**, *102*(10), 3615-3640.
       [http://dx.doi.org/10.1021/cr010368u] [PMID: 12371896]

[41]   Zhu, Y-P.; Ren, T-Z.; Yuan, Z-Y. Insights into mesoporous metal phosphonate hybrid materials for
       catalysis. *Catal. Sci. Technol.,* **2015**, *5*(9), 4258-4279.
       [http://dx.doi.org/10.1039/C5CY00107B]

[42]   Takeshita, T.; Yamaji, K. Important roles of Fischer–Tropsch synfuels in the global energy future.
       Energy Policy, **2008**, *36*(8), 2773-2784.
       [http://dx.doi.org/10.1016/j.enpol.2008.02.044]

[43]   Wang, Y.; Wang, X.; Antonietti, M. Polymeric graphitic carbon nitride as a heterogeneous
       organocatalyst: from photochemistry to multipurpose catalysis to sustainable chemistry. *Angew. Chem.
       Int. Ed. Engl.,* **2012**, *51*(1), 68-89.
       [http://dx.doi.org/10.1002/anie.201101182] [PMID: 22109976]

[44]   (a). Chng, L.L.; Erathodiyil, N.; Ying, J.Y. Nanostructured catalysts for organic transformations. *Acc.
       Chem. Res.,* **2013**, *46*(8), 1825-1837.
       [http://dx.doi.org/10.1021/ar300197s] [PMID: 23350747]
       (b). Roduner, E. Size matters: why nanomaterials are different. *Chem. Soc. Rev.,* **2006**, *35*(7), 583-592.
       [http://dx.doi.org/10.1039/b502142c] [PMID: 16791330]
       (c). Huang, W.; Liu, J.H-C.; Alayoglu, P.; Li, Y.; Witham, C.A.; Tsung, C-K.; Toste, F.D.; Somorjai,
       G.A. Highly active heterogeneous palladium nanoparticle catalysts for homogeneous electrophilic
       reactions in solution and the utilization of a continuous flow reactor. *J. Am. Chem. Soc.,* **2010**,
       132(47), 16771-16773.
       [http://dx.doi.org/10.1021/ja108898t] [PMID: 21062037]
       (d). Reynhardt, J.P.; Yang, Y.; Sayari, A.; Alper, H. Periodic mesoporous silica-supported recyclable
       rhodium-complexed dendrimer catalysts. *Chem. Mater.,* **2004**, *16*(21), 4095-4102.
       [http://dx.doi.org/10.1021/cm0493142]

[45]   (a). Li, Y.; Liu, J.H-C.; Witham, C.A.; Huang, W.; Marcus, M.A.; Fakra, S.C.; Alayoglu, P.; Zhu, Z.;
       Thompson, C.M.; Arjun, A.; Lee, K.; Gross, E.; Toste, F.D.; Somorjai, G.A. A Pt-cluster-based
       heterogeneous catalyst for homogeneous catalytic reactions: X-ray absorption spectroscopy and
       reaction kinetic studies of their activity and stability against leaching. *J. Am. Chem. Soc.,* **2011**,
       133(34), 13527-13533.
       [http://dx.doi.org/10.1021/ja204191t] [PMID: 21721543]
       (b). Andrés, R.; de Jesús, E.; Flores, J.C. Catalysts based on palladium dendrimers. *New J. Chem.,*
       2007, *31*(7), 1161-1191.
       [http://dx.doi.org/10.1039/b615761k]

[46]　Gawande, M. Sustainable Nanocatalysts for Organic Synthetic Transformations. *Organic. Chem. Curr. Res.,* **2014**, *1*, e137.

[47]　(a). Liu, B.; Zhang, Z. Catalytic conversion of biomass into chemicals and fuels over magnetic catalysts. *ACS Catal.,* **2015**, *6*(1), 326-338.
　　　[http://dx.doi.org/10.1021/acscatal.5b02094]
　　　(b). Zhang, Z.; Deng, K. Recent advances in the catalytic synthesis of 2, 5-furandicarboxylic acid and its derivatives. *ACS Catal.,* **2015**, *5*(11), 6529-6544.
　　　[http://dx.doi.org/10.1021/acscatal.5b01491]

[48]　Larsen, S.C. Nanocrystalline zeolites and zeolite structures: synthesis, characterization, and applications. *J. Phys. Chem. C,* **2007**, *111*(50), 18464-18474.
　　　[http://dx.doi.org/10.1021/jp074980m]

[49]　(a). Somorjai, G.A.; McCrea, K. Roadmap for catalysis science in the 21ˢᵗ century: a personal view of building the future on past and present accomplishments. *Appl. Catal. A Gen.,* **2001**, *222*(1), 3-18.
　　　[http://dx.doi.org/10.1016/S0926-860X(01)00825-0]
　　　(b). Panov, A.; Larsen, R.; Totah, N.; Larsen, S.; Grassian, V. Photooxidation of toluene and p-xylene in cation-exchanged zeolites X, Y, ZSM-5, and Beta: The role of zeolite physicochemical properties in product yield and selectivity. *J. Phys. Chem. B,* **2000**, *104*(24), 5706-5714.
　　　[http://dx.doi.org/10.1021/jp000831r]

[50]　Sohni, S.; Norulaini, N.A.N.; Hashim, R.; Khan, S.B.; Fadhullah, W.; Mohd Omar, A.K. Physicochemical characterization of Malaysian crop and agro-industrial biomass residues as renewable energy resources. *Ind. Crops Prod.,* **2018**, *111*, 642-650.
　　　[http://dx.doi.org/10.1016/j.indcrop.2017.11.031]

[51]　Wakerley, D.W.; Kuehnel, M.F.; Orchard, K.L.; Ly, K.H.; Rosser, T.E.; Reisner, E. Solar-driven reforming of lignocellulose to H 2 with a CdS/CdO x photocatalyst. *Nat. Energy,* **2017**, *2*(4), 17021.
　　　[http://dx.doi.org/10.1038/nenergy.2017.21]

[52]　Kim, S.H.; Ahn, S-Y.; Kwak, S-Y. Suppression of dioxin emission in incineration of poly (vinyl chloride)(PVC) as hybridized with titanium dioxide (TiO₂) nanoparticles. *Appl. Catal. B,* **2008**, *79*(3), 296-305.
　　　[http://dx.doi.org/10.1016/j.apcatb.2007.10.035]

[53]　Zhang, P.; Hu, Y.; Li, B.; Zhang, Q.; Zhou, C.; Yu, H.; Zhang, X.; Chen, L.; Eichhorn, B.; Zhou, S. Kinetically Stabilized Pd@ Pt Core–Shell Octahedral Nanoparticles with Thin Pt Layers for Enhanced Catalytic Hydrogenation Performance. *ACS Catal.,* **2015**, *5*(2), 1335-1343.
　　　[http://dx.doi.org/10.1021/cs501612g]

[54]　Bet-Moushoul, E.; Farhadi, K.; Mansourpanah, Y.; Nikbakht, A.M.; Molaei, R.; Forough, M. Application of CaO-based/Au nanoparticles as heterogeneous nanocatalysts in biodiesel production. *Fuel,* **2016**, *164*, 119-127.
　　　[http://dx.doi.org/10.1016/j.fuel.2015.09.067]

[55]　Gharib, A.; Jahangir, M.; Roshani, M.; Pesyan, N.N.; Scheeren, J.W.; Mohadesazadeh, S.; Lagzian, S. Synthesis of (s)-(-)-propranolol by using cs2. 5h0. 5pw12o40 nanocatalyst as green, eco-friendly, reusable, and recyclable catalyst. *Synth. React. Inorg. Met.-Org. Nano-Met. Chem.,* **2015**, *45*(3), 350-355.
　　　[http://dx.doi.org/10.1080/15533174.2013.832323]

[56]　Zhang, D.; Chen, L.; Ge, G. A green approach for efficient p-nitrophenol hydrogenation catalyzed by a Pd-based nanocatalyst. *Catal. Commun.,* **2015**, *66*, 95-99.
　　　[http://dx.doi.org/10.1016/j.catcom.2015.03.021]

[57]　Chen, B-H.; Liu, W.; Li, A.; Liu, Y-J.; Chao, Z-S. A simple and convenient approach for preparing core-shell-like silica@nickel species nanoparticles: highly efficient and stable catalyst for the dehydrogenation of 1,2-cyclohexanediol to catechol. *Dalton Trans.,* **2015**, *44*(3), 1023-1038.
　　　[http://dx.doi.org/10.1039/C4DT01476F] [PMID: 25407395]

[58] Das, S.K.; El-Safty, S.A. Development of mesoscopically assembled sulfated zirconia nanoparticles as promising heterogeneous and recyclable biodiesel catalysts. *ChemCatChem,* **2013**, *5*(10), 3050-3059.
[http://dx.doi.org/10.1002/cctc.201300192]

[59] Lueangchaichaweng, W.; Li, L.; Wang, Q-Y.; Su, B-L.; Aprile, C.; Pescarmona, P.P. Novel mesoporous composites of gallia nanoparticles and silica as catalysts for the epoxidation of alkenes with hydrogen peroxide. *Catal. Today,* **2013**, *203*, 66-75.
[http://dx.doi.org/10.1016/j.cattod.2012.02.037]

[60] Javidi, J.; Esmaeilpour, M.; Dodeji, F.N. Immobilization of phosphomolybdic acid nanoparticles on imidazole functionalized $Fe_3O_4@ SiO_2$: a novel and reusable nanocatalyst for one-pot synthesis of Biginelli-type 3, 4-dihydro-pyrimidine-2-(1 H)-ones/thiones under solvent-free conditions. *RSC Advances,* **2015**, *5*(1), 308-315.
[http://dx.doi.org/10.1039/C4RA09929J]

[61] Zhang, S.; Shen, X.; Zheng, Z.; Ma, Y.; Qu, Y. 3D graphene/nylon rope as a skeleton for noble metal nanocatalysts for highly efficient heterogeneous continuous-flow reactions. *J. Mater. Chem. A Mater. Energy Sustain.,* **2015**, *3*(19), 10504-10511.
[http://dx.doi.org/10.1039/C5TA00409H]

[62] Maleki, A.; Paydar, R. Graphene oxide–chitosan bionanocomposite: a highly efficient nanocatalyst for the one-pot three-component synthesis of trisubstituted imidazoles under solvent-free conditions. *RSC Advances,* **2015**, *5*(42), 33177-33184.
[http://dx.doi.org/10.1039/C5RA03355A]

[63] *World Population Prospects, The 2015 Revision, Key Findings and Advance Tables ESA/P/WP.241*; Department of Economic and Social Affairs, Population Division, United Nations: New York, **2015**.

[64] *Building a common vision for sustainable food and agriculture, Principles and Approaches*; Food and Agriculture Organization of the United Nations: Rome, **2014**.

[65] Kah, M.; Hofmann, T. Nanopesticide research: current trends and future priorities. *Environ. Int.,* **2014**, *63*, 224-235.
[http://dx.doi.org/10.1016/j.envint.2013.11.015] [PMID: 24333990]

[66] (a). Joseph, T.; Morrison, M. *Nanotechnology in agriculture and food: a nanoforum report.,* **2006**.
(b). Kashyap, P.L.; Xiang, X.; Heiden, P. Chitosan nanoparticle based delivery systems for sustainable agriculture. *Int. J. Biol. Macromol.,* **2015**, *77*, 36-51.
[http://dx.doi.org/10.1016/j.ijbiomac.2015.02.039] [PMID: 25748851]

[67] Siddiqui, M.H.; Al-Whaibi, M.H.; Mohammad, F. *Nanotechnology and Plant Sciences: Nanoparticles and Their Impact on Plants*; Springer, **2015**.
[http://dx.doi.org/10.1007/978-3-319-14502-0]

[68] Chinnamuthu, C.; Boopathi, P.M. Nanotechnology and agroecosystem. *Madras Agric. J.,* **2009**, *96*(1-6), 17-31.

[69] Monreal, C.; McGill, W.; Nyborg, M. Spatial heterogeneity of substrates: effects on hydrolysis, immobilization and nitrification of urea-N. *Can. J. Soil Sci.,* **1986**, *66*(3), 499-511.
[http://dx.doi.org/10.4141/cjss86-050]

[70] (a). Kottegoda, N.; Munaweera, I.; Madusanka, N.; Karunaratne, V. A green slow-release fertilizer composition based on urea-modified hydroxyapatite nanoparticles encapsulated wood. *Curr. Sci (Bangalore),* **2011**, *101*(1), 73-78.
(b). DeRosa, M.C.; Monreal, C.; Schnitzer, M.; Walsh, R.; Sultan, Y. Nanotechnology in fertilizers. *Nat. Nanotechnol.,* **2010**, *5*(2), 91-91.
[http://dx.doi.org/10.1038/nnano.2010.2] [PMID: 20130583]

[71] Rai, V.; Acharya, S.; Dey, N. Implications of nanobiosensors in agriculture. *J. Biomater. Nanobiotechnol.,* **2012**, *3*(2A), 315.

[http://dx.doi.org/10.4236/jbnb.2012.322039]

[72]   Kookana, R.S.; Boxall, A.B.; Reeves, P.T.; Ashauer, R.; Beulke, S.; Chaudhry, Q.; Cornelis, G.; Fernandes, T.F.; Gan, J.; Kah, M.; Lynch, I.; Ranville, J.; Sinclair, C.; Spurgeon, D.; Tiede, K.; Van den Brink, P.J. Nanopesticides: guiding principles for regulatory evaluation of environmental risks. *J. Agric. Food Chem.,* **2014**, *62*(19), 4227-4240.
[http://dx.doi.org/10.1021/jf500232f] [PMID: 24754346]

[73]   (a). Ditta, A.; Arshad, M.; Ibrahim, M. Nanoparticles in Sustainable Agricultural Crop Production: Applications and Perspectives.*Nanotechnology and Plant Sciences*; Springer, **2015**, pp. 55-75.
[http://dx.doi.org/10.1007/978-3-319-14502-0_4]
(b). Shende, S.; Ingle, A.P.; Gade, A.; Rai, M. Green synthesis of copper nanoparticles by Citrus medica Linn. (Idilimbu) juice and its antimicrobial activity. *World J. Microbiol. Biotechnol.,* **2015**, *31*(6), 865-873.
[http://dx.doi.org/10.1007/s11274-015-1840-3] [PMID: 25761857]
(c). Gajbhiye, M.; Kesharwani, J.; Ingle, A.; Gade, A.; Rai, M. Fungus-mediated synthesis of silver nanoparticles and their activity against pathogenic fungi in combination with fluconazole. *Nanomedicine (Lond.),* **2009**, *5*(4), 382-386.
[http://dx.doi.org/10.1016/j.nano.2009.06.005] [PMID: 19616127]
(d). Cioffi, N.; Torsi, L.; Ditaranto, N.; Sabbatini, L.; Zambonin, P.G.; Tantillo, G.; Ghibelli, L.; D'Alessio, M.; Bleve-Zacheo, T.; Traversa, E. Antifungal activity of polymer-based copper nanocomposite coatings. *Appl. Phys. Lett.,* **2004**, *85*(12), 2417-2419.
[http://dx.doi.org/10.1063/1.1794381]

[74]   November, R. Photochemical Purification of wastewater from the Fungicides and Pesticides Using Advanced Oxidation Processes. *Aust. J. Basic Appl. Sci.,* **2014**, *8*(1), 434-441.

[75]   Raveendran, P.; Fu, J.; Wallen, S.L. Completely "green" synthesis and stabilization of metal nanoparticles. *J. Am. Chem. Soc.,* **2003**, *125*(46), 13940-13941.
[http://dx.doi.org/10.1021/ja029267j] [PMID: 14611213]

[76]   Meena Kumari, M.; Jacob, J.; Philip, D. Green synthesis and applications of Au-Ag bimetallic nanoparticles. *Spectrochim. Acta A Mol. Biomol. Spectrosc.,* **2015**, *137*, 185-192.
[http://dx.doi.org/10.1016/j.saa.2014.08.079] [PMID: 25218228]

[77]   Bar, H.; Bhui, D.K.; Sahoo, G.P.; Sarkar, P.; De, S.P.; Misra, A. Green synthesis of silver nanoparticles using latex of Jatropha curcas. *Colloids Surf. A Physicochem. Eng. Asp.,* **2009**, *339*(1), 134-139.
[http://dx.doi.org/10.1016/j.colsurfa.2009.02.008]

[78]   Li, S.; Shen, Y.; Xie, A.; Yu, X.; Qiu, L.; Zhang, L.; Zhang, Q. Green synthesis of silver nanoparticles using Capsicum annuum L. extract. *Green Chem.,* **2007**, *9*(8), 852-858.
[http://dx.doi.org/10.1039/b615357g]

[79]   (a). Philip, D. Green synthesis of gold and silver nanoparticles using Hibiscus rosa sinensis. *Physica E,* **2010**, *42*(5), 1417-1424.
[http://dx.doi.org/10.1016/j.physe.2009.11.081]
(b). Thovhogi, N.; Diallo, A.; Gurib-Fakim, A.; Maaza, M. Nanoparticles green synthesis by Hibiscus Sabdariffa flower extract: Main physical properties. *J. Alloys Compd.,* **2015**, *647*, 392-396.
[http://dx.doi.org/10.1016/j.jallcom.2015.06.076]

[80]   Saxena, A.; Tripathi, R.; Zafar, F.; Singh, P. Green synthesis of silver nanoparticles using aqueous solution of Ficus benghalensis leaf extract and characterization of their antibacterial activity. *Mater. Lett.,* **2012**, *67*(1), 91-94.
[http://dx.doi.org/10.1016/j.matlet.2011.09.038]

[81]   Jagtap, U.B.; Bapat, V.A. Green synthesis of silver nanoparticles using Artocarpus heterophyllus Lam. seed extract and its antibacterial activity. *Ind. Crops Prod.,* **2013**, *46*, 132-137.
[http://dx.doi.org/10.1016/j.indcrop.2013.01.019]

[82]   Sulaiman, G.M.; Mohammed, W.H.; Marzoog, T.R.; Al-Amiery, A.A.A.; Kadhum, A.A.H.;

Mohamad, A.B. Green synthesis, antimicrobial and cytotoxic effects of silver nanoparticles using Eucalyptus chapmaniana leaves extract. *Asian Pac. J. Trop. Biomed.,* **2013**, *3*(1), 58-63.
[http://dx.doi.org/10.1016/S2221-1691(13)60024-6] [PMID: 23570018]

[83]	Edison, T.J.I.; Sethuraman, M. Instant green synthesis of silver nanoparticles using Terminalia chebula fruit extract and evaluation of their catalytic activity on reduction of methylene blue. *Process Biochem.,* **2012**, *47*(9), 1351-1357.
[http://dx.doi.org/10.1016/j.procbio.2012.04.025]

[84]	Dipankar, C.; Murugan, S. The green synthesis, characterization and evaluation of the biological activities of silver nanoparticles synthesized from Iresine herbstii leaf aqueous extracts. *Colloids Surf. B Biointerfaces,* **2012**, *98*, 112-119.
[http://dx.doi.org/10.1016/j.colsurfb.2012.04.006] [PMID: 22705935]

[85]	Basu, S.; Maji, P.; Ganguly, J. Rapid green synthesis of silver nanoparticles by aqueous extract of seeds of Nyctanthes arbor-tristis. *Appl. Nanosci.,* **2016**, *6*(1), 1-5.
[http://dx.doi.org/10.1007/s13204-015-0407-9]

[86]	Rupiasih, N.N.; Aher, A.; Gosavi, S.; Vidyasagar, P. Green synthesis of silver nanoparticles using latex extract of Thevetia peruviana: a novel approach towards poisonous plant utilization.*Recent Trends in Physics of Material Science and Technology*; Springer, **2015**, pp. 1-10.
[http://dx.doi.org/10.1007/978-981-287-128-2_1]

[87]	Ghaedi, M.; Yousefinejad, M.; Safarpoor, M.; Khafri, H.Z.; Purkait, M. Rosmarinus officinalis leaf extract mediated green synthesis of silver nanoparticles and investigation of its antimicrobial properties. *J. Ind. Eng. Chem.,* **2015**, *31*, 167-172.
[http://dx.doi.org/10.1016/j.jiec.2015.06.020]

[88]	Korbekandi, H.; Chitsazi, M.R.; Asghari, G.; Bahri Najafi, R.; Badii, A.; Iravani, S. Green biosynthesis of silver nanoparticles using Quercus brantii (oak) leaves hydroalcoholic extract. *Pharm. Biol.,* **2015**, *53*(6), 807-812.
[http://dx.doi.org/10.3109/13880209.2014.942868] [PMID: 25697607]

[89]	Nayak, D.; Ashe, S.; Rauta, P.R.; Kumari, M.; Nayak, B. Bark extract mediated green synthesis of silver nanoparticles: Evaluation of antimicrobial activity and antiproliferative response against osteosarcoma. *Mater. Sci. Eng. C,* **2016**, *58*, 44-52.
[http://dx.doi.org/10.1016/j.msec.2015.08.022] [PMID: 26478285]

[90]	Zahir, A.A.; Chauhan, I.S.; Bagavan, A.; Kamaraj, C.; Elango, G.; Shankar, J.; Arjaria, N.; Roopan, S.M.; Rahuman, A.A.; Singh, N. Green Synthesis of Silver and Titanium Dioxide Nanoparticles Using Euphorbia prostrata Extract Shows Shift from Apoptosis to G0/G1 Arrest followed by Necrotic Cell Death in Leishmania donovani. *Antimicrob. Agents Chemother.,* **2015**, *59*(8), 4782-4799.
[http://dx.doi.org/10.1128/AAC.00098-15] [PMID: 26033724]

[91]	Sadeghi, B.; Gholamhoseinpoor, F. A study on the stability and green synthesis of silver nanoparticles using Ziziphora tenuior (Zt) extract at room temperature. *Spectrochim. Acta A Mol. Biomol. Spectrosc.,* **2015**, *134*, 310-315.
[http://dx.doi.org/10.1016/j.saa.2014.06.046] [PMID: 25022503]

[92]	Ahmed, M.J.; Murtaza, G.; Mehmood, A.; Bhatti, T.M. Green synthesis of silver nanoparticles using leaves extract of Skimmia laureola: Characterization and antibacterial activity. *Mater. Lett.,* **2015**, *153*, 10-13.
[http://dx.doi.org/10.1016/j.matlet.2015.03.143]

[93]	Pawar, O.; Deshpande, N.; Dagade, S.; Waghmode, S.; Nigam Joshi, P. Green synthesis of silver nanoparticles from purple acid phosphatase apoenzyme isolated from a new source *Limonia acidissima. J. Exp. Nanosci.,* **2016**, *11*(1), 28-37.
[http://dx.doi.org/10.1080/17458080.2015.1025300]

[94]	Nasrollahzadeh, M.; Sajadi, S.M.; Maham, M. Green synthesis of palladium nanoparticles using Hippophae rhamnoides Linn leaf extract and their catalytic activity for the Suzuki–Miyaura coupling

in water. *J. Mol. Catal. Chem.,* **2015**, *396*, 297-303.
[http://dx.doi.org/10.1016/j.molcata.2014.10.019]

[95]   Nasrollahzadeh, M.; Sajadi, S.M.; Rostami-Vartooni, A.; Bagherzadeh, M. Green synthesis of Pd/CuO nanoparticles by *Theobroma cacao* L. seeds extract and their catalytic performance for the reduction of 4-nitrophenol and phosphine-free Heck coupling reaction under aerobic conditions. *J. Colloid Interface Sci.,* **2015**, *448*, 106-113.
[http://dx.doi.org/10.1016/j.jcis.2015.02.009] [PMID: 25721860]

[96]   Nasrollahzadeh, M.; Sajadi, S.M.; Rostami-Vartooni, A.; Khalaj, M. Green synthesis of Pd/Fe$_3$O$_4$ nanoparticles using Euphorbia condylocarpa M. bieb root extract and their catalytic applications as magnetically recoverable and stable recyclable catalysts for the phosphine-free Sonogashira and Suzuki coupling reactions. *J. Mol. Catal. Chem.,* **2015**, *396*, 31-39.
[http://dx.doi.org/10.1016/j.molcata.2014.09.029]

[97]   Thema, F.; Beukes, P.; Gurib-Fakim, A.; Maaza, M. Green synthesis of Monteponite CdO nanoparticles by *Agathosma betulina* natural extract. *J. Alloys Compd.,* **2015**, *646*, 1043-1048.
[http://dx.doi.org/10.1016/j.jallcom.2015.05.279]

[98]   (a). Nasrollahzadeh, M.; Sajadi, S.M. Green synthesis of copper nanoparticles using Ginkgo biloba L. leaf extract and their catalytic activity for the Huisgen [3+2] cycloaddition of azides and alkynes at room temperature. *J. Colloid Interface Sci.,* **2015**, *457*, 141-147.
[http://dx.doi.org/10.1016/j.jcis.2015.07.004] [PMID: 26164245]
(b). Nasrollahzadeh, M.; Sajadi, S.M.; Hatamifard, A. Anthemis xylopoda flowers aqueous extract assisted *in situ* green synthesis of Cu nanoparticles supported on natural Natrolite zeolite for N-formylation of amines at room temperature under environmentally benign reaction conditions. *J. Colloid Interface Sci.,* **2015**, *460*, 146-153.
[http://dx.doi.org/10.1016/j.jcis.2015.08.040] [PMID: 26319331]

[99]   (a). Naika, H.R.; Lingaraju, K.; Manjunath, K.; Kumar, D.; Nagaraju, G.; Suresh, D.; Nagabhushana, H. Green synthesis of CuO nanoparticles using Gloriosa superba L. extract and their antibacterial activity. *J. Taibah. Univ. Sci.,* **2015**, *9*(1), 7-12.
[http://dx.doi.org/10.1016/j.jtusci.2014.04.006]
(b). Nasrollahzadeh, M.; Maham, M.; Sajadi, S.M. Green synthesis of CuO nanoparticles by aqueous extract of Gundelia tournefortii and evaluation of their catalytic activity for the synthesis of N-monosubstituted ureas and reduction of 4-nitrophenol. *J. Colloid Interface Sci.,* **2015**, *455*, 245-253.
[http://dx.doi.org/10.1016/j.jcis.2015.05.045] [PMID: 26073846]
(c). Nethravathi, P.; Kumar, M.P.; Suresh, D.; Lingaraju, K.; Rajanaika, H.; Nagabhushana, H.; Sharma, S. Tinospora cordifolia mediated facile green synthesis of cupric oxide nanoparticles and their photocatalytic, antioxidant and antibacterial properties. *Mater. Sci. Semicond. Process.,* **2015**, *33*, 81-88.
[http://dx.doi.org/10.1016/j.mssp.2015.01.034]

[100]   (a). Paul, B.; Bhuyan, B.; Purkayastha, D.D.; Dey, M.; Dhar, S.S. Green synthesis of gold nanoparticles using Pogestemon benghalensis (B) O. Ktz. leaf extract and studies of their photocatalytic activity in degradation of methylene blue. *Mater. Lett.,* **2015**, *148*, 37-40.
[http://dx.doi.org/10.1016/j.matlet.2015.02.054]
(b). Aswathy, S.A.; Philip, D. Green synthesis of gold nanoparticles using Trigonella foenum-graecum and its size-dependent catalytic activity. *Spectrochim. Acta A Mol. Biomol. Spectrosc.,* **2012**, *97*, 1-5.
[http://dx.doi.org/10.1016/j.saa.2012.05.083] [PMID: 22743607]
(c). Ateeq, M.; Shah, M.R.; Ain, N.U.; Bano, S.; Anis, I.; Lubna, ; Faizi, S.; Bertino, M.F.; Sohaila Naz, S. Green synthesis and molecular recognition ability of patuletin coated gold nanoparticles. *Biosens. Bioelectron.,* **2015**, *63*, 499-505.
[http://dx.doi.org/10.1016/j.bios.2014.07.076] [PMID: 25129513]
(d). Annadhasan, M.; Kasthuri, J.; Rajendiran, N. Green synthesis of gold nanoparticles under sunlight irradiation and their colorimetric detection of Ni^{2+} and Co^{2+} ions. *RSC Advances,* **2015**, *5*(15), 11458-11468.

[http://dx.doi.org/10.1039/C4RA14034F]

[101] Davar, F.; Majedi, A.; Mirzaei, A. Green Synthesis of ZnO Nanoparticles and Its Application in the Degradation of Some Dyes. *J. Am. Ceram. Soc.,* **2015**, *98*(6), 1739-1746.
[http://dx.doi.org/10.1111/jace.13467]

[102] Ansari, M.A.; Khan, H.M.; Alzohairy, M.A.; Jalal, M.; Ali, S.G.; Pal, R.; Musarrat, J. Green synthesis of Al2O3 nanoparticles and their bactericidal potential against clinical isolates of multi-drug resistant Pseudomonas aeruginosa. *World J. Microbiol. Biotechnol.,* **2015**, *31*(1), 153-164.
[http://dx.doi.org/10.1007/s11274-014-1757-2] [PMID: 25304025]

[103] Nasrollahzadeh, M.; Maham, M.; Rostami-Vartooni, A.; Bagherzadeh, M.; Sajadi, S.M. Barberry fruit extract assisted *in situ* green synthesis of Cu nanoparticles supported on a reduced graphene oxide–Fe$_3$O$_4$ nanocomposite as a magnetically separable and reusable catalyst for the O-arylation of phenols with aryl halides under ligand-free conditions. *RSC Advances,* **2015**, *5*(79), 64769-64780.
[http://dx.doi.org/10.1039/C5RA10037B]

[104] Dhanasekar, N.N.; Rahul, G.R.; Narayanan, K.B.; Raman, G.; Sakthivel, N. Green chemistry approach for the synthesis of gold nanoparticles using the *Fungus Alternaria* sp. *J. Microbiol. Biotechnol.,* **2015**, *25*(7), 1129-1135.
[http://dx.doi.org/10.4014/jmb.1410.10036] [PMID: 25737119]

[105] Schmidt, K. *Green nanotechnology: it's easier than you think.,* **2007**.

[106] Nguyen, V.H.; Shim, J-J. Green synthesis and characterization of carbon nanotubes/polyaniline nanocomposites. *J. Spectrosc.,* **2015**.
[http://dx.doi.org/10.1155/2015/297804]

[107] Zhou, H.; Mao, Y.; Wong, S.S. Probing structure-parameter correlations in the molten salt synthesis of BaZrO$_3$ perovskite submicrometer-sized particles. *Chem. Mater.,* **2007**, *19*(22), 5238-5249.
[http://dx.doi.org/10.1021/cm071456j]

[108] (a). Wu, C.; Mosher, B.P.; Zeng, T. One-step green route to narrowly dispersed copper nanocrystals. *J. Nanopart. Res.,* **2006**, *8*(6), 965-969.
[http://dx.doi.org/10.1007/s11051-005-9065-2]
(b). Vigneshwaran, N.; Nachane, R.P.; Balasubramanya, R.H.; Varadarajan, P.V. A novel one-pot 'green' synthesis of stable silver nanoparticles using soluble starch. *Carbohydr. Res.,* **2006**, *341*(12), 2012-2018.
[http://dx.doi.org/10.1016/j.carres.2006.04.042] [PMID: 16716274]

[109] Lorestani, F.; Shahnavaz, Z.; Mn, P.; Alias, Y.; Manan, N.S. One-step hydrothermal green synthesis of silver nanoparticle-carbon nanotube reduced-graphene oxide composite and its application as hydrogen peroxide sensor. *Sens. Actuators B Chem.,* **2015**, *208*, 389-398.
[http://dx.doi.org/10.1016/j.snb.2014.11.074]

[110] Zhang, Y.; Liu, S.; Wang, L.; Qin, X.; Tian, J.; Lu, W.; Chang, G.; Sun, X. One-pot green synthesis of Ag nanoparticles-graphene nanocomposites and their applications in SERS, H$_2$O$_2$, and glucose sensing. *RSC Advances,* **2012**, *2*(2), 538-545.
[http://dx.doi.org/10.1039/C1RA00641J]

[111] Shankar, S.; Prasad, R.; Selvakannan, P.; Jaiswal, L.; Laxman, R. Green synthesis of silver nanoribbons from waste X-ray films using alkaline protease. *Mater. Express,* **2015**, *5*(2), 165-170.
[http://dx.doi.org/10.1166/mex.2015.1221]

[112] Kiple, K.F. *The Cambridge world history of food. 2*; Cambridge University Press, **2000**, Vol. 2, .

[113] Hartley, C. W. S. *oil palm (Elaeis guincenis Jacq.)*; , **1967**.

[114] Hussein, A.K. Applications of nanotechnology in renewable energies—A comprehensive overview and understanding. *Renew. Sustain. Energy Rev.,* **2015**, *42*, 460-476.
[http://dx.doi.org/10.1016/j.rser.2014.10.027]

[115] Material innovativ 2016. . Materials and production technologies for tomorrow's mobility. e.V., G. d.

A., Ed. Würzburg, 2016

[116] VDV Academy Conference. . Electric buses - market of the future! and ElekBu 2016, Berlin, Berlin, **2016**

[117] In Zero-emission ships with batteries and fuel cells [H2BZ-Hesse] Electrification of ships and ferries on inland waterways, Frankfurt am Main, HA Hessen Agentur GmbH, transfer Agent Bingen (TSB), H2BZ initiative Hesse, H2BZ -Kooperationsnetzwerk Rheinland-Pfalz eV, energy Agency Rheinland-Pfalz: Frankfurt am Main, 2016.

[118] Pathak, D.; Wagner, T.; Adhikari, T.; Nunzi, J. AgInSe 2. PCBM. P3HT inorganic organic blends for hybrid bulk heterojunction photovoltaics. *Synth. Met.,* **2015**, *200*, 102-108.
[http://dx.doi.org/10.1016/j.synthmet.2015.01.001]

[119] Tian, B.; Zheng, X.; Kempa, T.J.; Fang, Y.; Yu, N.; Yu, G.; Huang, J.; Lieber, C.M. Coaxial silicon nanowires as solar cells and nanoelectronic power sources. *Nature,* **2007**, *449*(7164), 885-889.
[http://dx.doi.org/10.1038/nature06181] [PMID: 17943126]

[120] Serrano, E.; Rus, G.; Garcia-Martinez, J. Nanotechnology for sustainable energy. *Renew. Sustain. Energy Rev.,* **2009**, *13*(9), 2373-2384.
[http://dx.doi.org/10.1016/j.rser.2009.06.003]

[121] Pathak, D.; Wagner, T.; Adhikari, T.; Nunzi, J. Photovoltaic performance of AgInSe$_2$-conjugated polymer hybrid system bulk heterojunction solar cells. *Synth. Met.,* **2015**, *199*, 87-92.
[http://dx.doi.org/10.1016/j.synthmet.2014.11.015]

[122] Pathak, D.; Wagner, T.; Šubrt, J.; Kupcik, J. Characterization of mechanically synthesized AgInSe$_2$ nanostructures 1. . *Can. J. Phys.,* **2014**, *92*(7/8), 789-796.
[http://dx.doi.org/10.1139/cjp-2013-0546]

[123] Whittaker-Brooks, L.; Gao, J.; Hailey, A.K.; Thomas, C.R.; Yao, N.; Loo, Y-L. Bi$_2$S$_3$ nanowire networks as electron acceptor layers in solution-processed hybrid solar cells. *J. Mater. Chem. C Mater. Opt. Electron. Devices,* **2015**, *3*(11), 2686-2692.
[http://dx.doi.org/10.1039/C4TC02534B]

[124] *Welcome to the hydrogen and fuel cell technology in Hessen!,* In *H2BZ Knowledge Encyclopedia, 2016*; Vol. **2016**.

[125] Liao, C-H.; Huang, C-W.; Wu, J. Hydrogen production from semiconductor-based photocatalysis *via* water splitting. *Catalysts,* **2012**, *2*(4), 490-516.
[http://dx.doi.org/10.3390/catal2040490]

[126] principle of operation. In H2BZ *Knowledge Encyclopedia,* **2016**.

[127] (a). Mao, S.S.; Shen, S.; Guo, L. Nanomaterials for renewable hydrogen production, storage and utilization. *Progress in Natural Science: Materials International,* **2012**, *22*(6), 522-534.
[http://dx.doi.org/10.1016/j.pnsc.2012.12.003]
(b). Fujishima, A.; Honda, K. Electrochemical photolysis of water at a semiconductor electrode. *Nature,* **1972**, *238*(5358), 37-38.
[http://dx.doi.org/10.1038/238037a0] [PMID: 12635268]

[128] Yin, Y.; Jin, Z.; Hou, F. Enhanced solar water-splitting efficiency using core/sheath heterostructure CdS/TiO$_2$ nanotube arrays. *Nanotechnology,* **2007**, *18*(49), 495608.
[http://dx.doi.org/10.1088/0957-4484/18/49/495608] [PMID: 20442481]

[129] Maeda, K.; Domen, K. Photocatalytic water splitting: recent progress and future challenges. *J. Phys. Chem. Lett.,* **2010**, *1*(18), 2655-2661.
[http://dx.doi.org/10.1021/jz1007966]

[130] Erogbogbo, F.; Lin, T.; Tucciarone, P.M.; LaJoie, K.M.; Lai, L.; Patki, G.D.; Prasad, P.N.; Swihart, M.T. On-demand hydrogen generation using nanosilicon: splitting water without light, heat, or electricity. *Nano Lett.,* **2013**, *13*(2), 451-456.
[http://dx.doi.org/10.1021/nl304680w] [PMID: 23317111]

[131] H2BZ technology. In *H2BZ Knowledge Encyclopedia*, **2016**

[132] Innovation & Sustainability Projects > Hydrogen & Fuel Cell Technology *Available at: http://www.hessen-agentur.de/dynasite.cfm?dsmid=16043&newsid=22376&dsnocache=1.*, 2008 [Accessed on: Jan 20 2016];

[133] Zhou, L. Progress and problems in hydrogen storage methods. *Renew. Sustain. Energy Rev.*, **2005**, *9*(4), 395-408.
[http://dx.doi.org/10.1016/j.rser.2004.05.005]

[134] Yürüm, Y.; Taralp, A.; Veziroglu, T.N. Storage of hydrogen in nanostructured carbon materials. *Int. J. Hydrogen Energy*, **2009**, *34*(9), 3784-3798.
[http://dx.doi.org/10.1016/j.ijhydene.2009.03.001]

[135] (a). Li, J.; Furuta, T.; Goto, H.; Ohashi, T.; Fujiwara, Y.; Yip, S. Theoretical evaluation of hydrogen storage capacity in pure carbon nanostructures. *J. Chem. Phys.*, **2003**, *119*(4), 2376-2385.
[http://dx.doi.org/10.1063/1.1582831]
(b). Yoon, M.; Yang, S.; Hicke, C.; Wang, E.; Geohegan, D.; Zhang, Z. Calcium as the superior coating metal in functionalization of carbon fullerenes for high-capacity hydrogen storage. *Phys. Rev. Lett.*, **2008**, *100*(20), 206806.
[http://dx.doi.org/10.1103/PhysRevLett.100.206806] [PMID: 18518569]
(c). Muniz, A.R.; Meyyappan, M.; Maroudas, D. On the hydrogen storage capacity of carbon nanotube bundles. *Appl. Phys. Lett.*, **2009**, *95*(16), 163111.
[http://dx.doi.org/10.1063/1.3253711]
(d). Okati, A.; Zolfaghari, A.; Sadat Hashemi, F.; Anousheh, N.; Jooya, H. Hydrogen Physisorption on Stone-Wales Defect-embedded Single-walled Carbon Nanotubes. *Fuller. Nanotub. Carbon Nanostruct.*, **2009**, *17*(3), 324-335.
[http://dx.doi.org/10.1080/15363830902776599]
(e). Chang, J-K.; Chen, C-Y.; Tsai, W-T. Decorating carbon nanotubes with nanoparticles using a facile redox displacement reaction and an evaluation of synergistic hydrogen storage performance. *Nanotechnology*, **2009**, *20*(49), 495603.
[http://dx.doi.org/10.1088/0957-4484/20/49/495603] [PMID: 19893152]
(f). Darkrim, F.; Levesque, D. Monte Carlo simulations of hydrogen adsorption in single-walled carbon nanotubes. *J. Chem. Phys.*, **1998**, *109*(12), 4981-4984.
[http://dx.doi.org/10.1063/1.477109]
(g). Shao, H.; Xin, G.; Zheng, J.; Li, X.; Akiba, E. Nanotechnology in Mg-based materials for hydrogen storage. *Nano Energy*, **2012**, *1*(4), 590-601.
[http://dx.doi.org/10.1016/j.nanoen.2012.05.005]

[136] Silambarasan, D.; Surya, V.; Vasu, V.; Iyakutti, K. One-step process of hydrogen storage in single walled carbon nanotubes-tin oxide nano composite. *Int. J. Hydrogen Energy*, **2013**, *38*(10), 4011-4016.
[http://dx.doi.org/10.1016/j.ijhydene.2013.01.129]

[137] De Jongh, P.E.; Allendorf, M.; Vajo, J.J.; Zlotea, C. Nanoconfined light metal hydrides for reversible hydrogen storage. *MRS Bull.*, **2013**, *38*(06), 488-494.
[http://dx.doi.org/10.1557/mrs.2013.108]

[138] Manekkathodi, A.; Lu, M.Y.; Wang, C.W.; Chen, L.J. Direct growth of aligned zinc oxide nanorods on paper substrates for low-cost flexible electronics. *Adv. Mater.*, **2010**, *22*(36), 4059-4063.
[http://dx.doi.org/10.1002/adma.201001289] [PMID: 20512820]

[139] Tobjörk, D.; Österbacka, R. Paper electronics. *Adv. Mater.*, **2011**, *23*(17), 1935-1961.
[http://dx.doi.org/10.1002/adma.201004692] [PMID: 21433116]

[140] (a). Takushi, S. Nanotechnology Incorporated Into Paper Electronics. *Available at: http://guardianlv.com/2014/03/nanotechnology-incorporated-into-paper-electronics/.*, 2016 [Accessed on: Feb 26 2016];
(b). Martins, R.; Nathan, A.; Barros, R.; Pereira, L.; Barquinha, P.; Correia, N.; Costa, R.; Ahnood, A.;

Ferreira, I.; Fortunato, E. Complementary metal oxide semiconductor technology with and on paper. *Adv. Mater.,* **2011**, *23*(39), 4491-4496.
[http://dx.doi.org/10.1002/adma.201102232] [PMID: 21898609]
(c). Russo, A.; Ahn, B.Y.; Adams, J.J.; Duoss, E.B.; Bernhard, J.T.; Lewis, J.A. Pen-on-paper flexible electronics. *Adv. Mater.,* **2011**, *23*(30), 3426-3430.
[http://dx.doi.org/10.1002/adma.201101328] [PMID: 21688330]

[141]　Paper Electronics is Successful: Where Next? *Available at: http://www.nanotech-now.com/ news.cgi?story_id=44546.,* 2016 [Accessed on: Feb 2016];

[142]　Morales-Narváez, E.; Golmohammadi, H.; Naghdi, T.; Yousefi, H.; Kostiv, U.; Horák, D.; Pourreza, N.; Merkoçi, A. Nanopaper as an optical sensing platform. *ACS Nano,* **2015**, *9*(7), 7296-7305.
[http://dx.doi.org/10.1021/acsnano.5b03097] [PMID: 26135050]

[143]　Zhong, J.; Zhu, H.; Zhong, Q.; Dai, J.; Li, W.; Jang, S-H.; Yao, Y.; Henderson, D.; Hu, Q.; Hu, L.; Zhou, J. Self-Powered Human-Interactive Transparent Nanopaper Systems. *ACS Nano,* **2015**, *9*(7), 7399-7406.
[http://dx.doi.org/10.1021/acsnano.5b02414] [PMID: 26118467]

[144]　(a). Choi, W.; Chung, D.; Kang, J.; Kim, H.; Jin, Y.; Han, I.; Lee, Y.; Jung, J.; Lee, N.; Park, G. Fully sealed, high-brightness carbon-nanotube field-emission display. *Appl. Phys. Lett.,* **1999**, *75*(20), 3129-3131.
[http://dx.doi.org/10.1063/1.125253]
(b). Chang, P-L.; Wu, C-C.; Leu, H-J. Using patent analyses to monitor the technological trends in an emerging field of technology: a case of carbon nanotube field emission display. *Scientometrics,* **2010**, *82*(1), 5-19.
[http://dx.doi.org/10.1007/s11192-009-0033-y]

[145]　Saito, Y. *Carbon Nanotube and Related Field Emitters: Fundamentals and Applications*; Wiley, **2010**.
[http://dx.doi.org/10.1002/9783527630615]

[146]　Sharma, P. K.; Miao, W.; Giri, A.; Raghunathan, S. *Dekker encyclopedia of nanoscience and nanotechnology.,* **2005**.

[147]　(a). Jang, E.; Jun, S.; Jang, H.; Lim, J.; Kim, B.; Kim, Y. White-light-emitting diodes with quantum dot color converters for display backlights. *Adv. Mater.,* **2010**, *22*(28), 3076-3080.
[http://dx.doi.org/10.1002/adma.201000525] [PMID: 20517873]
(b). Xu, J.; Liu, J.; Cui, D.; Gerhold, M.; Wang, A.Y.; Nagel, M.; Lippert, T.K. Laser-assisted forward transfer of multi-spectral nanocrystal quantum dot emitters. *Nanotechnology,* **2006**, *18*(2), 025403.
[http://dx.doi.org/10.1088/0957-4484/18/2/025403]

[148]　Berger, M. Nanotechnology drives electronic paper displays. *Available at: http://www.nanowerk.com/spotlight/spotid=14546.php.,* 2016 [Accessed on: Feb 2016];

[149]　Oh, S.W.; Kim, C.W.; Cha, H.J.; Pal, U.; Kang, Y.S. Encapsulated-dye all-organic charged colored ink nanoparticles for electrophoretic image display. *Adv. Mater.,* **2009**, *21*(48), 4987-4991.
[http://dx.doi.org/10.1002/adma.200901595] [PMID: 25377075]

[150]　(a). Pan, J.; Wang, J.; Shaddock, D.M. Lead-free solder joint reliability-state of the art and perspectives. *J. Microelectronics and Electronic Packaging.,* **2005**, *2*(1), 72-83.
[http://dx.doi.org/10.4071/1551-4897-2.1.72]
(b). Abtew, M.; Selvaduray, G. Lead-free solders in microelectronics. *Mater. Sci. Eng. Rep.,* **2000**, *27*(5), 95-141.
(c). Gibson, A.; Choi, S.; Bieler, T.; Subramanian, K. Environmental concerns and materials issues in manufactured solder joints. *Proceedings of the 1997 IEEE International Symposium on Electronics and the Environment,* **1997**, pp. 246-251.

[151]　Zinn, A.A.; Fried, A.; Stachowiak, T.; Chang, J.; Stoltenberg, R.M. *Nanoparticle paste formulations and methods for production and use thereof*; Google Patents, **2015**.

[152]　Nai, S.; Wei, J.; Gupta, M. Lead-free solder reinforced with multiwalled carbon nanotubes. *J. Electron. Mater.,* **2006**, *35*(7), 1518-1522.

[http://dx.doi.org/10.1007/s11664-006-0142-9]

[153] Lee, J-G.; Subramanian, K.; Lee, J-G. Development of nanocomposite lead-free electronic solders, Advanced Packaging Materials: Processes, Properties and Interfaces *Proceedings. International Symposium on, IEEE: 2005,* **2005**, 276-281.

[154] Jiang, H.; Moon, K-s.; Hua, F.; Wong, C. Synthesis and thermal and wetting properties of tin/silver alloy nanoparticles for low melting point lead-free solders. . *Chem. Mater.,* **2007**, *19*(18), 4482-4485.
[http://dx.doi.org/10.1021/cm0709976]

[155] Shu, Y.; Rajathurai, K.; Gao, F.; Cui, Q.; Gu, Z. Synthesis and thermal properties of low melting temperature tin/indium (Sn/In) lead-free nanosolders and their melting behavior in a vapor flux. *J. Alloys Compd.,* **2015**, *626*, 391-400.
[http://dx.doi.org/10.1016/j.jallcom.2014.11.173]

[156] Wadud, M.; Gafur, M.; Qadir, M.; Rahman, M. Thermal and Electrical Properties of Sn-Zn-Bi Ternary Soldering Alloys. *Mater. Sci. Appl.,* **2015**, *6*(11), 1008.
[http://dx.doi.org/10.4236/msa.2015.611100]

[157] Zinn, A.; Stoltenberg, R.; Fried, A.; Chang, J.; Elhawary, A.; Beddow, J.; Chiu, F. Nano copper based solder-free electronic assembly material. *Nanotech.,* **2012**, *2*, 71-74.

[158] Gadde, J.R.; Giramkar, V.D.; Joseph, S.; Phatak, G.J. In *CNT-lead free solder composite electrodeposition for obtaining high speed interconnect for memsapplication,* 2nd International Symposium on Physics and Technology of Sensors (ISPTS): 2015; pp 120-124.

[159] Wang, Z.L.; Song, J. Piezoelectric nanogenerators based on zinc oxide nanowire arrays. *Science,* **2006**, *312*(5771), 242-246.
[http://dx.doi.org/10.1126/science.1124005] [PMID: 16614215]

[160] Wang, S.; Lin, L.; Wang, Z.L. Nanoscale triboelectric-effect-enabled energy conversion for sustainably powering portable electronics. *Nano Lett.,* **2012**, *12*(12), 6339-6346.
[http://dx.doi.org/10.1021/nl303573d] [PMID: 23130843]

[161] Wang, Z.L. Self-powered nanosensors and nanosystems. *Adv. Mater.,* **2012**, *24*(2), 280-285.
[http://dx.doi.org/10.1002/adma.201102958] [PMID: 22329002]

[162] Wang, Z.L.; Zhu, G.; Yang, Y.; Wang, S.; Pan, C. Progress in nanogenerators for portable electronics. *Mater. Today,* **2012**, *15*(12), 532-543.
[http://dx.doi.org/10.1016/S1369-7021(13)70011-7]

[163] Wang, Z.L.; Wu, W. Nanotechnology-enabled energy harvesting for self-powered micro-/nanosystems. *Angew. Chem. Int. Ed. Engl.,* **2012**, *51*(47), 11700-11721.
[http://dx.doi.org/10.1002/anie.201201656] [PMID: 23124936]

[164] Herbert, G.J.; Iniyan, S.; Sreevalsan, E.; Rajapandian, S. A review of wind energy technologies. *Renew. Sustain. Energy Rev.,* **2007**, *11*(6), 1117-1145.
[http://dx.doi.org/10.1016/j.rser.2005.08.004]

[165] (a).Kalsin, A.M.; Fialkowski, M.; Paszewski, M.; Smoukov, S.K.; Bishop, K.J.; Grzybowski, B.A. Electrostatic self-assembly of binary nanoparticle crystals with a diamond-like lattice. *Science,* **2006**, *312*(5772), 420-424.
[http://dx.doi.org/10.1126/science.1125124] [PMID: 16497885]
(b).Fan, F-R.; Tian, Z-Q.; Wang, Z.L. Flexible triboelectric generator. *Nano Energy,* **2012**, *1*(2), 328-334.
[http://dx.doi.org/10.1016/j.nanoen.2012.01.004]
(c).Grzybowski, B.A.; Winkleman, A.; Wiles, J.A.; Brumer, Y.; Whitesides, G.M. Electrostatic self-assembly of macroscopic crystals using contact electrification. *Nat. Mater.,* **2003**, *2*(4), 241-245.
[http://dx.doi.org/10.1038/nmat860] [PMID: 12690397]

[166] Yang, Y.; Zhu, G.; Zhang, H.; Chen, J.; Zhong, X.; Lin, Z-H.; Su, Y.; Bai, P.; Wen, X.; Wang, Z.L. Triboelectric nanogenerator for harvesting wind energy and as self-powered wind vector sensor

system. *ACS Nano,* **2013**, *7*(10), 9461-9468.
[http://dx.doi.org/10.1021/nn4043157] [PMID: 24044652]

[167] Xie, Y.; Wang, S.; Lin, L.; Jing, Q.; Lin, Z-H.; Niu, S.; Wu, Z.; Wang, Z.L. Rotary triboelectric nanogenerator based on a hybridized mechanism for harvesting wind energy. *ACS Nano,* **2013**, *7*(8), 7119-7125.
[http://dx.doi.org/10.1021/nn402477h] [PMID: 23768179]

[168] Fan, F-R.; Lin, L.; Zhu, G.; Wu, W.; Zhang, R.; Wang, Z.L. Transparent triboelectric nanogenerators and self-powered pressure sensors based on micropatterned plastic films. *Nano Lett.,* **2012**, *12*(6), 3109-3114.
[http://dx.doi.org/10.1021/nl300988z] [PMID: 22577731]

[169] Werner, M.; Kohly, W.; Šimić, M. *Nanotechnologies in Automobiles – Innovation Potentials in Hesse for the Automotive Industry and its Subcontractors*; Hessen Nanotech, **2008**.

[170] Malani, A.S.; Chaudhari, A.D.; Sambhe, R.U. A Review on Applications of Nanotechnology in Automotive Industry. *World Academy of Science, Engineering and Technology, International Journal of Mechanical, Aerospace, Industrial. Mechatronic and Manufacturing Engineering.,* **2015**, *10*(1), 36-40.

[171] Wong, K.V.; Paddon, P.A. Nanotechnology impact on the automotive industry. *Recent Pat. Nanotechnol.,* **2014**, *8*(3), 181-199.
[http://dx.doi.org/10.2174/187221050803141027101058] [PMID: 25360613]

[172] America on Wheels: Safe and Green with the Help of Nanotechnology. *Available at: https://www.crcpress.com/authors/news/i2129-america-on-wheels-safe-and-green-with-the-help-of-nanotechnology.,* 2016 [Accessed on: Jan 2016];

[173] chugh, K. Green Nanotechnology: It's easier than you think. *Available at: http://followgreenliving.com/green-nanotechnology/.,* 2016 [Accessed on: Jan 2016];

[174] Arivalagan, K.; Ravichandran, S.; Rangasamy, K.; Karthikeyan, E. Nanomaterials and its potential applications. *Int. J. Chemtech Res.,* **2011**, *3*(2), 534-538.

[175] Soutter, W. , Nanotechnology in Electric Vehicle Batteries. *Available at: http://www.azonano.com/article.aspx?ArticleID=3157.,* 2016 [Accessed on: Jan 2016];

[176] (a). Gupta, H.; Agrawal, G.; Mathur, J. An overview of Nanofluids: A new media towards green environment. *Int. J. Environ. Sci.,* **2012**, *3*(1), 433-440.(b). Mishra, P. C.; Nayak, S. K.; Mitra, P.; Mukherjee, S.; Paria, S. Application and Future of Nanofluids in Automobiles: An Overview on Current Research.

[177] Bartos, P.; Chemistry, R.S.o. *Nanotechnology in Construction*; Royal Society of Chemistry, **2004**.
[http://dx.doi.org/10.1039/9781847551528]

[178] Elvin, D.G. Nanotechnology for Green Building Green Technology Forum: 2007.

[179] Greßler, S.; Nentwich, M. *Nano and the environment – Part I: Potential environmental benefits and sustainability effects (NanoTrust Dossier No. 026en – March 2012);* Wien, 2012-03-03, 2012; p 4.

[180] Grieger, K. D.; Hjorth, R.; Rice, J.; Kumar, N.; Bang, J. *Nano-remediation: tiny particles cleaning up big environmental problems.,* **2015**.

[181] *Water for a sustainable World, The United Nations World Water Development Report,* United Nations 2015.

[182] Fryxell, G.E.; Lin, Y.; Fiskum, S.; Birnbaum, J.C.; Wu, H.; Kemner, K.; Kelly, S. Actinide sequestration using self-assembled monolayers on mesoporous supports. *Environ. Sci. Technol.,* **2005**, *39*(5), 1324-1331.
[http://dx.doi.org/10.1021/es049201j] [PMID: 15787373]

[183] Tratnyek, P.G.; Johnson, R.L. Nanotechnologies for environmental cleanup. *Nano Today,* **2006**, *1*(2), 44-48.

[http://dx.doi.org/10.1016/S1748-0132(06)70048-2]

[184] Qu, X.; Brame, J.; Li, Q.; Alvarez, P.J. Nanotechnology for a safe and sustainable water supply: enabling integrated water treatment and reuse. *Acc. Chem. Res.*, **2013**, *46*(3), 834-843.
[http://dx.doi.org/10.1021/ar300029v] [PMID: 22738389]

[185] Li, Q.; Mahendra, S.; Lyon, D.Y.; Brunet, L.; Liga, M.V.; Li, D.; Alvarez, P.J. Antimicrobial nanomaterials for water disinfection and microbial control: potential applications and implications. *Water Res.*, **2008**, *42*(18), 4591-4602.
[http://dx.doi.org/10.1016/j.watres.2008.08.015] [PMID: 18804836]

[186] Tchobanoglous, G.; Burton, F.L.; Stensel, H.D. *Metcalf; Eddy, Wastewater Engineering: Treatment and Reuse*; McGraw-Hill Education, **2003**.

[187] Qu, X.; Alvarez, P.J.; Li, Q. Applications of nanotechnology in water and wastewater treatment. *Water Res.*, **2013**, *47*(12), 3931-3946.
[http://dx.doi.org/10.1016/j.watres.2012.09.058] [PMID: 23571110]

[188] Yavuz, C.T.; Mayo, J.T.; Yu, W.W.; Prakash, A.; Falkner, J.C.; Yean, S.; Cong, L.; Shipley, H.J.; Kan, A.; Tomson, M.; Natelson, D.; Colvin, V.L. Low-field magnetic separation of monodisperse Fe3O4 nanocrystals. *Science*, **2006**, *314*(5801), 964-967.
[http://dx.doi.org/10.1126/science.1131475] [PMID: 17095696]

[189] Brame, J.; Li, Q.; Alvarez, P.J. Nanotechnology-enabled water treatment and reuse: emerging opportunities and challenges for developing countries. *Trends Food Sci. Technol.*, **2011**, *22*(11), 618-624.
[http://dx.doi.org/10.1016/j.tifs.2011.01.004]

[190] Hossain, F.; Perales-Perez, O.J.; Hwang, S.; Román, F. Antimicrobial nanomaterials as water disinfectant: applications, limitations and future perspectives. *Sci. Total Environ.*, **2014**, *466-467*, 1047-1059.
[http://dx.doi.org/10.1016/j.scitotenv.2013.08.009] [PMID: 23994736]

[191] (a). Dong, S.; Huang, G.; Wang, X.; Hu, Q.; Huang, T. Magnetically mixed hemimicelles solid-phase extraction based on ionic liquid-coated Fe3O4 nanoparticles for the analysis of trace organic contaminants in water. *Anal. Methods*, **2014**, *6*(17), 6783-6788.
[http://dx.doi.org/10.1039/C4AY01185F]
(b). Sun, L.; Sun, X.; Du, X.; Yue, Y.; Chen, L.; Xu, H.; Zeng, Q.; Wang, H.; Ding, L. Determination of sulfonamides in soil samples based on alumina-coated magnetite nanoparticles as adsorbents. *Anal. Chim. Acta*, **2010**, *665*(2), 185-192.
[http://dx.doi.org/10.1016/j.aca.2010.03.044] [PMID: 20417329]
(c). Zhu, L.; Pan, D.; Ding, L.; Tang, F.; Zhang, Q.; Liu, Q.; Yao, S. Mixed hemimicelles SPE based on CTAB-coated Fe3O4/SiO2 NPs for the determination of herbal bioactive constituents from biological samples. *Talanta*, **2010**, *80*(5), 1873-1880.
[http://dx.doi.org/10.1016/j.talanta.2009.10.037] [PMID: 20152426]

[192] Yao, Y.; Miao, S.; Liu, S.; Ma, L.P.; Sun, H.; Wang, S. Synthesis, characterization, and adsorption properties of magnetic Fe3O4@ graphene nanocomposite. *Chem. Eng. J.*, **2012**, *184*, 326-332.
[http://dx.doi.org/10.1016/j.cej.2011.12.017]

[193] Zhao, G.; Song, S.; Wang, C.; Wu, Q.; Wang, Z. Determination of triazine herbicides in environmental water samples by high-performance liquid chromatography using graphene-coated magnetic nanoparticles as adsorbent. *Anal. Chim. Acta*, **2011**, *708*(1-2), 155-159.
[http://dx.doi.org/10.1016/j.aca.2011.10.006] [PMID: 22093359]

[194] Wiles, P.; Abrahamson, J. Carbon fibre layers on arc electrodes—I: their properties and cool-down behaviour. *Carbon*, **1978**, *16*(5), 341-349.
[http://dx.doi.org/10.1016/0008-6223(78)90072-6]

[195] Iijima, S. Helical microtubules of graphitic carbon. *Nature*, **1991**, *354*(6348), 56-58.
[http://dx.doi.org/10.1038/354056a0]

[196]   Jung, C.; Heo, J.; Han, J.; Her, N.; Lee, S-J.; Oh, J.; Ryu, J.; Yoon, Y. Hexavalent chromium removal by various adsorbents: powdered activated carbon, chitosan, and single/multi-walled carbon nanotubes. *Separ. Purif. Tech.,* **2013,** *106,* 63-71.
[http://dx.doi.org/10.1016/j.seppur.2012.12.028]

[197]   Upadhyayula, V.K.; Deng, S.; Mitchell, M.C.; Smith, G.B. Application of carbon nanotube technology for removal of contaminants in drinking water: a review. *Sci. Total Environ.,* **2009,** *408*(1), 1-13.
[http://dx.doi.org/10.1016/j.scitotenv.2009.09.027] [PMID: 19819525]

[198]   aGupta, V.K.; Kumar, R.; Nayak, A.; Saleh, T.A.; Barakat, M.A. Adsorptive removal of dyes from aqueous solution onto carbon nanotubes: a review. *Adv. Colloid Interface Sci.,* **2013,** *193-194,* 24-34.
[http://dx.doi.org/10.1016/j.cis.2013.03.003] [PMID: 23579224] bYu, J-G.; Zhao, X-H.; Yang, H.; Chen, X-H.; Yang, Q.; Yu, L-Y.; Jiang, J-H.; Chen, X-Q. Aqueous adsorption and removal of organic contaminants by carbon nanotubes. *Sci. Total Environ.,* **2014,** *482-483,* 241-251.
[http://dx.doi.org/10.1016/j.scitotenv.2014.02.129] [PMID: 24657369]

[199]   Wang, H.; Lin, K-Y.; Jing, B.; Krylova, G.; Sigmon, G.E.; McGinn, P.; Zhu, Y.; Na, C. Removal of oil droplets from contaminated water using magnetic carbon nanotubes. *Water Res.,* **2013,** *47*(12), 4198-4205.
[http://dx.doi.org/10.1016/j.watres.2013.02.056] [PMID: 23582309]

[200]   Brodie, B. Note sur un nouveau procédé pour la purification et la désagrégation du graphite. *Ann. Chim. Phys.,* **1855,** *45,* 351-353.

[201]   Novoselov, K.S.; Geim, A.K.; Morozov, S.V.; Jiang, D.; Zhang, Y.; Dubonos, S.V.; Grigorieva, I.V.; Firsov, A.A. Electric field effect in atomically thin carbon films. *Science,* **2004,** *306*(5696), 666-669.
[http://dx.doi.org/10.1126/science.1102896] [PMID: 15499015]

[202]   Sitko, R.; Zawisza, B.; Malicka, E. Graphene as a new sorbent in analytical chemistry. *TrAC Trends in Analytical Chemistry.,* **2013,** *51,* 33-43.
[http://dx.doi.org/10.1016/j.trac.2013.05.011]

[203]   Kyzas, G.Z.; Deliyanni, E.A.; Matis, K.A. Graphene oxide and its application as an adsorbent for wastewater treatment. *J. Chem. Technol. Biotechnol.,* **2014,** *89*(2), 196-205.
[http://dx.doi.org/10.1002/jctb.4220]

[204]   Sedaghat, S. Synthesis of clay-CNTs nanocomposite. *J. Nanostruct. Chem.,* **2013,** *3*(1), 1-4.
[http://dx.doi.org/10.1186/2193-8865-3-24]

[205]   Singh, V.; Joung, D.; Zhai, L.; Das, S.; Khondaker, S.I.; Seal, S. Graphene based materials: past, present and future. *Prog. Mater. Sci.,* **2011,** *56*(8), 1178-1271.
[http://dx.doi.org/10.1016/j.pmatsci.2011.03.003]

[206]   Perreault, F.; Fonseca de Faria, A.; Elimelech, M. Environmental applications of graphene-based nanomaterials. *Chem. Soc. Rev.,* **2015,** *44*(16), 5861-5896.
[http://dx.doi.org/10.1039/C5CS00021A] [PMID: 25812036]

[207]   Dimiev, A.M.; Tour, J.M. Mechanism of graphene oxide formation. *ACS Nano,* **2014,** *8*(3), 3060-3068.
[http://dx.doi.org/10.1021/nn500606a] [PMID: 24568241]

[208]   Apul, O.G.; Wang, Q.; Zhou, Y.; Karanfil, T. Adsorption of aromatic organic contaminants by graphene nanosheets: comparison with carbon nanotubes and activated carbon. *Water Res.,* **2013,** *47*(4), 1648-1654.
[http://dx.doi.org/10.1016/j.watres.2012.12.031] [PMID: 23313232]

[209]   Zhao, J.; Wang, Z.; White, J.C.; Xing, B. Graphene in the aquatic environment: adsorption, dispersion, toxicity and transformation. *Environ. Sci. Technol.,* **2014,** *48*(17), 9995-10009.
[http://dx.doi.org/10.1021/es5022679] [PMID: 25122195]

[210]   Jacoby, M. Graphene moves toward applications: as composites and inks become commercial products, advanced electronics remain a long way off. *Chem. Eng. News,* **2011,** *89,* 10-15.

[211] Sohni, S.; Gul, K.; Ahmad, F.; Ahmad, I.; Khan, A.; Khan, N.; Bahadar Khan, S. Highly efficient removal of acid red☐17 and bromophenol blue dyes from industrial wastewater using graphene oxide functionalized magnetic chitosan composite. *Polym. Compos.,* **2017**.
[http://dx.doi.org/10.1002/pc.24349]

[212] Zhang, Y.; Chen, B.; Zhang, L.; Huang, J.; Chen, F.; Yang, Z.; Yao, J.; Zhang, Z. Controlled assembly of Fe_3O_4 magnetic nanoparticles on graphene oxide. *Nanoscale,* **2011**, *3*(4), 1446-1450.
[http://dx.doi.org/10.1039/c0nr00776e] [PMID: 21301708]

[213] Gul, K.; Sohni, S.; Ahmad, I.; Ullah Khattak, N.; Zada, R.; Akhtar, N.; Bahisht, N. Synthesis and Characterization of Graphene/Fe_3O_4 Nanocomposite as an Effective Adsorbent for Removal of Acid Red-17 and Remazol Brilliant Blue R from Aqueous Solutions. *Curr. Nanosci.,* **2016**, *12*(5), 554-563.
[http://dx.doi.org/10.2174/1573413712666160224004953]

[214] (a). Cong, H-P.; Chen, J-F.; Yu, S-H. Graphene-based macroscopic assemblies and architectures: an emerging material system. *Chem. Soc. Rev.,* **2014**, *43*(21), 7295-7325.
[http://dx.doi.org/10.1039/C4CS00181H] [PMID: 25065466]
(b). Bai, H.; Li, C.; Shi, G. Functional composite materials based on chemically converted graphene. *Adv. Mater.,* **2011**, *23*(9), 1089-1115.
[http://dx.doi.org/10.1002/adma.201003753] [PMID: 21360763]

[215] Sheet, I.; Kabbani, A.; Holail, H. Removal of heavy metals using nanostructured graphite oxide, silica nanoparticles and silica/graphite oxide composite. *Energy Procedia,* **2014**, *50*, 130-138.
[http://dx.doi.org/10.1016/j.egypro.2014.06.016]

[216] Luo, Y-B.; Shi, Z-G.; Gao, Q.; Feng, Y-Q. Magnetic retrieval of graphene: extraction of sulfonamide antibiotics from environmental water samples. *J. Chromatogr. A,* **2011**, *1218*(10), 1353-1358.
[http://dx.doi.org/10.1016/j.chroma.2011.01.022] [PMID: 21288529]

[217] Mohmood, I.; Lopes, C.B.; Lopes, I.; Ahmad, I.; Duarte, A.C.; Pereira, E. Nanoscale materials and their use in water contaminants removal-a review. *Environ. Sci. Pollut. Res. Int.,* **2013**, *20*(3), 1239-1260.
[http://dx.doi.org/10.1007/s11356-012-1415-x] [PMID: 23292223]

[218] Kostal, J.; Prabhukumar, G.; Lao, U.L.; Chen, A.; Matsumoto, M.; Mulchandani, A.; Chen, W. Customizable biopolymers for heavy metal remediation. *J. Nanopart. Res.,* **2005**, *7*(4-5), 517-523.
[http://dx.doi.org/10.1007/s11051-005-5132-y]

[219] Watlington, K. *Emerging nanotechnologies for site remediation and wastewater treatment. Report*; US Environmental Protection Agency, Office of Solid Waste and Emergency Response, Office of Superfund Remediation and Technology Innovation, Technology Innovation and Field Services Division: Washington, DC, **2005**.

[220] Diallo, M.S.; Christie, S.; Swaminathan, P.; Johnson, J.H., Jr; Goddard, W.A., III Dendrimer enhanced ultrafiltration. 1. Recovery of Cu(II) from aqueous solutions using PAMAM dendrimers with ethylene diamine core and terminal NH2 groups. *Environ. Sci. Technol.,* **2005**, *39*(5), 1366-1377.
[http://dx.doi.org/10.1021/es048961r] [PMID: 15787379]

[221] Knecht, M.R.; Garcia-Martinez, J.C.; Crooks, R.M. Synthesis, characterization, and magnetic properties of dendrimer-encapsulated nickel nanoparticles containing< 150 atoms. *Chem. Mater.,* **2006**, *18*(21), 5039-5044.
[http://dx.doi.org/10.1021/cm061272p]

[222] (a). Ottaviani, M.F.; Montalti, F.; Romanelli, M.; Turro, N.J.; Tomalia, D.A. Characterization of starburst dendrimers by EPR. 4. Mn (II) as a probe of interphase properties. *J. Phys. Chem.,* **1996**, *100*(26), 11033-11042.
[http://dx.doi.org/10.1021/jp953261h]
(b). Hedden, R.C.; Bauer, B.J. Structure and dimensions of PAMAM/PEG dendrimer-star polymers. *Macromolecules,* **2003**, *36*(6), 1829-1835.
[http://dx.doi.org/10.1021/ma025752n]

[223] Diallo, M.S.; Balogh, L.; Shafagati, A.; Johnson, J.H.; Goddard, W.A.; Tomalia, D.A. Poly (amidoamine) dendrimers: a new class of high capacity chelating agents for Cu (II) ions. *Environ. Sci. Technol.,* **1999**, *33*(5), 820-824.
[http://dx.doi.org/10.1021/es980521a]

[224] Xu, Y.; Zhao, D. Removal of copper from contaminated soil by use of poly(amidoamine) dendrimers. *Environ. Sci. Technol.,* **2005**, *39*(7), 2369-2375.
[http://dx.doi.org/10.1021/es040380e] [PMID: 15871278]

[225] Barakat, M.A.; Ramadan, M.H.; Alghamdi, M.A.; Algarny, S.S.; Woodcock, H.L.; Kuhn, J.N. Remediation of Cu(II), Ni(II), and Cr(III) ions from simulated wastewater by dendrimer/titania composites. *J. Environ. Manage.,* **2013**, *117*, 50-57.
[http://dx.doi.org/10.1016/j.jenvman.2012.12.025] [PMID: 23353877]

[226] Chou, C-M.; Lien, H-L. Dendrimer-conjugated magnetic nanoparticles for removal of zinc (II) from aqueous solutions. . *J. Nanopart. Res.,* **2011**, *13*(5), 2099-2107.
[http://dx.doi.org/10.1007/s11051-010-9967-5]

[227] Fryxell, G.E.; Wu, H.; Lin, Y.; Shaw, W.J.; Birnbaum, J.C.; Linehan, J.C.; Nie, Z.; Kemner, K.; Kelly, S. Lanthanide selective sorbents: self-assembled monolayers on mesoporous supports (SAMMS). *J. Mater. Chem.,* **2004**, *14*(22), 3356-3363.
[http://dx.doi.org/10.1039/b408181a]

[228] (a). Mattigod, S.V.; Fryxell, G.E.; Parker, K.E. Anion binding in self-assembled monolayers in mesoporous supports (SAMMS). *Inorg. Chem. Commun.,* **2007**, *10*(6), 646-648.
[http://dx.doi.org/10.1016/j.inoche.2007.02.014]
(b). Mattigod, S.V.; Feng, X.; Fryxell, G.E.; Liu, J.; Gong, M. Separation of complexed mercury from aqueous wastes using self-assembled mercaptan on mesoporous silica. *Sep. Sci. Technol.,* **1999**, *34*(12), 2329-2345.
[http://dx.doi.org/10.1081/SS-100100775]
(c). Lin, Y.; Fryxell, G.E.; Wu, H.; Engelhard, M. Selective sorption of cesium using self-assembled monolayers on mesoporous supports. *Environ. Sci. Technol.,* **2001**, *35*(19), 3962-3966.
[http://dx.doi.org/10.1021/es010710k] [PMID: 11642461]
(d). Yantasee, W.; Lin, Y.; Fryxell, G.E.; Busche, B.J.; Birnbaum, J.C. Removal of heavy metals from aqueous solution using novel nanoengineered sorbents: self-assembled carbamoylphosphonic acids on mesoporous silica. *Sep. Sci. Technol.,* **2003**, *38*(15), 3809-3825.
[http://dx.doi.org/10.1081/SS-120024232]

[229] Mansoori, G.A. T. R. B., A. Ahmadpour, Z. and Eshaghi Environmental Application of Nanotechnology.*Annual Review of Nano Research*; World Scientific, **2008**, Vol. 2, pp. 439-493.
[http://dx.doi.org/10.1142/9789812790248_0010]

[230] Fryxell, G.E.; Lin, Y.; Wu, H.; Kemner, K.M. Environmental applications of self-assembled monolayers on mesoporous supports (SAMMS). *Stud. Surf. Sci. Catal.,* **2002**, *141*, 583-590.
[http://dx.doi.org/10.1016/S0167-2991(02)80593-6]

[231] Bois, L.; Bonhommé, A.; Ribes, A.; Pais, B.; Raffin, G.; Tessier, F. Functionalized silica for heavy metal ions adsorption. *Colloids Surf. A Physicochem. Eng. Asp.,* **2003**, *221*(1), 221-230.
[http://dx.doi.org/10.1016/S0927-7757(03)00138-9]

[232] Fryxell, G.E.; Mattigod, S.V.; Lin, Y.; Wu, H.; Fiskum, S.; Parker, K.; Zheng, F.; Yantasee, W.; Zemanian, T.S.; Addleman, R.S. Design and synthesis of self-assembled monolayers on mesoporous supports (SAMMS): The importance of ligand posture in functional nanomaterials. *J. Mater. Chem.,* **2007**, *17*(28), 2863-2874.
[http://dx.doi.org/10.1039/b702422c]

[233] (a). Chen, X.; Feng, X.; Liu, J.; Fryxell, G.E.; Gong, M. Mercury separation and immobilization using self-assembled monolayers on mesoporous supports (SAMMS). *Sep. Sci. Technol.,* **1999**, *34*(6-7), 1121-1132.
[http://dx.doi.org/10.1080/01496399908951084]

(b). Watlington, K. *Emerging nanotechnologies for site remediation and wastewater treatment,* **2005**.
(c). Kwon, S.; Thomas, J.; Reed, B.E.; Levine, L.; Magar, V.S.; Farrar, D.; Bridges, T.S.; Ghosh, U. Evaluation of sorbent amendments for *in situ* remediation of metal-contaminated sediments. *Environ. Toxicol. Chem.,* **2010**, *29*(9), 1883-1892.
[PMID: 20821645]

[234] (a). Fryxell, G.E.; Liu, J.; Hauser, T.A.; Nie, Z.; Ferris, K.F.; Mattigod, S.; Gong, M.; Hallen, R.T. Design and synthesis of selective mesoporous anion traps. *Chem. Mater.,* **1999**, *11*(8), 2148-2154.
[http://dx.doi.org/10.1021/cm990104c]
(b). Kelly, S.; Kemner, K.; Fryxell, G.E.; Liu, J.; Mattigod, S.V.; Ferris, K. X-ray-absorption fine-structure spectroscopy study of the interactions between contaminant tetrahedral anions and self-assembled monolayers on mesoporous supports. *J. Phys. Chem. B,* **2001**, *105*(27), 6337-6346.
[http://dx.doi.org/10.1021/jp0045890]

[235] Mattigod, S. V.; Fryxell, G. E.; Serne, R. J.; Parker, K. E. **2003**.

[236] Kemner, K.M.; Feng, X.; Liu, J.; Fryxell, G.E.; Wang, L-Q.; Kim, A.Y.; Gong, M.; Mattigod, S. Investigation of the local chemical interactions between Hg and self-assembled monolayers on mesoporous supports. *J. Synchrotron Radiat.,* **1999**, *6*(Pt 3), 633-635.
[http://dx.doi.org/10.1107/S090904959801560X] [PMID: 15263405]

[237] Fryxell, G.E.; Liu, J.; Papirer, E. *Designing surface chemistry in mesoporous silica*; Pacific Northwest National Lab: Richland, WA, US, **2000**.

[238] Lin, Y.; Fiskum, S.K.; Yantasee, W.; Wu, H.; Mattigod, S.V.; Vorpagel, E.; Fryxell, G.E.; Raymond, K.N.; Xu, J. Incorporation of hydroxypyridinone ligands into self-assembled monolayers on mesoporous supports for selective actinide sequestration. *Environ. Sci. Technol.,* **2005**, *39*(5), 1332-1337.
[http://dx.doi.org/10.1021/es049169t] [PMID: 15787374]

[239] Park, Y.; Shin, W.S.; Choi, S-J. Removal of Co, Sr and Cs from aqueous solution using self-assembled monolayers on mesoporous supports. *Korean J. Chem. Eng.,* **2012**, *29*(11), 1556-1566.
[http://dx.doi.org/10.1007/s11814-012-0035-y]

[240] (a). Balachandran, K.; Venckatesh, R.; Sivaraj, R.; Rajiv, P. TiO_2 nanoparticles *versus* TiO_2-SiO_2 nanocomposites: a comparative study of photo catalysis on acid red 88. *Spectrochim. Acta A Mol. Biomol. Spectrosc.,* **2014**, *128*, 468-474.
[http://dx.doi.org/10.1016/j.saa.2014.02.127] [PMID: 24682063]
(b). Li, D.; Qu, J. The progress of catalytic technologies in water purification: a review. *J. Environ. Sci. (China),* **2009**, *21*(6), 713-719.
[http://dx.doi.org/10.1016/S1001-0742(08)62329-3] [PMID: 19803071]

[241] Herrmann, J-M. Heterogeneous photocatalysis: an emerging discipline involving multiphase systems. *Catal. Today,* **1995**, *24*(1), 157-164.
[http://dx.doi.org/10.1016/0920-5861(95)00005-Z]

[242] Tang, J.; Zou, Z.; Ye, J. Efficient photocatalytic decomposition of organic contaminants over CaBi2O4 under visible-light irradiation. *Angew. Chem. Int. Ed. Engl.,* **2004**, *43*(34), 4463-4466.
[http://dx.doi.org/10.1002/anie.200353594] [PMID: 15340944]

[243] Zhu, H.; Jiang, R.; Fu, Y.; Guan, Y.; Yao, J.; Xiao, L.; Zeng, G. Effective photocatalytic decolorization of methyl orange utilizing TiO_2/ZnO/chitosan nanocomposite films under simulated solar irradiation. *Desalination,* **2012**, *286*, 41-48.
[http://dx.doi.org/10.1016/j.desal.2011.10.036]

[244] Bahnemann, D.W. Ultrasmall metal oxide particles: preparation, photophysical characterization, and photocatalytic properties. *Isr. J. Chem.,* **1993**, *33*(1), 115-136.
[http://dx.doi.org/10.1002/ijch.199300017]

[245] Nadejde, C.; Neamtu, M.; Schneider, R.; Hodoroaba, V-D.; Ababei, G.; Panne, U. Catalytical degradation of relevant pollutants from waters using magnetic nanocatalysts. *Appl. Surf. Sci.,* **2015**,

352, 42-48.
[http://dx.doi.org/10.1016/j.apsusc.2015.01.036]

[246] (a). Pouretedal, H.; Shafeie, A.; Keshavarz, M. Preparation, characterization and catalytic activity of tin dioxide and zero-valent tin nanoparticles. *J. Korean Chem. Society.,* **2012**, *56*(4), 484-490.
[http://dx.doi.org/10.5012/jkcs.2012.56.4.484]
(b). Mizukoshi, Y.; Sato, K.; Konno, T.J.; Masahashi, N. Dependence of photocatalytic activities upon the structures of Au/Pd bimetallic nanoparticles immobilized on TiO_2 surface. *Appl. Catal. B,* **2010**, *94*(3), 248-253.
[http://dx.doi.org/10.1016/j.apcatb.2009.11.015]

[247] Fu, Y.; Chen, Q.; He, M.; Wan, Y.; Sun, X.; Xia, H.; Wang, X. Copper ferrite-graphene hybrid: a multifunctional heteroarchitecture for photocatalysis and energy storage. *Ind. Eng. Chem. Res.,* **2012**, *51*(36), 11700-11709.
[http://dx.doi.org/10.1021/ie301347j]

[248] Hoang, S.; Berglund, S.P.; Hahn, N.T.; Bard, A.J.; Mullins, C.B. Enhancing visible light photo-oxidation of water with TiO_2 nanowire arrays *via* cotreatment with H_2 and NH_3: synergistic effects between Ti^{3+} and N. *J. Am. Chem. Soc.,* **2012**, *134*(8), 3659-3662.
[http://dx.doi.org/10.1021/ja211369s] [PMID: 22316385]

[249] Pelaez, M.; Nolan, N.T.; Pillai, S.C.; Seery, M.K.; Falaras, P.; Kontos, A.G.; Dunlop, P.S.; Hamilton, J.W.; Byrne, J.A.; O'shea, K. A review on the visible light active titanium dioxide photocatalysts for environmental applications. *Appl. Catal. B,* **2012**, *125*, 331-349.
[http://dx.doi.org/10.1016/j.apcatb.2012.05.036]

[250] (a). Macak, J. M.; Zlamal, M.; Krysa, J.; Schmuki, P. **2007**.
(b). Liang, H. C.; Li, X. Z.; Nowotny, J. **2010**.

[251] Zhao, D.; Sheng, G.; Chen, C.; Wang, X. Enhanced photocatalytic degradation of methylene blue under visible irradiation on graphene@ TiO_2 dyade structure. *Appl. Catal. B,* **2012**, *111*, 303-308.
[http://dx.doi.org/10.1016/j.apcatb.2011.10.012]

[252] Zhang, X.; Sun, Y.; Cui, X.; Jiang, Z. A green and facile synthesis of TiO_2/graphene nanocomposites and their photocatalytic activity for hydrogen evolution. *Int. J. Hydrogen Energy,* **2012**, *37*(1), 811-815.
[http://dx.doi.org/10.1016/j.ijhydene.2011.04.053]

[253] Moradi, S.; Aberomand, A. P.; Khodadadi, B.; Abedini, K. S.; Givian, R. M. H. Preparation, characterization, and investigation of photocatalytic activity of TiO_2/SiO_2/Co nanocomposite using additives. **2013**.

[254] Ruzmanova, I.; Stoller, M.; Chianese, A. Photocatalytic treatment of olive mill waste water by magnetic core titanium dioxide nanoparticles. *Chem. Eng. Trans.,* **2013**, *32*, 2269-2274.

[255] Xie, J.; Li, Y.; Zhao, W.; Bian, L.; Wei, Y. Simple fabrication and photocatalytic activity of ZnO particles with different morphologies. *Powder Technol.,* **2011**, *207*(1), 140-144.
[http://dx.doi.org/10.1016/j.powtec.2010.10.019]

[256] Zhang, Z.; Shao, C.; Li, X.; Wang, C.; Zhang, M.; Liu, Y. Electrospun nanofibers of p-type NiO/n-type ZnO heterojunctions with enhanced photocatalytic activity. *ACS Appl. Mater. Interfaces,* **2010**, *2*(10), 2915-2923.
[http://dx.doi.org/10.1021/am100618h] [PMID: 20936796]

[257] Fu, Y.; Chen, H.; Sun, X.; Wang, X. Graphene-supported nickel ferrite: A magnetically separable photocatalyst with high activity under visible light. *AIChE J.,* **2012**, *58*(11), 3298-3305.
[http://dx.doi.org/10.1002/aic.13716]

[258] Wang, C.; Feng, C.; Gao, Y.; Ma, X.; Wu, Q.; Wang, Z. Preparation of a graphene-based magnetic nanocomposite for the removal of an organic dye from aqueous solution. *Chem. Eng. J.,* **2011**, *173*(1), 92-97.
[http://dx.doi.org/10.1016/j.cej.2011.07.041]

[259] (a). Yuan, Z-h. Synthesis, characterization and photocatalyticactivity of $ZnFe_2O_4/TiO_2$ nanocomposite. *J. Mater. Chem.,* **2001**, *11*(4), 1265-1268.
[http://dx.doi.org/10.1039/b006994i]
(b). Zhu, X.; Zhang, F.; Wang, M.; Ding, J.; Sun, S.; Bao, J.; Gao, C. Facile synthesis, structure and visible light photocatalytic activity of recyclable $ZnFe_2O_4/TiO_2$. *Appl. Surf. Sci.,* **2014**, *319*, 83-89.
[http://dx.doi.org/10.1016/j.apsusc.2014.07.051]

[260] Xiong, P.; Fu, Y.; Wang, L.; Wang, X. Multi-walled carbon nanotubes supported nickel ferrite: A magnetically recyclable photocatalyst with high photocatalytic activity on degradation of phenols. *Chem. Eng. J.,* **2012**, *195*, 149-157.
[http://dx.doi.org/10.1016/j.cej.2012.05.007]

[261] Fan, L.; Luo, C.; Sun, M.; Li, X.; Lu, F.; Qiu, H. Preparation of novel magnetic chitosan/graphene oxide composite as effective adsorbents toward methylene blue. *Bioresour. Technol.,* **2012**, *114*, 703-706.
[http://dx.doi.org/10.1016/j.biortech.2012.02.067] [PMID: 22464421]

[262] Bai, S.; Shen, X.; Zhong, X.; Liu, Y.; Zhu, G.; Xu, X.; Chen, K. One-pot solvothermal preparation of magnetic reduced graphene oxide-ferrite hybrids for organic dye removal. *Carbon,* **2012**, *50*(6), 2337-2346.
[http://dx.doi.org/10.1016/j.carbon.2012.01.057]

[263] Min, Y.; Zhang, F-J.; Zhao, W.; Zheng, F.; Chen, Y.; Zhang, Y. Hydrothermal synthesis of nanosized bismuth niobate and enhanced photocatalytic activity by coupling of graphene sheets. *Chem. Eng. J.,* **2012**, *209*, 215-222.
[http://dx.doi.org/10.1016/j.cej.2012.07.109]

[264] Sun, L.; Shao, R.; Tang, L.; Chen, Z. Synthesis of $ZnFe_2O_4/ZnO$ nanocomposites immobilized on graphene with enhanced photocatalytic activity under solar light irradiation. *J. Alloys Compd.,* **2013**, *564*, 55-62.
[http://dx.doi.org/10.1016/j.jallcom.2013.02.147]

[265] Fu, Y.; Wang, X. Magnetically separable $ZnFe_2O_4$–graphene catalyst and its high photocatalytic performance under visible light irradiation. *Ind. Eng. Chem. Res.,* **2011**, *50*(12), 7210-7218.
[http://dx.doi.org/10.1021/ie200162a]

[266] Fu, Y.; Xiong, P.; Chen, H.; Sun, X.; Wang, X. High photocatalytic activity of magnetically separable manganese ferrite–graphene heteroarchitectures. *Ind. Eng. Chem. Res.,* **2012**, *51*(2), 725-731.
[http://dx.doi.org/10.1021/ie2026212]

[267] Zhang, Y.; Zhang, N.; Tang, Z-R.; Xu, Y-J. Graphene transforms wide band gap ZnS to a visible light photocatalyst. The new role of graphene as a macromolecular photosensitizer. *ACS Nano,* **2012**, *6*(11), 9777-9789.
[http://dx.doi.org/10.1021/nn304154s] [PMID: 23106763]

[268] (a). Zhao, X.; Lv, L.; Pan, B.; Zhang, W.; Zhang, S.; Zhang, Q. Polymer-supported nanocomposites for environmental application: a review. *Chem. Eng. J.,* **2011**, *170*(2), 381-394.
[http://dx.doi.org/10.1016/j.cej.2011.02.071]
(b). Jiang, R.; Zhu, H.; Yao, J.; Fu, Y.; Guan, Y. Chitosan hydrogel films as a template for mild biosynthesis of CdS quantum dots with highly efficient photocatalytic activity. *Appl. Surf. Sci.,* **2012**, *258*(8), 3513-3518.
[http://dx.doi.org/10.1016/j.apsusc.2011.11.105]

[269] (a). Zhu, H.; Jiang, R.; Xiao, L.; Chang, Y.; Guan, Y.; Li, X.; Zeng, G. Photocatalytic decolorization and degradation of Congo Red on innovative crosslinked chitosan/nano-CdS composite catalyst under visible light irradiation. *J. Hazard. Mater.,* **2009**, *169*(1-3), 933-940.
[http://dx.doi.org/10.1016/j.jhazmat.2009.04.037] [PMID: 19477069]
(b). Jiang, R.; Zhu, H.; Li, X.; Xiao, L. Visible light photocatalytic decolourization of CI Acid Red 66 by chitosan capped CdS composite nanoparticles. *Chem. Eng. J.,* **2009**, *152*(2), 537-542.
[http://dx.doi.org/10.1016/j.cej.2009.05.037]

[270] Heiligtag, F.J.; Cheng, W.; de Mendonça, V.R.; Süess, M.J.; Hametner, K.; Günther, D.; Ribeiro, C.; Niederberger, M. Self-assembly of metal and metal oxide nanoparticles and nanowires into a macroscopic ternary aerogel monolith with tailored photocatalytic properties. *Chem. Mater.,* **2014**, *26*(19), 5576-5584.
[http://dx.doi.org/10.1021/cm502063f]

[271] Szabó, T.; Németh, J.; Dékány, I. Zinc oxide nanoparticles incorporated in ultrathin layer silicate films and their photocatalytic properties. *Colloids Surf. A Physicochem. Eng. Asp.,* **2003**, *230*(1), 23-35.
[http://dx.doi.org/10.1016/j.colsurfa.2003.09.010]

[272] (a). Kotov, N.A.; Meldrum, F.C.; Fendler, J.H. Monoparticulate layers of titanium dioxide nanocrystallites with controllable interparticle distances. *J. Phys. Chem.,* **1994**, *98*(36), 8827-8830.
[http://dx.doi.org/10.1021/j100087a002]
(b). De, I. Layered solid particles as self-assembled films. *Colloids Surf. A Physicochem. Eng. Asp.,* **1997**, *123*, 391-401.
(c). Kotov, N.; Meldrum, F.; Fendler, J.; Tombacz, E.; Dekany, I. Spreading of clay organocomplexes on aqueous solutions: construction of Langmuir-Blodgett clay organocomplex multilayer films. *Langmuir,* **1994**, *10*(10), 3797-3804.
[http://dx.doi.org/10.1021/la00022a066]

[273] Xiao, F-X.; Miao, J.; Liu, B. Layer-by-layer self-assembly of CdS quantum dots/graphene nanosheets hybrid films for photoelectrochemical and photocatalytic applications. *J. Am. Chem. Soc.,* **2014**, *136*(4), 1559-1569.
[http://dx.doi.org/10.1021/ja411651e] [PMID: 24392972]

[274] Zhang, Z.; Xiao, F.; Guo, Y.; Wang, S.; Liu, Y. One-pot self-assembled three-dimensional TiO$_2$-graphene hydrogel with improved adsorption capacities and photocatalytic and electrochemical activities. *ACS Appl. Mater. Interfaces,* **2013**, *5*(6), 2227-2233.
[http://dx.doi.org/10.1021/am303299r] [PMID: 23429833]

[275] Xiao, F-X.; Miao, J.; Wang, H-Y.; Liu, B. Self-assembly of hierarchically ordered CdS quantum dots-TiO$_2$ nanotube array heterostructures as efficient visible light photocatalysts for photoredox applications. *J. Mater. Chem. A Mater. Energy Sustain.,* **2013**, *1*(39), 12229-12238.
[http://dx.doi.org/10.1039/c3ta12856c]

[276] Liu, S.; Chen, Z.; Zhang, N.; Tang, Z-R.; Xu, Y-J. An Efficient Self-Assembly of CdS Nanowires–Reduced Graphene Oxide Nanocomposites for Selective Reduction of Nitro Organics under Visible Light Irradiation. *J. Phys. Chem. C,* **2013**, *117*(16), 8251-8261.
[http://dx.doi.org/10.1021/jp400550t]

[277] Das, S.; Sen, B.; Debnath, N. Recent trends in nanomaterials applications in environmental monitoring and remediation. *Environ. Sci. Pollut. Res. Int.,* **2015**, *22*(23), 18333-18344.
[http://dx.doi.org/10.1007/s11356-015-5491-6] [PMID: 26490920]

[278] Zhang, W-x. Nanoscale iron particles for environmental remediation: an overview. *J. Nanopart. Res.,* **2003**, *5*(3-4), 323-332.
[http://dx.doi.org/10.1023/A:1025520116015]

[279] Patil, S.S.; Shedbalkar, U.U.; Truskewycz, A.; Chopade, B.A.; Ball, A.S. Nanoparticles for environmental clean-up: A review of potential risks and emerging solutions. *Environmental Technology & Innovation,* **2016**, *5*, 10-21.
[http://dx.doi.org/10.1016/j.eti.2015.11.001]

[280] Lee, C-H.; Lin, T-S.; Mou, C-Y. Mesoporous materials for encapsulating enzymes. *Nano Today,* **2009**, *4*(2), 165-179.
[http://dx.doi.org/10.1016/j.nantod.2009.02.001]

[281] Wang, M.; Abad, D.; Kickhoefer, V.A.; Rome, L.H.; Mahendra, S. Vault Nanoparticles Packaged with Enzymes as an Efficient Pollutant Biodegradation Technology. *ACS Nano,* **2015**, *9*(11), 10931-10940.
[http://dx.doi.org/10.1021/acsnano.5b04073] [PMID: 26493711]

[282] Wang, P. Nanoscale biocatalyst systems. *Curr. Opin. Biotechnol.,* **2006**, *17*(6), 574-579.
[http://dx.doi.org/10.1016/j.copbio.2006.10.009] [PMID: 17084611]

[283] Kim, J.; Grate, J.W.; Wang, P. Nanostructures for enzyme stabilization. *Chem. Eng. Sci.,* **2006**, *61*(3), 1017-1026.
[http://dx.doi.org/10.1016/j.ces.2005.05.067]

[284] (a). Eslamian, S. *Handbook of Engineering Hydrology: Environmental Hydrology and Water Management*; CRC Press, **2014**.
(b). Watlington, K. *Emerging Nanotechnologies for Site Remediation and Wastewater Treatment*; US Environmental Protection Agency, Office of Solid Waste and Emergency Response, Office of Superfund Remediation and Technology Innovation, Technology Innovation and Field Services Division: Washington, DC, **2005**.

[285] Hu, A.; Apblett, A. *Nanotechnology for water treatment and purification*; Springer, **2014**, Vol. 22, .

[286] Mishra, A.; Clark, J.H.; Kraus, G.A.; Seidl, P.R.; Stankiewicz, A.; Kou, Y.; Sharma, R.; Dwivedi, S.; Hristovski, K.; Wu, Y. *Green Materials for Sustainable Water Remediation and Treatment*; Royal Society of Chemistry, **2013**.
[http://dx.doi.org/10.1039/9781849735001]

[287] Yan, M.; Ge, J.; Liu, Z.; Ouyang, P. Encapsulation of single enzyme in nanogel with enhanced biocatalytic activity and stability. *J. Am. Chem. Soc.,* **2006**, *128*(34), 11008-11009.
[http://dx.doi.org/10.1021/ja064126t] [PMID: 16925402]

[288] Yang, Z.; Si, S.; Zhang, C. Magnetic single-enzyme nanoparticles with high activity and stability. *Biochem. Biophys. Res. Commun.,* **2008**, *367*(1), 169-175.
[http://dx.doi.org/10.1016/j.bbrc.2007.12.113] [PMID: 18158913]

[289] Jay, W. *Grate, D. J. K., Prof. Jon Dordick*; Armored Enzyme Nanoparticles for Remediation of Subsurface, **2005**.

[290] WHO. *Water sanitation and hygiene for accelerating and sustaining progress on neglected tropical diseases, A global strategy 2015-2020,* **2015**, 38.

[291] Fan, L.; Luo, C.; Sun, M.; Qiu, H.; Li, X. Synthesis of magnetic β-cyclodextrin-chitosan/graphene oxide as nanoadsorbent and its application in dye adsorption and removal. *Colloids Surf. B Biointerfaces,* **2013**, *103*, 601-607.
[http://dx.doi.org/10.1016/j.colsurfb.2012.11.023] [PMID: 23261586]

[292] Lens, P.N.; Virkutyte, J.; Jegatheesan, V.; Kim, S-H.; Al-Abed, S. Nanotechnology for water and wastewater treatment. *Water Intelligence Online.,* **2013**, *12*, 9781780404592.
[http://dx.doi.org/10.2166/9781780404592]

[293] Yunus, I.S. Harwin; Kurniawan, A.; Adityawarman, D.; Indarto, A., Nanotechnologies in water and air pollution treatment. *Environ. Technol. Rev.,* **2012**, *1*(1), 136-148.
[http://dx.doi.org/10.1080/21622515.2012.733966]

[294] Klaine, S.J.; Alvarez, P.J.; Batley, G.E.; Fernandes, T.F.; Handy, R.D.; Lyon, D.Y.; Mahendra, S.; McLaughlin, M.J.; Lead, J.R. Nanomaterials in the environment: behavior, fate, bioavailability, and effects. *Environ. Toxicol. Chem.,* **2008**, *27*(9), 1825-1851.
[http://dx.doi.org/10.1897/08-090.1] [PMID: 19086204]

[295] Baruah, S.; Khan, M.N.; Dutta, J. Perspectives and applications of nanotechnology in water treatment. *Environ. Chem. Lett.,* **2015**, •••, 1-14.

[296] Alborzfar, M.; Jonsson, G. GrØn, C., Removal of natural organic matter from two types of humic ground waters by nanofiltration. *Water Res.,* **1998**, *32*(10), 2983-2994.
[http://dx.doi.org/10.1016/S0043-1354(98)00063-3]

[297] Otto, M.; Floyd, M.; Bajpai, S. Nanotechnology for site remediation. *Rem. J.,* **2008**, *19*(1), 99-108.
[http://dx.doi.org/10.1002/rem.20194]

[298] Goh, P.; Ismail, A.; Hilal, N. Nano-enabled membranes technology: Sustainable and revolutionary solutions for membrane desalination? *Desalination,* **2016**, *380*, 100-104.
[http://dx.doi.org/10.1016/j.desal.2015.06.002]

[299] Baruah, S.; Khan, M.N.; Dutta, J. Nanotechnology in water treatment. *Pollutants in Buildings, Water and Living Organisms*; Springer, **2015**, pp. 51-84.
[http://dx.doi.org/10.1007/978-3-319-19276-5_2]

[300] (a). Chin, S.S.; Chiang, K.; Fane, A.G. The stability of polymeric membranes in a TiO_2 photocatalysis process. *J. Membr. Sci.,* **2006**, *275*(1), 202-211.
[http://dx.doi.org/10.1016/j.memsci.2005.09.033]
(b). Tiraferri, A.; Vecitis, C.D.; Elimelech, M. Covalent binding of single-walled carbon nanotubes to polyamide membranes for antimicrobial surface properties. *ACS Appl. Mater. Interfaces,* **2011**, *3*(8), 2869-2877.
[http://dx.doi.org/10.1021/am200536p] [PMID: 21714565]
(c). Salta, M.; Wharton, J.A.; Stoodley, P.; Dennington, S.P.; Goodes, L.R.; Werwinski, S.; Mart, U.; Wood, R.J.; Stokes, K.R. Designing biomimetic antifouling surfaces. *Philosophical Transactions of the Royal Society of London A: Mathematical. Physical and Engineering Sciences,* **1929**, *2010*(368), 4729-4754.

[301] Surwade, S.P.; Smirnov, S.N.; Vlassiouk, I.V.; Unocic, R.R.; Veith, G.M.; Dai, S.; Mahurin, S.M. Water desalination using nanoporous single-layer graphene. *Nat. Nanotechnol.,* **2015**, *10*(5), 459-464.
[http://dx.doi.org/10.1038/nnano.2015.37] [PMID: 25799521]

[302] Nic, M. J. J., B. Kosata, Compendium of Chemical Terminology.*IUPAC,* 2nd ed; Blackwell Scientific Publications: Oxford, **2014**.

[303] Stefaniak, S.; Twardowska, I.; Allen, H.E.; Häggblom, M.M. *Viable methods of soil and water pollution monitoring, protection and remediation*; Springer Science & Business Media, **2007**, 69.

[304] Acevedo, M.F. *Real-Time Environmental Monitoring: Sensors and Systems*; CRC Press, **2015**.
[http://dx.doi.org/10.1201/b19209]

[305] Artiola, J.; Pepper, I.L.; Brusseau, M.L. *Environmental monitoring and characterization*; Academic Press, **2004**.

[306] Khopkar, S. *Environmental pollution monitoring and control*; New Age International, **2007**.

[307] Scognamiglio, V.; Arduini, F.; Palleschi, G.; Rea, G. Biosensing technology for sustainable food safety. *TrAC. Trends Analyt. Chem.,* **2014**, *62*, 1-10.
[http://dx.doi.org/10.1016/j.trac.2014.07.007]

[308] Duarte, K.; Justino, C.I.; Freitas, A.C.; Gomes, A.M.; Duarte, A.C.; Rocha-Santos, T.A. Disposable sensors for environmental monitoring of lead, cadmium and mercury. *TrAC. Trends Analyt. Chem.,* **2015**, *64*, 183-190.
[http://dx.doi.org/10.1016/j.trac.2014.07.006]

[309] De Silva, C.W. *Sensors and Actuators: Engineering System Instrumentation*; CRC Press, **2015**.

[310] Pumera, M. Graphene in biosensing. *Mater. Today,* **2011**, *14*(7), 308-315.
[http://dx.doi.org/10.1016/S1369-7021(11)70160-2]

[311] Segev-Bar, M.; Haick, H. Flexible sensors based on nanoparticles. *ACS Nano,* **2013**, *7*(10), 8366-8378.
[http://dx.doi.org/10.1021/nn402728g] [PMID: 23998193]

[312] Broza, Y.Y.; Haick, H. Nanomaterial-based sensors for detection of disease by volatile organic compounds. *Nanomedicine (Lond.),* **2013**, *8*(5), 785-806.
[http://dx.doi.org/10.2217/nnm.13.64] [PMID: 23656265]

[313] (a). Su, S.; Wu, W.; Gao, J.; Lu, J.; Fan, C. Nanomaterials-based sensors for applications in environmental monitoring. *J. Mater. Chem.,* **2012**, *22*(35), 18101-18110.
[http://dx.doi.org/10.1039/c2jm33284a]

(b). Kong, J.; Franklin, N.R.; Zhou, C.; Chapline, M.G.; Peng, S.; Cho, K.; Dai, H. Nanotube molecular wires as chemical sensors. *Science,* **2000**, *287*(5453), 622-625.
[http://dx.doi.org/10.1126/science.287.5453.622] [PMID: 10649989]
(c). Zhang, L.; Chang, H.; Hirata, A.; Wu, H.; Xue, Q-K.; Chen, M. Nanoporous gold based optical sensor for sub-ppt detection of mercury ions. *ACS Nano,* **2013**, *7*(5), 4595-4600.
[http://dx.doi.org/10.1021/nn4013737] [PMID: 23590120]
(d). Frasco, M.F.; Chaniotakis, N. Semiconductor quantum dots in chemical sensors and biosensors. *Sensors (Basel),* **2009**, *9*(9), 7266-7286.
[http://dx.doi.org/10.3390/s90907266] [PMID: 22423206]

[314] Arafat, M.M.; Dinan, B.; Akbar, S.A.; Haseeb, A.S. Gas sensors based on one dimensional nanostructured metal-oxides: a review. *Sensors (Basel),* **2012**, *12*(6), 7207-7258.
[http://dx.doi.org/10.3390/s120607207] [PMID: 22969344]

[315] Hu, N.; Yang, Z.; Wang, Y.; Zhang, L.; Wang, Y.; Huang, X.; Wei, H.; Wei, L.; Zhang, Y. Ultrafast and sensitive room temperature NH^3 gas sensors based on chemically reduced graphene oxide. *Nanotechnology,* **2014**, *25*(2), 025502.
[http://dx.doi.org/10.1088/0957-4484/25/2/025502] [PMID: 24334417]

[316] (a). Novikov, S.; Lebedeva, N.; Satrapinski, A.; Walden, J. Graphene Based Sensor for Environmental Monitoring of NO$_2$. *Procedia Eng.,* **2015** ,*120*, 586-589.
[http://dx.doi.org/10.1016/j.proeng.2015.08.731]
(b). Liu, Y.; Dong, X.; Chen, P. Biological and chemical sensors based on graphene materials. *Chem. Soc. Rev.,* **2012**, *41*(6), 2283-2307.
[http://dx.doi.org/10.1039/C1CS15270J] [PMID: 22143223]

[317] Ates, M. A review study of (bio)sensor systems based on conducting polymers. *Mater. Sci. Eng. C,* **2013**, *33*(4), 1853-1859.
[http://dx.doi.org/10.1016/j.msec.2013.01.035] [PMID: 23498205]

[318] Zhang, W.; Asiri, A.M.; Liu, D.; Du, D.; Lin, Y. Nanomaterial-based biosensors for environmental and biological monitoring of organophosphorus pesticides and nerve agents. *TrAC. Trends Analyt. Chem.,* **2014**, *54*, 1-10.
[http://dx.doi.org/10.1016/j.trac.2013.10.007]

[319] Kuila, T.; Bose, S.; Khanra, P.; Mishra, A.K.; Kim, N.H.; Lee, J.H. Recent advances in graphene-based biosensors. *Biosens. Bioelectron.,* **2011**, *26*(12), 4637-4648.
[http://dx.doi.org/10.1016/j.bios.2011.05.039] [PMID: 21683572]

[320] Radhakrishnan, S.; Krishnamoorthy, K.; Sekar, C.; Wilson, J.; Kim, S.J. A highly sensitive electrochemical sensor for nitrite detection based on Fe$_2$O$_3$ nanoparticles decorated reduced graphene oxide nanosheets. *Appl. Catal. B,* **2014**, *148*, 22-28.
[http://dx.doi.org/10.1016/j.apcatb.2013.10.044]

[321] Kaushik, A.; Kumar, R.; Arya, S.K.; Nair, M.; Malhotra, B.D.; Bhansali, S. Organic-inorganic hybrid nanocomposite-based gas sensors for environmental monitoring. *Chem. Rev.,* **2015**, *115*(11), 4571-4606.
[http://dx.doi.org/10.1021/cr400659h] [PMID: 25933130]

[322] Amjadi, M.; Pichitpajongkit, A.; Lee, S.; Ryu, S.; Park, I. Highly stretchable and sensitive strain sensor based on silver nanowire-elastomer nanocomposite. *ACS Nano,* **2014**, *8*(5), 5154-5163.
[http://dx.doi.org/10.1021/nn501204t] [PMID: 24749972]

[323] Li, L.; Tang, S.; Ding, D.; Hu, N.; Yang, S.; He, S.; Wang, Y.; Tan, Y.; Sun, J. A core-shell structured nanocomposite material for detection, adsorption and removal of Hg(II) ions in water. *J. Nanosci. Nanotechnol.,* **2012**, *12*(11), 8407-8414.
[http://dx.doi.org/10.1166/jnn.2012.6668] [PMID: 23421223]

[324] Lopez-Roldan, R.; Tusell, P.; Cortina, J.L.; Courtois, S. On-line bacteriological detection in water. *TrAC. Trends Analyt. Chem.,* **2013**, *44*, 46-57.
[http://dx.doi.org/10.1016/j.trac.2012.10.010]

[325] H.; Omar, A. M.; Rosma, A.; Huda, N.; Sohni, S., Analysis of Salmonella Contamination in Poultry Meat at Various Retailing, Different Storage Temperatures and Carcass Cuts-A Literature Survey. *Int. J. Poult. Sci.,* **2016**, *15*(3), 111-120.
[http://dx.doi.org/10.3923/ijps.2016.111.120]

[326] Leonard, P.; Hearty, S.; Brennan, J.; Dunne, L.; Quinn, J.; Chakraborty, T.; O'Kennedy, R. Advances in biosensors for detection of pathogens in food and water. *Enzyme Microb. Technol.,* **2003**, *32*(1), 3-13.
[http://dx.doi.org/10.1016/S0141-0229(02)00232-6]

[327] Joung, C-K.; Kim, H-N.; Lim, M-C.; Jeon, T-J.; Kim, H-Y.; Kim, Y-R. A nanoporous membrane-based impedimetric immunosensor for label-free detection of pathogenic bacteria in whole milk. *Biosens. Bioelectron.,* **2013**, *44*, 210-215.
[http://dx.doi.org/10.1016/j.bios.2013.01.024] [PMID: 23428735]

[328] Farahi, R.H.; Passian, A.; Tetard, L.; Thundat, T. Critical issues in sensor science to aid food and water safety. *ACS Nano,* **2012**, *6*(6), 4548-4556.
[http://dx.doi.org/10.1021/nn204999j] [PMID: 22564109]

CHAPTER 3

Role of Metal Based Nanomaterials in Photocatalysis

S. Sajjad[1,*], **S.A.K. Leghari**[2] and **A. Iqbal**[1]

[1] *International Islamic University, Islamabad, Pakistan*

[2] *Pakistan Institute of Engineering and Applied Sciences, Islamabad, Pakistan*

Abstract: The basic concepts of photocatalysis are explored in this chapter. Various parameters which control and influence the photocatalytic process are studied in relation to the mechanistic approach. Metal oxides like titanium dioxide, zinc oxide, cerium oxide, tungsten oxide and bismuth oxides play an important role in photocatalysis for environmental remediation, water splitting and solar cells. The phenomenon of photocatalysis is dependent on wavelength of incident light. However, the efficiency of most of the metal oxides is limited in the UV range. It is needed to modify their band gap energy levels. These modifications can be achieved by doping or coupling of metals, non-metals, metal oxides and carbon based materials. Metal oxides morphologies also affect the photocatalytic process due to enhanced surface area and surface defects providing more accessible sites for the diffusion of organics. Some new types of materials like perovskite and metal organic framework (MOF) are used as efficient photocatalysts. The role and mechanism of these materials have been discussed. All these nanomaterials are used for the environmental remediation, dye sensitized solar cells, air purifications, hydrogen production and self-cleaning process.

Keywords: Carbon materials, Metal organic frameworks, Metal oxides, Nanostructures, Quantum dots.

INTRODUCTION

Currently, there are various existing energy sources like fossil fuels which are being used by the human population. Fossil fuels are mainly hydrocarbon deposits such as petroleum, coal or natural gas. Because of these hydrocarbons, the serious disadvantage associated with these energy sources is environmental pollution. Pollution, associated with the use of fossil fuels is mainly referred as air pollution. Some water and land pollution also arise during the use of such fossil fuels. Besides the fossil fuels, there are various other activities which cause harmful effects on our natural water reservoirs and generate the water pollution. These

---

* **Corresponding author S. Sajjad:** International Islamic University, Islamabad, Pakistan; Tel/Fax: 0092519019813; E-mail: shalisajad@yahoo.com

energy and environmental challenges can be monitored by the use of solar energy [1]. Photocatalysis is the most investigated phenomenon as it is a promising technology to convert solar energy into chemical fuel, electricity and the degradation of organic pollutant. Likewise, photocatalysis is now an intensively researched field due to practical interest in self-cleaning process, self-sterilizing surfaces and hydrogen generation [1]. The photocatalyst is of great importance in all sort of photocatalytic reactions [2]. The role of metal oxide functional materials for an efficient photocatalyst is under investigation for many years [3 - 6]. In this category, the most promising photocatalysts are nanomaterials. Nanomaterials possess remarkable features as photocatalyst due to enlarged surface area, small lateral diffusion length and low reflectivity. Moreover their size dependent characteristics provide greater number of adsorption sites, maximum photon absorption ability, reduced electron hole pair annihilation as well as suppressed carrier scattering effects [7 - 10]. All these properties make the nanomaterials a hot topic of interest for researchers in light harvesting field [7].

DEFINITION AND MECHANISM OF PHOTOCATALYSIS

The word photocatalysis is composed of two words. The word photo means light, and catalysis means decomposition.

Photocatalytic reactions have been classified into two categories:

1. Uphill reaction. The photon energy is converted into chemical energy in this reaction.
2. Downhill reaction. A huge positive change in the Gibbs free energy accompanies water splitting where hydrogen and oxygen are the final products (Fig. **1**) [11].

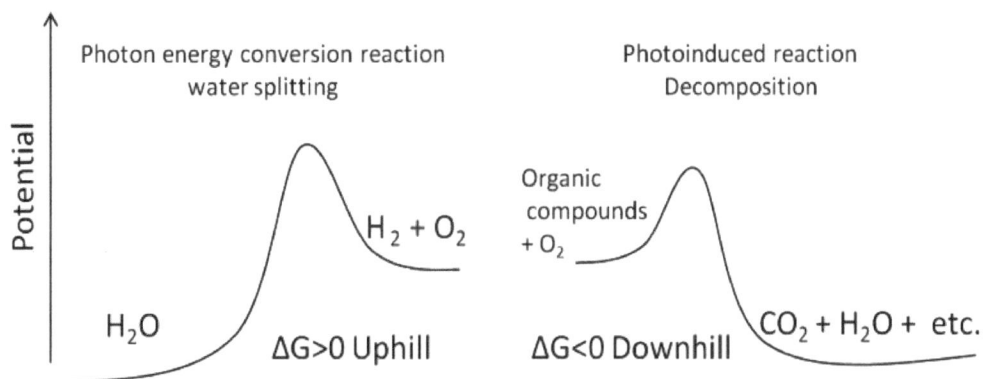

Fig. (1). Classification of photocatalysis [11].

In photocatalysis, light is used to activate the catalyst to initiate redox reaction between the photo generated electrons and holes and the adsorbed species on the surface of the catalyst.

Following are the steps involved in photocatalysis;

Photon Absorption

The incident photons strike with the semiconductor material are recognized as photo catalyst. If the interacting photons have energy greater than or equal to the band gap energy of the photo catalyst, then this light energy will be absorbed by the photo catalyst.

Generation of Electron-hole pair

Absorbed photons then shifted the electrons from the valence band (VB) to the conduction band (CB) leaving free holes in the VB and thus generate the electron-hole pair.

Charge Transport

After the formation of photo generated electron hole pair, two possibilities can arise. One possibility is the recombination of these species which creates heat and second possibility is the migration of these species towards the surface to initiate the redox reaction.

Mainly, two phenomena are responsible for the transport of charge carriers to the surface. Migration of charges can occur through thermal diffusion or *via* the presence of near-surface electric fields that causes the field-driven "drift" migration of charges.

Formation of Free Radicals or Redox Reaction

At the surface, photo generated electrons initiate the reduction reaction with adsorbed species whereas holes prompt strong oxidizing agents like hydroxyl radical on interacting with surface hydroxyl groups or oxidizing adsorbed species (Fig. **2**) [12].

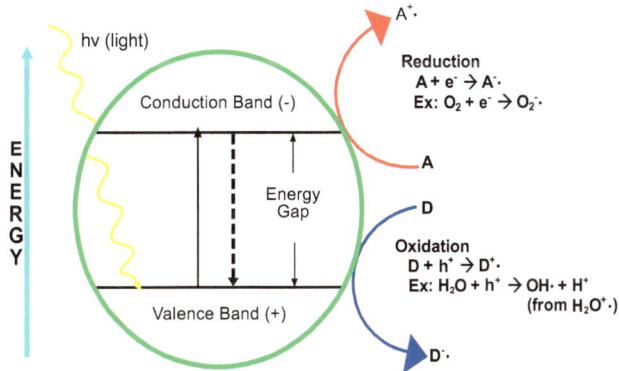

Fig. (2). Schematic diagram showing the chemical reactions occurring in photocatalysis [12].

Chemical Reactions Involved in Photocatalysis

• Absorption of photons and generation of electron-hole pair:

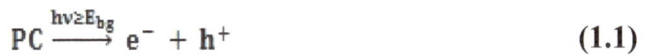

$$PC \xrightarrow{h\nu \geq E_{bg}} e^- + h^+ \tag{1.1}$$

• Reduction reactions or role of photo generated e⁻ to form free radicals:

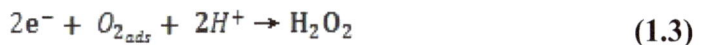

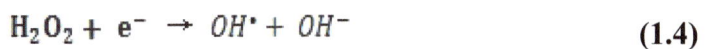

$$e^- + O_{2_{ads}} \rightarrow O_2^{\cdot -} \tag{1.2}$$

$$2e^- + O_{2_{ads}} + 2H^+ \rightarrow H_2O_2 \tag{1.3}$$

$$H_2O_2 + e^- \rightarrow OH^\cdot + OH^- \tag{1.4}$$

• Oxidation reaction or role of holes to form free radicals:

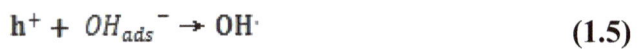

$$h^+ + OH_{ads}^- \rightarrow OH^\cdot \tag{1.5}$$

• Role of hydroxide free radicals for the decomposition of organic compound:

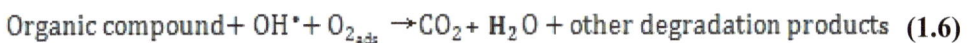

$$\text{Organic compound} + OH^\cdot + O_{2_{ads}} \rightarrow CO_2 + H_2O + \text{other degradation products} \tag{1.6}$$

TYPES OF PHOTOCATALYSIS

Photocatalysis is mainly divided into two categories [13]

- Homogeneous photocatalysis
- Heterogeneous photocatalysis

Homogeneous Photocatalysis

Homogeneous photocatalysis is based on the catalyst and the reactants that exist in the identical phase in chemical system [12].

Heterogeneous Photocatalysis

In heterogeneous catalysis, the catalyst and reactants are of distinct phase. Mainly an interface develops between a solid photocatalyst and a fluid consisting of the reactants and the products of the reaction [13]. Moreover, heterogeneous photocatalysis can be carried out in different media: pure organic liquid phases or aqueous solutions [14]. Furthermore heterogeneous photocatalysis is further divided into following two categories:

- Direct heterogeneous photocatalysis
- Indirect heterogeneous photocatalysis

Direct Heterogeneous Photocatalysis

In direct heterogeneous photocatalysis, adsorbed species mainly interact with incident light. Adsorbed molecules onto the catalyst surface absorb light of suitable wavelength and come into an excited state. After it, the excited molecules inject electrons into the catalyst substrate to initiate the redox reaction.

Indirect Heterogeneous Photocatalysis

Indirect photocatalysis makes use of electronic transitions in the catalyst substrate to initiate the surface reactions. In this process, catalyst mainly interacts with incident light [12].

PARAMETERS AFFECTING THE RATE OF PHOTOCATALYTIC REACTION

Light Intensity

Speed of a photochemical reaction is influenced by the photon absorption potential of a photocatalyst. Estimation of absorbed light quanta (ϕ) by any photocatalyst and the rate of reaction are given as:

$$(\Phi)_{Overall} = (\text{rate of reaction})/(\text{rate of absorption of radiation})$$

Reaction Temperature

Increased temperature leads to the decrease photocatalytic response of a photocatalyst as high temperature multiplies the rate of electron hole pair annihilation as well as elevates the desorption process of adsorbed molecules [15].

Influence of pH on Photocatalytic Activity of Photocatalyst

The role of pH on the effectiveness of a photocatalyst is quite difficult to understand because of its multiple functions. Primarily, pH is associated with the ionization state of the surface such as;

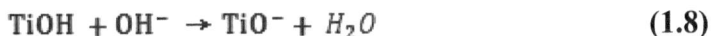

$$TiOH + H^+ \rightarrow TiOH_2^+ \tag{1.7}$$

$$TiOH + OH^- \rightarrow TiO^- + H_2O \tag{1.8}$$

Thus, it affects the adsorption behavior of dye or pollutant molecules on the photocatalyst, which is the crucial step for the photocatalytic process [16]. Secondly, positive holes are examined as the major oxidation species at low pH, while hydroxyl radicals are treated as the prevalent specie at neutral or high pH values [16]. As in case of titania, hydroxyl free radicals are greatly generated in alkaline solution through oxidation of hydroxide ions existing on titania surface, thus the effectiveness of the process is significantly increased. But the degradation behavior of organic pollutant is also largely affected by their nature [16, 17]. Some are photo catalytically degraded at lower pH values while some are effectively degraded at higher pH values. pH behavior also affects the size of the photocatalyst as it deals with the surface charges. In case of TiO_2 photocatalyst, under acidic environment particles frequently accumulate that do not only reduce the accessible surface area for molecule adsorption but additionally photon absorption behavior of a photocatalyst is also significantly decreased [16].

Nature of Photocatalyst

The nature of photocatalyst affects the photocatalytic activity. The surface morphology and particle size are the main parameters in this regard [15].

Effect of Oxidizing Agent on Photocatalytic Degradation of Organic Pollutant

Oxygen serves as electrons scavenger to maintain the photocatalytic performance at optimum rate. Thus the quantity of O_2 passing into the system is an essential parameter. So to have a better supply of oxygen, some oxidizing agents can also be introduced. Few reports support the addition of hydrogen peroxide to enhance the rate of photocatalytic reactions with adequate oxygen supply [19].

Effect of Catalyst Concentration

It has been found that the initial rate of the photochemical reaction is directly proportional to the catalyst concentration. An enhanced catalyst concentration supports the elevated amount of reactive sites on the photocatalyst surface. Subsequently number of hydroxyl and superoxide radicals are intensified. But the exceeded concentration of the catalyst than the optimum limit is found to be suppressed the decaying rate because the catalyst surface becomes unavailable for photon absorption [18].

Effect of Calcination Temperature on the Activity of the Photocatalysts

It has been observed that calcination temperature greatly influences the photocatalytic behavior of nanomaterials as it transforms the crystal phase of nanomaterials [19]. In case of titania nanoparticles, the increased rate of photocatalytic activity was observed with the increase in temperature from 300 to 500°C. But further increase in calcination temperature from 500 to 900°C decreased the photocatalytic activity of photocatalyst [19]. Since the increase in calcination temperature beyond 500°C can cause the transformation of crystal phase from anatase to rutile which has little photocatalytic activity [19, 20].

ROLE OF METAL OXIDE NANOSTUCTURES IN PHOTOCATLYSIS

Nano-crystalline photocatalysts are ultra-small semiconductor particles which are a few nanometers in sizes. Interest in such small semiconductor particles originates from their unique photophysical and photocatalytic properties [21]. An ideal photocatalyst must have photostability, chemically and biologically inert nature (non-toxic), high availability, low cost and highly active photocatalyst [15, 21]. In case of direct photocatalysis, photocatalyst adsorb species like organic dye which mainly act as a photosensitizer. Numerous organic molecules such as Methyl orange, Rhodamine B, Porphyrins and Phthalocyanines can be used as photosensitizers [7, 22 - 24]. These organic compounds showed an important function in the photosensitization of dye synthesized solar cells [25 - 27].

The search for new semi-conducting materials or the engineering of the existing ones for efficient photocatalyst has been extremely changing. This search meets a wide range of compounds of different compositions, morphology and electronic structure. A wide range of materials have been studied for this purpose that includes the variety of binary compounds like Nb_2O_5, TiO_2, Fe_2O_3, ZnO, WO_3 and Cu_2O, ternary oxides ($AgPO_4$, $SrTiO_3$, $BaTiO_3$, and $CaTiO_3$) as well as quaternary oxides. However, few drawbacks are associated with these compounds. Binary metal sulphide semiconductors like CdS, CdSe and PbS are regarded as insufficiently stable for catalysis at least in aqueous media because they readily undergo photo cathodic corrosion [28]. For instance, iron oxides are not suitable semiconductors as they readily undergo photo cathodic corrosion [28]. Zinc oxide behaves as a good photocatalyst in UV light irradiation but it is also unstable in H_2O with $Zn(OH)_2$ being formed on the particle surface. This results in catalyst deactivation. Among all other semiconductor photocatalyst, titania is most widely investigated material as a photocatalyst because of its high photocatalytic activity, large chemical stability and robustness against photo corrosion, low price and nontoxicity [8, 17, 29 - 34].

Titanium Dioxide (TiO$_2$)

Structural Properties and its Effect on Energy Band Structure of TiO$_2$

Physical and chemical properties of a material are highly dependent upon its phase structure. The phase structure of titania is one of the most important parameter to determine its photocatalytic performance. Titania has mainly four polymorphic forms: anatase, rutile, brookite and $TiO_2(B)$ [35]. All these types of material consist of TiO_6 octahedra in which titanium (Ti^{4+}) atoms are coordinated with six oxygen atoms. The different polymorphic structures of TiO_2 exist due to difference in the arrangement of octahedron units, share edges and corners in a variety of manners. In anatase phase, octahedra exhibit zigzag array and share four edges [35]. In rutile, octahedra are allocated only two edges and attached linearly, while in brookite both corners and edges are associated. The structure of $TiO_2(B)$ resembles to the layered structure, consisting of corrugated sheets in which both edges and corners are shared, as shown in Fig. (**3**).

Differences in a lattice structure emerge different electronic band structures. Thus for anatase, the band gap value was found to be 3.2 eV, for rutile 3.0 eV and for brookite approximately 3.2 eV [34]. Particularly, two forms of titania, rutile and anatase are the most extensively researched as photocatalyst. Anatase having 3.2 eV responds to 384nm, and rutile having 3.02 eV responds to 410 nm [35]. Anatase has drawn much attention of researchers as a photocatalyst because anatase has a slightly higher redox driving force than rutile, as well as anatase has much higher surface area as compared to rutile which leads to enhanced adsorption capability and huge greater generation of active sites.

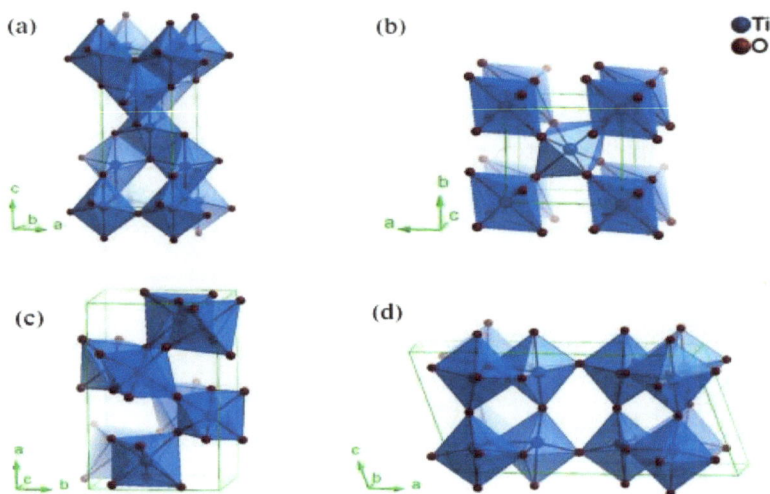

Fig. (3). (**a**) anatase phase (**b**) rutile (**c**) brookite (**d**) TiO_2 (B) [35].

Drawback of Titania Photocatalyst

In case of photocatalysis, the main drawback that is associated with TiO_2 is its large band gap. Therefore, it only absorbs light in ultraviolet region [35]. In order to make the titania efficient to absorb visible light, two most efficient strategies are developed. The first one is band gap engineering, which refers to narrowing the band gap of TiO_2 to make it active under visible light by doping the titania with other elements. The second one is the surface sensitization, which refers to the application of other visible light active materials as a light harvester to sensitize TiO_2. Doping the titania with other elements causes the introduction of an impurity level in the forbidden band. This intermediate energy level helps acts as either an electron acceptor or donor which makes the titania to absorb visible light efficiently. Metals and non-metals are both revealed as a promising dopant to make the titania a visible light active photocatalyst [35].

Non-metal Doping in Titania

For this purpose, nitrogen acts as a good dopant in the titania structure due to its comparable atomic size with oxygen, small ionization energy and high stability. Similarly, there are many non-metals like F, C, S which make the titania to absorb the visible light effectively. Fluorine dopant does not shift the titania band gap but it modifies the surface acidity. It favors the transition of Ti^{+4} into reduced Ti^{+3} ions due to the electron withdrawing nature of fluorine ions. Consequently, charge separation is intensified and the efficiency of photo induced process is improved. Phosphorous and sulphur also act as dopants showing the positive response for visible light activity in titania. Band gap of titania is relatively short with the aid of non-metals dopants due to alteration in lattice parameters along with the existence of trapped sites inside the conduction and the valence bands [36].

Nitrogen Fluorine co-doped titania was explored in visible light photocatalysis due to the identical structural affinity of two dopants. The union of two elements nitrogen and fluorine in titania magnifies its visible light response due to nitrogen and the fluorine doping. These elements played appreciable role in charge separation [36].

Transition Metal as Dopants in Titania

Modification of titania with transition metals such as Cr, Co, V, Mn, Mo, Nb, W, Ru, Pt and Ag makes it able to absorb visible light. Doping the titania with metal ions causes decreased photo corrosion effects and suppressed the charge recombination at metal sites [37]. Surface modification of titania with noble metals like Ag, Au, Pt and Pd remarkably enhanced the photocatalytic activity of titania in visible light. These metals function as electron sink and promote the interfacial charge shift to enhance the lifetime of photogenerated electron hole pair [38]. Silver based titania composites are eagerly investigated due to astonishing electrical, optical and catalytic features [39, 40]. In addition, silver nanoparticles exhibited high bactericidal activity and biocompatibility in comparison with other nanoparticles. In silver-titania composite, Ag nanoparticles work as an electron sink to suppress the annihilation of photo-generated electron pairs, hence the photocatalytic efficiency of the composite is significantly enhanced [39].

In Cr doped titania, Cr dopant only capture one type of charge carrier. Where as Fe ions can efficiently decrease the chances of charge recombination by trapping both electrons and holes and thus considered as better doping candidates than Cr, Co and Ni ions [35]. Metal co-doping of titania exhibited higher photocatalytic activity than single cation doping. It has been noticed that co-doping of two cations with different charges can enhance the stability of the photocatalyst due to charge balancing effect. Fe and Ni co-doped titania showed excellent photocatalytic activity under visible light [35]. Anatase TiO_2 coated nickle ferrite ($NiFe_2O_4$) nanoparticles displayed photocatalytic performance that can be directed towards the antimicrobial activity. Diverse metal ions (*e.g* Nd^{+3}, Fe^{+3} and W^{+4}) doped titania coated nickel ferrite nanocomposites, offer outstanding antimicrobial activity by blocking the electron hole annihilation and shorten the band gap of titania [41]. The ferrite magnetic nanoparticles encapsulated with the photocatalytic shell retain super-paramagnetic characteristics and magnetic strength so that it can act as removable antimicrobial photocatalytic composite nanoparticles [41].

Impact of Morphology on the Photocatalytic Efficiency of TiO$_2$ Photocatalyst

Crystallographic behavior and morphology of a photocatalyst are two important parameters that determine the quality of its photocatalytic performance. Therefore, the photocatalytic deterioration potential of titania is highly influenced by its morphological features and crystallization ratio of anatase/rutile phase [42]. Phan and co-worker prepared different samples of titania nanoparticles *via* fluctuating the quantity of used HCl. These synthesized samples exhibit different morphological features and distinct crystallization ratio of anatase/rutile phase. Then they labeled as DTC-2.15 that contained huge clusters of tiny anatase titania nanoparticles. Second DTC-5.0 specimen was exhibiting 3D features due to anisotropic crystal extension of prism like titania nanoparticles. Consequently, they attained uniform flower shaped titania particles in which rutile was dominant phase of crystal phase of titania [42]. Likewise, third specimen named as DTC-6.0 possessed imperfect cauliflower shaped structure having the decreased rutile titania phase behavior as compared to that found in DTC-5.0. Fourth one was DTC-9.0 based on the flawless assemblage of titania nanoprisms that appeared in a plump flower like morphology. The photocatalytic performance of aforementioned samples was assessed by the photocatalytic decay of methylene blue (MB) organic dyes in the presence of UV beam. Among all the synthesized samples, DTC-5.0 showed superior photocatalytic performance whereas DTC 9.0 showed lower photocatalytic efficiency [42]. All the prepared samples exhibited enhanced reactive behavior than the ball or rectangular shaped pure rutile titania particles. Moreover, an unusual aspect was taken into consideration that the decay rate constant for DTC-2.15 was much smaller and even inferior than the MB photolysis in spite of possessing pure anatase phase, tiny clustered nanoparticles and enhanced surface area [42]. Although this sample was composed of pure anatase phase but the agglomerated features of tiny nanoparticles became an obstacle for the penetration and oxidation of organic molecules under UV irradiations. So inefficiency of this sample was due to UV light shielding effects. Ball shaped sample showed slightly improved photo-efficiency having rutile nanorods and granules [42]. The partial sticking of anatase nanoparticles on rutile nano prism provided the more assessable cites for molecular oxygen [42, 43]. Since the conduction band edge of the anatase is about 0.2 eV more negative than that of rutile phase, thus facilitating the interfacial electron transfer from rutile phase acting as electron sink (Fig. **4**). These photo-generated electrons from the conduction band of the rutile phase were scavenged by oxygen molecules to form superoxide radicals. The rutile phase transferred these excited electrons to the lower energy anatase trapping sites, thus increasing the separation of charge carriers [42].

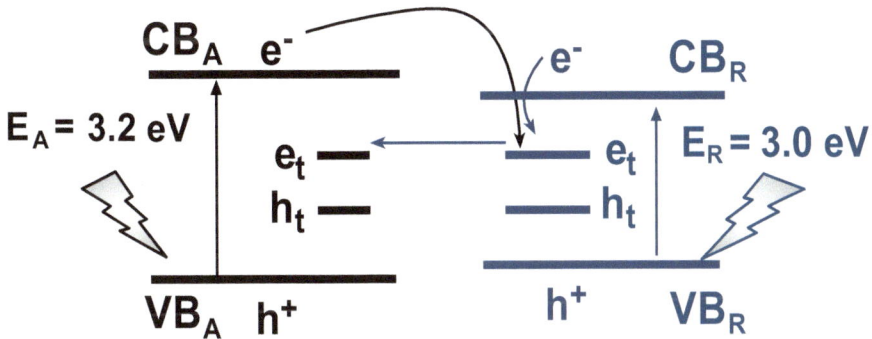

Fig. (4). Demonstrating the role of rutile phase as electron sink and the antenna effect in mixed phase nano-crystalline titania [42].

The cauliflower like nano structures had less photocatalytic activity due to interposition of larger numbers of anatase particles which adversely affected the scavenging process of photo-generated electrons by oxygen [42]. So finally it is confirmed that the photocatalytic efficiency are strongly influenced by morphological and crystalline structures of titania [41].

The electro spun titania and nanofibers acquire the following advantages for photocatalytic application:

• High surface area
• Controllable pore sizes
• Controlled thickness of nano-fibrous

Much literature has been studied on synthesis, structure, antimicrobial and photocatalytic activities of titania nanofibers. For example ZnO/TiO_2 composite nanofibers were fabricated by electro-spinning which showed excellent antimicrobial activity against *E. Coli* under UV irradiation [44]. CuO/TiO_2 nanofibers had excellent photocatalytic activity under visible light irradiations [45]. Co-doped titania nanofibers and NiO/TiO_2 nanocomposites nanofibers showed excellent antimicrobial activity. These nanostructures effectively interact with bacteria and disrupt their cell membranes and enzymes to kill the bacteria [46, 47]. The photocatalytic efficiency of TiO_2/Al_2O_3 nanofibers and porous nanofibers was compared. It was found that porous nanofibers demonstrated much higher photocatalytic activity than the other [41]. The enhanced photocatalytic activity of porous TiO_2/Al_2O_3 nanofibers was due to the large surface area and porous structure that offered the greater adsorption sites for different organic dye molecules [41].

Zinc Oxide as a Photocatalyst

Zinc oxide occurs as white powder and it is an amphoteric oxide. ZnO is 2-6 compound semi-conductor. Its iconicity lies at the borderline between ionic and covalent semiconductors. Zinc oxide is found in three crystal structures: cubic zinc blend, cubic rock salt and hexagonal wurtzite as shown in Fig. (5) [48].

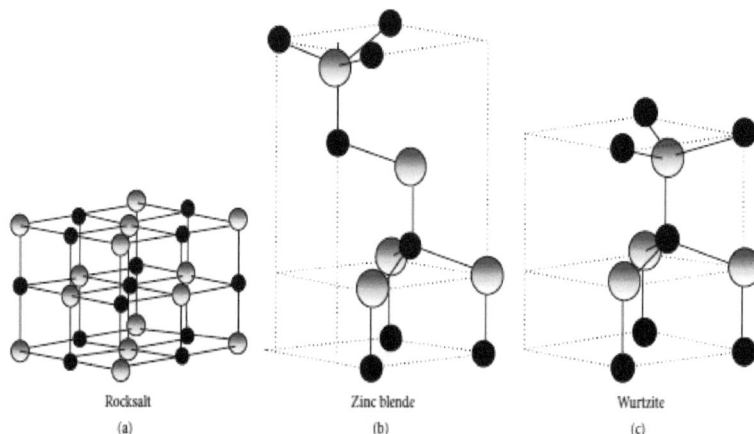

Fig. (5). Crystal structures of zinc oxide [49].

ZnO has a direct band gap having energy 3.37 eV. It is a biocompatible, biodegradable and bio safe for environmental applications [49]. Photocatalytic reactions occur at the surface of photocatalyst. The rate of photocatalytic reaction is enhanced with the increase in surface area. Photocatalytic properties of zinc oxide nanostructures are better than bulk zinc oxide because the band gap of zinc oxide nanostructure is lower than the bulk zinc oxide due to the presence of various defects [50]. Defects develop the sub bands inside the structure from where electron hole transitions are easier. These defects also reduce the electron hole recombination by trapping. Various nanostructures of ZnO having different morphologies are being investigated as photocatalyst [50].

Impact of Morphology on Photocatalytic Property of Zinc Oxide Photocatalyst

Kajbafvala and co-workers synthesized zinc oxide nano-spheres and flower like morphology through simple microwave irradiation methods. In their observation, they found that degradation of methylene blue under UV irradiation through zinc oxide nano spheres was much faster than in the presence of zinc oxide flower like morphology [50]. This observation was due to the fact that specific surface area in zinc oxide nano-spheres was greater than in the flower like morphology. This is due to the porous structure of ZnO nano spheres. Whereas in case of flower like morphology, more flat surfaces were generated resulting in less porosity [50]. Photocatalytic activity of zinc oxide nanoparticles with different morphologies was found to increase in order of plate- like ZnO < flower like ZnO < needle like ZnO < sphere like ZnO [51]. Moreover, the experimental results confirmed the zinc oxide film as good photocatalyst in comparison with zinc oxide nanorods and powder. Superior photocatalytic property of Zinc oxide film was due to the large surface to volume ratio, the effective electron hole separation of the shottcky barrier and thinness. Thus, zinc oxide film was able to adsorb and transport more dye molecules on surface [52]. ZnO dense nano-sheets built network that showed high photocatalytic activity due to increased surface area to volume ratio [53]. Zinc oxide nanorods grown on a paper substrate were used as photocatalytic papers. Photocatalytic activity of these papers was tested for the degradation of organic molecules and in the inactivity of *E. coli*. Results confirmed that it was good photocatalyst under visible light [54]. ZnO nanorods grown on glass substrate showed good photocatalytic property under visible light [55]. ZnO nanowires destroy the outer membrane of *E.coli* and make its nuclei inactive under light irradiation [56].

Zinc Oxide Nanocomposite

ZnO can be sensitized by doping it with low band gap metal oxides. Establishment of a hetero junction is the crucial step to determine the superficial electron transfer phenomena. An effective junction was formulated by the combination of ZnO and CdS [57] as shown in Fig. (6).

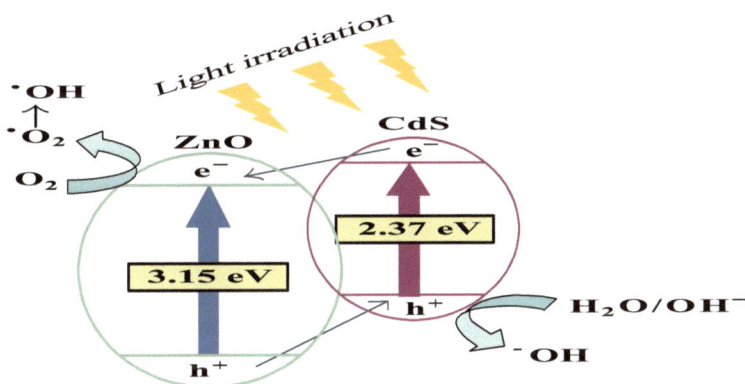

Fig. (6). Heterojunction formed by the combination of ZnO and CdS [58].

Visible light generates the electron hole pair in CdS and electron is then transferred into the conduction band of ZnO by ballistic diffusion [57]. The transfer of electrons from conduction band of CdS to the conduction band of ZnO occur in 18 picoseconds which is less than the lifetime of electron in CdS. CdS/ZnO nanowires hetero-structure absorbs light up to 550 nm and thus are highly photoactive under visible light [58]. Zinc oxide/titania nanocomposite is superior with respect to the light harvesting range and act as the efficient photocatalyst under UV-visible light irradiation. The increased quantum efficiency of the system is due to strong coupling effect of titania and ZnO in nanocomposites. Thus, the lifetime of electron hole pairs is increased. ZnO/SnO_2 nanocomposite also acts as good photocatalyst. Tin oxide is a wide direct band gap semiconductor having band gap energy 3.7 eV [59]. The conduction band of SnO_2 is located below as compared to the conduction band of ZnO. Therefore, photo-generated electrons diffuse from the conduction band of ZnO to the conduction band of SnO_2. While the opposite directed flow of holes is essential to suppress the chances of electron hole pair annihilation [59]. Thus, magnificent photocatalytic performance is attributed to the heterojunction between ZnO and SnO_2 that broadens the separation of photogenerated charge carriers [59]. It has also been noticed that increase in calcination temperature decreases the photocatalytic activity of ZnO/SnO_2 nanofibers because of the reduction in the surface area of ZnO/SnO_2 nanofibers [60].

Graphene Nanocomposite/ZnO

The unique morphology of composite in graphene/Zinc oxide (G-ZnO) nanocomposite results into magnificent photocatalytic efficiency [61]. These G-ZnO composite thin films were synthesized by using the electro-static spray deposition technique. The reason for its good photocatalytic activity was its enhanced specific surface area, increased light absorption and the lifetime of photo generated electron hole pairs. Mn doped ZnO/graphene nanocomposite showed more efficient photocatalytic activity under visible light [61, 62].

ZnO/Ag_2S Nanocomposite

ZnO/Ag_2S nanocomposite showed efficient photocatalytic activity for the degradation of Eriochrome Black dye as shown in Fig. (**7**). The band gap energy of Ag_2S is 1.1 eV [63]. Due to low band gap energy, Ag_2S can absorb a broad solar spectrum. Photoluminescence spectroscopy supported the UV-visible absorption behavior of ZnO/Ag_2S nanoparticle [63, 65] that had a main peak in visible range along with a peak in ultraviolet region. ZnO/Fe nanowires were grown on Fe doped ZnO seeding layer exhibited high photocatalytic activity under visible and UV light [66]. Cr doped zinc oxide nanowires, Co-doped zinc oxide nanorods [67] and Ag-doped zinc oxide nanowires showed good photocatalytic activity under visible light [48].

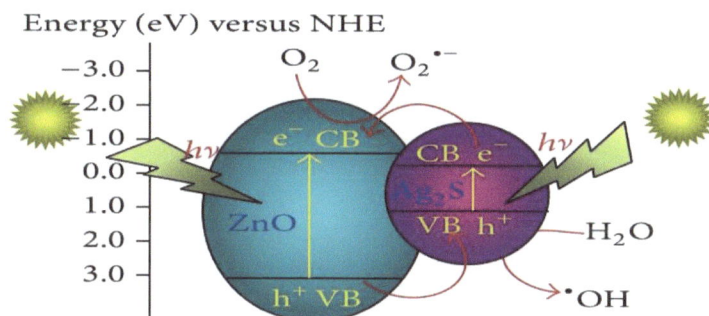

Fig. (7). Heterojunction of ZnO and Ag$_2$S nanocomposite [64].

Role of Complex Bismuth Oxide Compounds as a Photocatalyst

Bismuth based complex oxides are hot topic of interest for the researchers as a solution for the current environmental and energy crises due to their enhanced solar light collection ability. In bismuth oxide compounds, the deformation of lone pair in Bi 6s orbital lead the strong overlapping of O 2p and Bi 6s orbitals in the valence band [68]. Consequently advantageous movability of photogenerated charge species occurs that makes it a good photocatalyst [68 - 72]. Bismuth based oxides include BiVO$_4$, BiWO$_4$, Bi$_2$WO$_6$, BiMoO$_6$, Bi$_4$Ti$_3$O$_{12}$, BiFeO$_3$, Bi$_2$Fe$_4$O$_9$, Bi$_5$FeTiO$_{15}$, BiOCl and Bi$_5$O$_7$I. In bismuth based complex metal oxides Bi-M-O, M is considered as favorable site for the reduction reaction while oxidative phenomena is likely to be occurred at Bi or on O site [68].

Morphological Control of Bismuth Related Compounds to Enhance the Photocatalytic Efficiency

The size, morphology and crystal structure are the parameters affecting the photocatalytic activity. The average diffusion time of the charge carriers depend on the grain radius and is estimated by formula $t=r^2/\pi^3 D$ [68] where r is the grain radius and D is the diffusion constant of the carrier. Thus, if the grain radius is small, then large number of photogenerated charge species will be shifted to the surface to take part in photocatalytic action. Photocatalytic activity of different BiVO$_4$ samples having different morphologies was tested. Because of the layered structure, plate like morphology is the primary nanostructure for BiVO$_4$, BiWO$_6$ and BiMoO$_6$ [68]. Thus, two dimensional (disc and plate like) structure showed the highest photocatalytic activity due to large surface area and the thin thickness of the laminar structure. Other morphologies like dendrite, flower and olive like structures as shown in Fig. (**8**) have been synthesized. All such morphologies showed higher photo-catalytic activity [68].

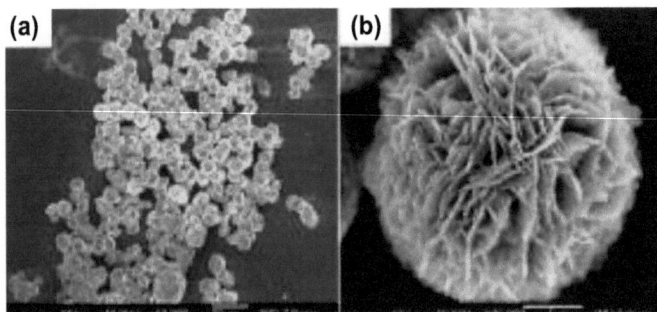

Fig. (8). Flower like Bi_2WO_6 hierarchical structure [68].

It was hypothesized that large surface area and small density of recombination centers are essential features for a high photocatalytic activity [68]. Thus, to utilize the features of large surface area, mesoporous structure of bismuth metal oxide photocatalyst was investigated. Such intrinsic characteristics provide greater amount of reactive sites for degradation reaction as well as efficiently decrease the rate of the electron hole pairs annihilation [68].

Surface Modification of Bismuth Related Oxides

Photocatalytic phenomena occur at the surface of the photocatalyst, therefore the surface features are considered as crucial among all other parameters [68]. While investigating the effects of structural features on the photocatalytic performance of Bi_2WO_6, it was notified that Bi_2WO_6 microspheres were particularly assigned to bismuth-rich hydrophilic surface. A detailed analysis of the impact of surface acidity on the photocatalytic degradation of Rhodamine B in BiOCl [69], Bi_2O_3 [70], $BiVO_4$ [71] and Bi_2WO_6 [72] materials revealed the fact that high photocatalytic performance was occurred in the presence of higher surface acidity environment [73]. Acid sites may induce strong interactions with the pollutant molecules, consequently distance between the pollutant and the photocatalyst surface is much decreased. As a result of it, the photogenerated electrons, holes and radicals easily interact with pollutant molecules, leading to an efficient degradation under visible light. In the same manner, surface fluorination is another efficient technique to modify the surface attributes of oxide photocatalysts to enhance their photocatalytic behavior. Remarkable photocatalytic efficiency of fluorinated Bi_2WO_6 [74] was due to the presence of fluorine on the catalyst surface which behaved as electron sink and increased the interfacial electron transfer rates by tightly holding trapped electrons. Similarly, higher photocatalytic activity was observed in Mo doped $BiVO_4$ than pure $BiVO_4$ due to the higher surface acidity of Mo doped $BiVO_4$. Various metals were also used as dopants like W, Fe, B, Cu, Zn, Ti, Nb, Sn, Co, Pb, Rb, Ru, Ag, Ga, Sr and Ir [68, 75].

Photosensitization of Quantum Dots

Quantum dots are fluorescent nanomaterials having magnitudes of discrete nanometers, composed of thousands of atoms of second and sixth group elements (*e.g.* CDs, CdSe, CdTe) or third and fifth group elements (InAs and InP) [7]. Due to charge quantum confinement, quantum dots possess valuable property of photoluminiscence. The presence of appreciative luminescence, quantum efficiency, broad continuous absorption power and very confined PL band makes the quantum dots a promising element for the photosensitization of wide band transition metal oxides [9]. CdS is a highly captivating material among all quantum dots having band gap energy (2.4eV) for visible light absorption. CdS/TiO$_2$ composite was highly active in visible light. As TiO$_2$ molecules could not be excited under visible light due to high band energy. But in CdS/TiO$_2$ composite [76], visible light generated the electron-hole pair in CdS and the photogenerated electrons were then conducted from the conduction level of the CdS to the conduction level of the adjacent TiO$_2$. Whereas the photogenerated holes were stayed in valence band of CdS [76]. The high efficiency of this photocatalyst under visible light irradiation was due to strong coupling and effective electron transfer between nano sized CdS and TiO$_2$ nanocrystal as shown in Fig. (**9**) [7].

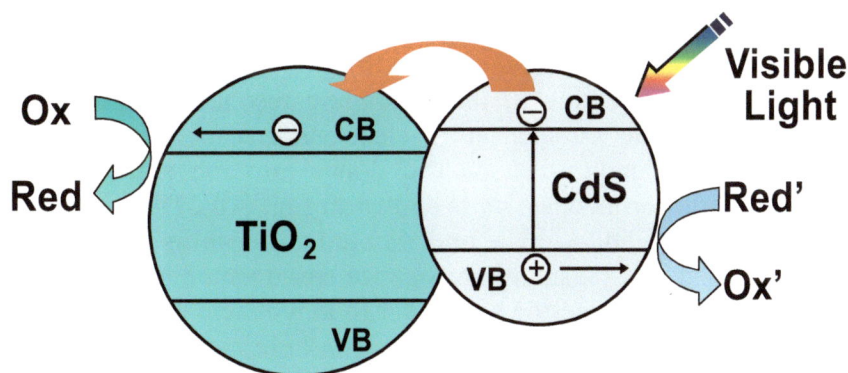

Fig. (9). Diagrammatical illustration of the principle of charge transfer between CdS and TiO$_2$ [7].

Similarly, ZnO disk-CdS nanocomposite hetero-structure showed 2.8 times higher photocatalytic activity than ZnO rod-CdS nanocomposite under visible light due to polar boundaries and remarkable carrier separation [77, 78]. Furthermore, quantum dots based transition metal oxides hetero structure composites such as CdTe/CdSe quantum dots on TiO$_2$ nanotube array, CdTe quantum dot monolayer sensitized ZnO nanowire and Cd /TiO$_2$ nanofibers hetero architectures are reported [79 - 81]. With the use of photoluminiscence property of quantum dots, the absorption behavior of wide band gap semiconductor can be made active in visible light and even in infra-red region. But few drawbacks of quantum dots have been noticed like toxicity of heavy metals that arise the possible environmental damages and photo-corrosion due to self-oxidation phenomena exhibited by the photogenerated holes in the valence band [7].

Plasmonic Metal Nanostructures as Photosensitizer

Surface plasmon resonance can be described as the induced collective vibrations of valence electrons due to resonant photon that occurs only if the frequency of interacting photons gets harmonized with the basic frequency of surface electron vibrations on specific metal resonance structures such as Cu, Ag and Au [7]. Electromagnetic radiations induce intense localized electromagnetic fields on the metal nanostructures. This can result into the increase of optical absorption and scattering at a peculiar wavelength. This phenomenon is not only affected by the nature of metal but also by the variations in size and shape of the metallic nanostructures. Such as, plasmonic response of silver can be tuned from UV towards the visible region by confining the size of silver nanoparticles in the low range of nanometers. In the same fashion, plasmonic behavior of gold nanoparticles can be altered from the visible to the infrared edge of the spectrum *via* modulating the aspect ratio of gold nanorods. The composite of titania with gold nanoparticles is highly active under visible light irradiation. Due to plasmon effect, gold nanoparticles are highly active under visible light irradiation [82, 83]. The excited electrons can be conducted from the gold nanoparticles surface to the conduction band of titania due to the plasmonic resonance behavior of gold nanoparticles. Meanwhile compensative electrons are injected into the gold nanoparticles by introducing the specific donor into the solution [82]. The suggested charge transfer mechanism is shown in Fig. (**10**). Distinct behavior of plasmonic resonance structure in UV and visible light beam is manifested. Upon interaction with UV beam, plasmonic resonance based nanomaterials behave like co-catalyst that work as an electron scavenger to move them apart from the holes. Thus amplifies the lifetime of generated electron hole pair. Whereas in case of visible light beam, plasmonic nanostructures function as photosensitizer by enhancing the solar light collection and improving the visible light energy conversion efficiency [7].

Carbon Based Nanostructures as Photocatalyst

Carbon nanostructures are considered as the promising candidates for the photocatalysis as they possess appreciative electrical conductance, chemical durability and high surface area. The photocatalytic performance of the carbon nanotube/transition metal oxide [84] composite is greatly intensified by increasing the number of adsorbed reactants *via* the presence of huge amount of reactive

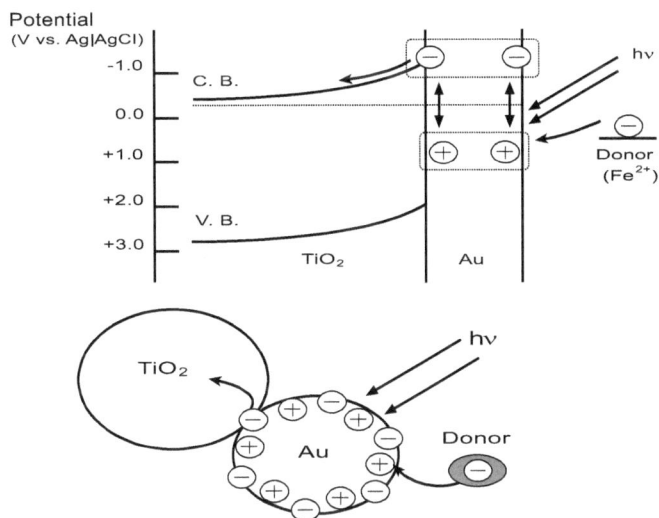

Fig. (10). Charge transfer mechanism in gold/titania nanocomposite [82].

sites, excellent charge separation and feasible visible light excitation in carbon nanotubes (CNT). The composites of CNT with titania and Ni particles are effectively operated in visible light beam. Multi-walled carbon nanotube functions as photo sensitizer. In these composites, firstly visible light beam activates the CNT and then the excited electrons are transferred from CNT to the conduction band of a transition metal oxide for the initiation of reduction reaction, meanwhile holes are in transition metal oxide [85] to initiate the redox reaction shown in Fig. (**11**) [7]. Graphene is another engineered carbon based nanomaterial that is based upon the hexagonal framework of sp^2 hybridized carbon atoms. Graphene owns outstanding features such as increased specific surface area, ballistic transport of charge carriers and superb optical transmittance [86].

$ZnWO_4$/graphene hybrid photocatalyst is functional in both UV and visible light beam [88]. Electron hole pairs can be induced in $ZnWO_4$ upon UV light exposure. The valence band of $ZnWO_4$ is located at the low position than the LUMO of graphene. So the holes can be transferred from the valence band of $ZnWO_4$ to the highest occupied molecular orbital (HOMO) of graphene. Consequently photogenerated electrons can reside in the conduction band of $ZnWO_4$ and favors the occurrence of surface reaction to produce radicals. Under visible light, primarily electrons are shifted from HOMO to LUMO of graphene and then directed towards the conduction band of $ZnWO_4$ to involve in the surface reduction reaction [7] as shown in Fig. (**12**). Therefore, the lifetime of photogenerated electron hole pair is effectively enhanced.

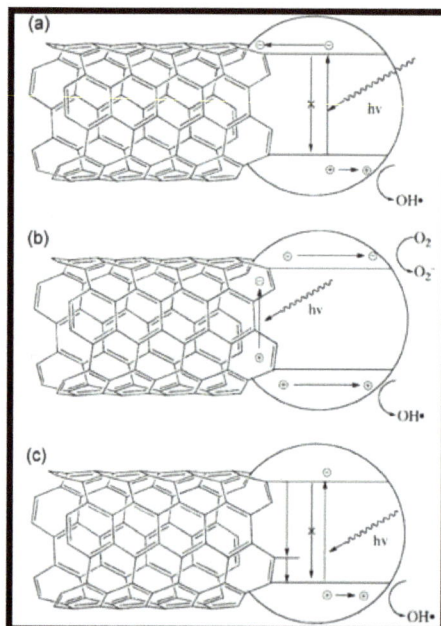

Fig. (11). (**a**) CNTs act as an electron sink to suppress the recombination of electron hole pair. (**b**) Carbon nanotube act as the photosensitizer. (**c**) CNT acts as an impurity by introducing additional energy level within the transition metal oxide band gap [87].

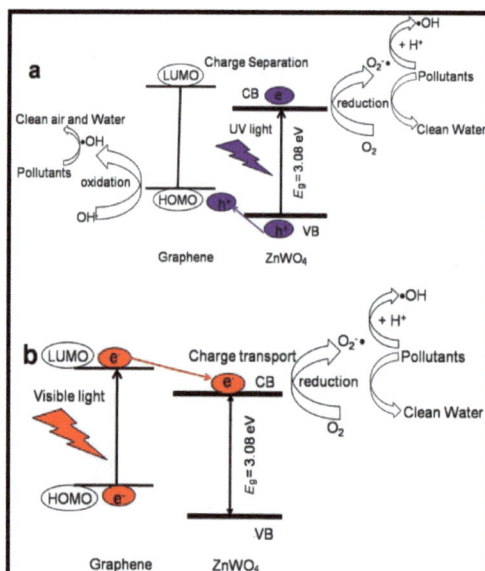

Fig. (12). Schematic representation of photocatalytic process in $ZnWO_4$/graphenephotocatalyst (**a**) under UV light irradiation (**b**) under visible light irradiation [88].

Carbon nanodots are another constitute of engineered carbon based nanomaterials that consist of discrete and quasi spherical nanoparticles having size within the range of 10 nm [89 - 92]. Carbon nanodots act as efficient photosensitizer under visible light. Photocatalytic mechanism (Fig. **13**) of TiO_2/carbon nanodots composite initiates with the absorbance of visible light followed by the emission of UV light due to the photolumiscence up conversion of carbon nanodots. Subsequently, electron hole pair is induced in titania nanoparticles with the interaction of above mentioned emitted UV light that is resulted in the generation of active oxygen radicals [93] for the decay of organic pollutants [7].

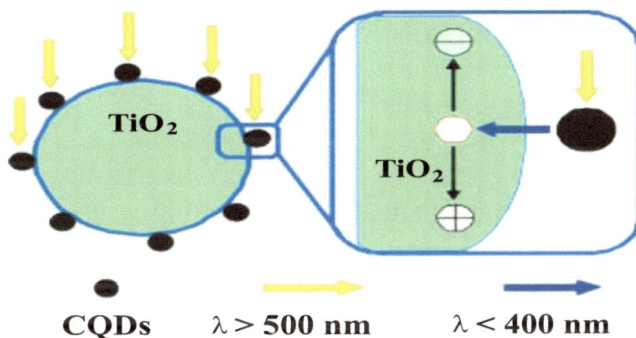

Fig. (13). Photocatalytic mechanism of TiO_2/carbon nanodots under visible light [94].

Likewise ZnO/carbon nanodots, carbon nanodots/$SrTiO_3$ film and carbon nanodots/ZnO nanorod composite arrays are also effective visible light photocatalyst [95 - 97]. In (m-$BiVO_4$/carbon nanodot) composite [98], carbon nanodot depicts the dual behavior. Primarily, carbon nanodots behave as an electron sink to capture the electrons and transfer electrons which are generated from m-$BiVO_4$ nanoparticles at wavelength shorter than the 520nm that efficiently intensifies the charge separation. Besides this, carbon nanodots correspond to longer wavelength and radiate the shorter ones (300nm to 530nm) which can further activate the m-$BiVO_4$ to generate electron hole pair for photocatalytic decay. Due to marvelous photocatalytic performance of carbon nanodots, they are fused with Cu_2O, Ag_3PO_4 and Fe_2O_3 as well [99 - 101].

Tungsten Trioxide as Photocatalyst

Tungsten oxide (WO_3) is an n-type semiconductor with a band gap of Eg = 2.6-2.8eV [102], which enables the absorption of light in the visible range. WO_3 has many interesting optical, electrical, structural and electrochromic properties [103]. It is also used in photocatalysis for efficient phenol degradation [104], photo electrolytic oxidation of several toxic compounds [105] and photoelectron chemical water decomposition [106, 107]. Nevertheless, WO_3 is unstable in basic environment. WO_3 can couple with TiO_2 or dope into its lattice [108 - 111]. The coupling of WO_3/TiO_2 allows better charge separation due to the difference of their CB levels. Photo-excited electrons are transferred from CB of TiO_2 to CB of WO_3. Therefore, electrons are separated from photo-induced holes.

WO_3 nanoparticles modified by using iron (III) or silver ions as the electron acceptor on the surface are capable of oxidizing water under visible light irradiation [112 - 115]. WO_3 has high efficiency in photocatalytic degradation of organic compounds, including a large fraction of environmentally hazardous materials [116, 117]. Hence, WO_3 is used as a cleaning agent which can be applied to energy renewal, energy storage and environmental cleanup. WO_3 modified or doped by Pt [118], Nb [119], and phthalo cyanine [120] can photodegrade organic compounds with much higher efficiency than that of the bare WO_3. It can also enhance the photoactivity of wide band gap semiconductors, such as, TiO_2, SnO_2 [121] and ZnO [122, 123]. The WO_3 doping or coupling onto the surfaces of above semiconducting metal oxides can encourage the charge separation, because the band gap of WO_3 is low and photogenerated carriers easily accumulate in its conductor or valence band. A $WO_3/MWCNT$ composite was fabricated for olefin skeletal isomerization [124]. WO_3 was also modified by MWCNTs in five different concentrations. The $MWCNTs/WO_3$ nanocomposites were studied for photocatalytic degradation of Rhodamine B dye in comparison with MWCNTs and bare WO_3. The results proved that 5.0 wt% $MWCNTs/WO_3$ composite had the highest photocatalytic activity [125].

Cerium Oxide Photocatalysts

The bulk ceria (CeO_2) has band gap value of 3.2 eV which makes it a photocatalyst sensitive to ultraviolet radiations. Under visible light irradiation, the photogenerated charge carriers cannot migrate easily to the surface and thus remain photo catalytically inefficient [126]. The ceria can absorb visible light by tuning the band gap. This can be achieved by a combination of modifying techniques such as producing intrinsic defects, doping, coupling, deposition of plasmonic resonant materials and developing intrinsic composites of ceria heterophases. CeO_2 based photocatalysts have high oxygen storage capacity and ability to release or uptake oxygen due to their Ce4p/Ce3p redox cycles [127]. Diffusion rate of photogenerated excitons on catalyst surface is fundamental step for enhancing the photocatalytic activity. The doping of CeO_2 with 3d transition metals can improve this diffusion rate [128]. The doping of CeO_2 with Fe, Mn, Ti, and Co transition-metal ions enhances the photocatalytic efficiencies as compare to bare CeO_2 [129]. It was reported that co-doping of ceria also influences the morphology of the nanostructures. Thus the surface area is increased which enhances the photocatalytic activity under UV light [130]. The N-doped CeO_2

[131] and Au-supported CeO_2 nanostructures [132] showed excellent photocatalytic activity under visible irradiation. The photo thermo catalyst of yttrium-doped CeO_2 was prepared which showed high photoactivity at 100 °C [133]. The different combination of CeO_2 with Ag_3PO_4 [134], $BiVO_4$ [135] and Cu_2O [136] showed enhanced visible light response. CdS/CeO_2 composites were prepared because of their matched band structures and were explored in the field of photocatalysis [137 - 141]. The photocatalytic performance of CdS/CeO_x nanowires and CeO_2/CdS nanospheres prepared by electrochemical process for hydrogen evolution under visible light illumination was investigated. One dimensional CdS/CeO_2 nanowires and nanoparticle composites were investigated under visible light irradiation for the photocatalytic reduction of nitroaromatics and water splitting to hydrogen [142].

Various morphologies of cerium oxides like nanotubes [143], nanowires [144], octahedra [145], nanocubes [146, 147] and nanorods [146 - 150] have been fabricated to enhance shape dependent properties. It is reported that CeO_2 acts as a photocatalytic material with gold (Au) nanoparticles which influence the strong localized surface plasmon resonance (LSPR) of Au at around 550 nm. The organic acids were stoichiometrically decomposed to carbon dioxide under irradiation of visible light by Au/CeO_2 nanocomposite which mineralize the organic compounds. The mechanistic approach is discussed in Fig. (**14**) [151].

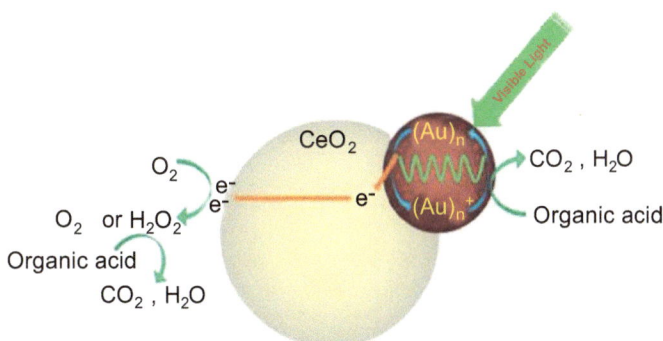

Fig. (14). Expected working mechanism for mineralization of organic acids in aqueous suspension of Au/CeO_2 under irradiation of visible light [151].

PEROVSKITES AS PHOTOCATALYST

Among hybrid crystalline compounds the organic-inorganic perovskites have prime importance, containing a variety of inorganic anions combined with organic cations of versatile properties [152, 153]. The three-dimensional framework of organometal halide perovskite materials like methylammonium lead iodide ($CH_3NH_3PbI_3$) supports a multitude of photovoltaic's functionalities, such as light sensitizers, absorbers and ambipolar electron-hole transporters [152, 153]. The organic-inorganic hybrid perovskites are generally expressed in ABX_3 stoichiometry, where, (A= small organic cation, B = divalent metal cation, X = halogen). The structure of hybrid perovskites shows that the BX^{6-} anion forms octahedral geometry with a component neutralizing the charge and fill the interstices [154, 155].

The superb conductive features of the inorganic semiconductor component can hybrid with the substantial light-matter interaction of the organic portion. The layered organic-inorganic perovskites contain lead halide as semiconductor layers that sandwiched between organic ammonium insulator layers [156 - 160]. Lead halide that possess strong exciton binding energy is well acclaimed as typical ionic crystals [161]. The organic layer has low value of dielectric constant and more band gap as compared to inorganic layer. This is the reason for high exciton binding energy for perovskites [162]. The peroveskites are proved to be used as nonlinear optical material [163, 164] and have its utiliaztion in luminescent devices [165, 166]. The nanostructured devices use organometallic halide perovskites [167]. The efforts have been carried out for organometallic trihalide perovskite absorbers to be used in extremely effective solar mobile phones [168 - 171]. In Fig. (15) the two-dimensional inorganic layers and an organic ammonium layer are assembled alternately. These layers are consist of two-dimensional sheet of $[MX_6]^{-4}$ octahedron which are coupled at the four corners with halide ions on the plane.

Organic -Inorganic Layered Perovskites

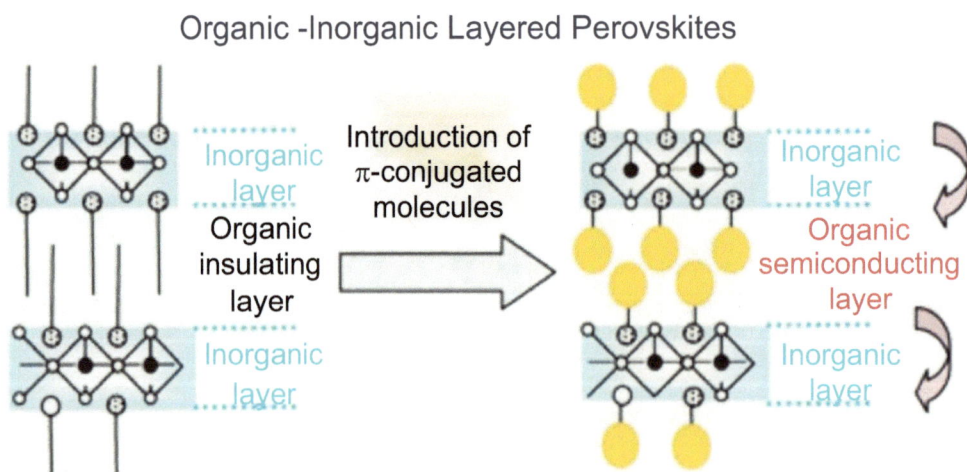

Fig. (15). Schematic structure of the organic-inorganic hybrid crystal [172].

The hybrid perovskite $CH_3NH_3PbX_3$ solar cells are unstable with the fact that they have shown good power conversion efficiency. The moisture and heat affect their stability. The hybrid perovskites undergo degradation, even in the presence of little moisture [173, 174]. Gratzel and co-workers have observed that hybrid perovskite solar cells fabrication should be carried out under a controlled environment with less than 1% humidity [171, 175, 176]. Similarly, Li-dan Tang and co-worker have found that $CH_3NH_3PbBr_3$ is more thermally stable as compared to other perovskites like $CH_3NH_3PbI_3$ [177]. The solar energy transfer process is shown in Fig. (**16**).

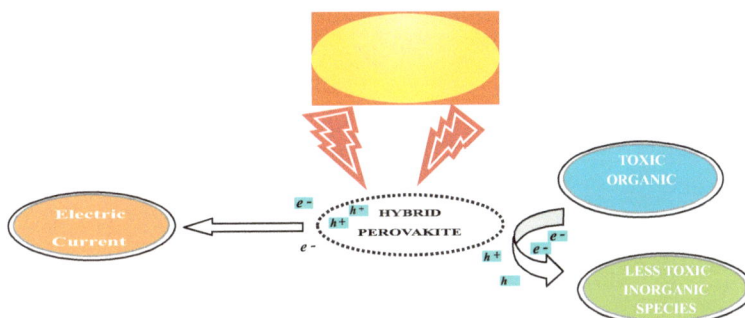

Fig. (16). Photo catalytic process of perovskites.

METAL ORGANIC FRAMEWORKS AS PHOTOCATALYSTS

The potentially most important porous material is the composite of organic and inorganic moieties called metal organic frameworks (MOFs). MOFs are encouraging candidates of porous crystalline solids due to enlarged surface area and the dominant pore volumes. Mostly, MOFs also display permanent porosity and high thermal durability to above 300 °C. Such intrinsic features of highly organized structures (surface area and pore size distribution) make them distinctive than other materials. The preparation of MOFs that have potential to absorb light for the advantageous photocatalytic deterioration of organic pollutants, is feasible due to the richness of metal nodes and organic linkers, as well as the controllability of different functional groups [178]. MIL-53 [179] displayed enhanced photocatalytic performance in the degradation of organic dyes. Cavka *et al.* synthesized stable photocatalytic zirconium metal MOF (UiO-66(Zr): $[Zr_6O_4(OH)_4(CO_2)_{12}]$) in 2008 [180].

It was reported that UiO-66 did not show any morphologic transition on suspension in water at 100 °C for 4 h without under UV light irradiation for photocatalytic hydrogen emergence [181]. In 2009 Serre and co-worker formulated the prospective stable photocatalyst $Ti_8O_8(OH)_4(O_2C-C_6H-$ (MIL125(Ti)) that resulted the concurrent titanium center reduction accompanied by the oxidation of adsorbed alcohol by irradiating with UV-visible beam [182].

One example of photo decay of Rhodamine 6G in aqueous solution in presence of visible light by iron(III)-based MOFs was first stated by Larurier *et al.* in 2013 [183], where Fe–O clusters initiate the photocatalytic reaction by absorbing visible light and the charge separation is accelerated by the organic linkers. Because of the well-ordered porous structure along with organic linker/metal clusters MOFs are applied in catalysis [184, 185], sorting [186], gas storage [187, 188] and carbon dioxide capture [189].

One of the most investigated class of ordered porous solids being inorganic-organic material with ultrahigh porosity, remarkable internal surface areas, dimension, size and shape is metal-organic frameworks (MOFs). Recently explored MOFs as a new nano photocatalyst become a hot topic of interest for the researchers interested in fields of chemistry, chemical engineering, materials science and others. Some MOFS such as Zn(II) [190 - 193], Co(II)/Co(III) [21, 23 - 28], Fe(II)/Fe(III) [179, 192, 193], Cd(II) [193, 194] and Cu(I)/Cu(II) [193, 194] that explored as photocatalyst to degrade organic pollutants under the irradiation of 200-600 nm light. The photocatalytic performances of various MOFs are evaluated for the deterioration of organic pollutants [200 - 203]. Various MOFs with band gaps between 1.0 and 5.5 eV calculated by optical properties indicates utilization in photocatalysis [195]. Alvaro *et al.* reported photocatalytic activities of MOF-5 [196]. The suggested mechanism of photocatalytic process is demonstrated in Fig. (**17**). The compound consists of 1,4-benzenedicarboxilic acid as ligand and metal is zinc. Moreover, it has absorbed visible light [178]. Herein organic compound phenol is degraded in aqueous media, the degradation perform-ance is compared with commercially available TiO_2(Degussa P-25) (Fig. **17b**) [178]. Moreover organic part favors the charge distribution that resulted in amplified efficiency [178]. To initiate the catalytic process (Fig. **17c**) electron is directed from phenol to the hole generated in MOF-5 or production of reactive oxygen radical by the electron transferred. The light source is used to determine the comparable efficiency of MOF-5 with other photocatalysts under visible irradiation suggest MOF-5 as better photocatalyst as compared to traditional metal oxides because of the less absorption of light with wavelength greater than 350 nm while MOF-5 absorbs more longer wavelength (>400nm) [178].

The photoactive MOFs show unique photocatalytic properties than other materials, especially in organic synthesis applications. MOFs create the opportunity to combine photocatalyst with organocatalyst. MOFs can be considered as semiconducting materials because many of them exhibit a broad UV-Vis absorption with an edge falling into the range of typical semiconductor band gap values [196]. These absorption bands can be assigned to n–π* transition of the aromatic ligand or a localized ligand-to metal charge transfer. Therefore, MOFs are expected to be used as photocatalysts in the wastewater treatment [206

- 208].

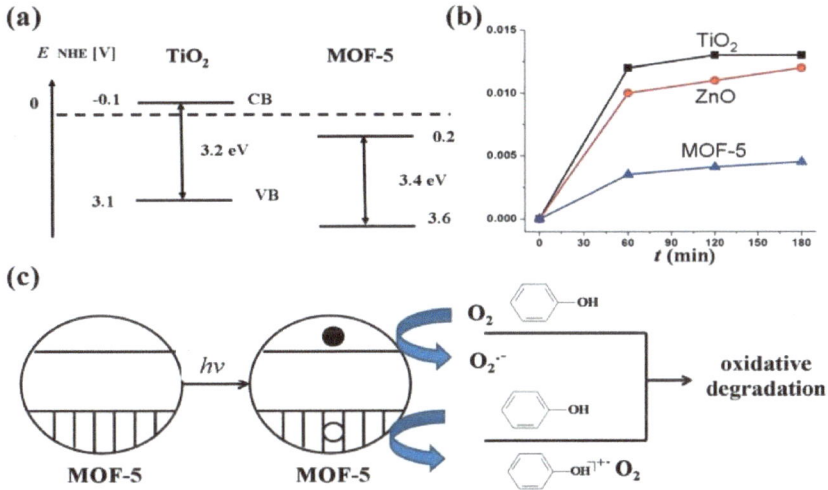

Fig. (17). (a) Theoretical band gap values of MOF-5 and TiO$_2$ **(b)** phenol degradation(g/mol) process with time **(c)** Purposed mechanism for degradation process of by MOF-5 [178].

APPLICATIONS OF PHOTOCATALYSIS

Under the exposure of light certain semi-conducting materials named as photocatalyst trigger a chemical reaction (photocatalysis) that results in decomposition of organic molecule [197]. The selection of applications for photocatalysts and photocatalytic process are described below:

Self-cleaning Process

Cleanliness and maintenance issues are the key points for applications of self-cleaning surfaces. Two major variants are directly associated with the treatment of surfaces by using the photocatalyst;

- Self-cleaning photocatalytic surfaces support the photon induced destruction of adherent organic molecules [198] and is of particular interest for anti-bacterial, anti-virus and anti-fungicidal applications. During antibacterial activity cell membrane is damaged resulting into oxidative attack of internal cellular components to kill the bacteria [199].
- Super hydrophilic surfaces: photo induced hydrophilicity is caused by the exposure of a TiO$_2$ [200] treated surface to intense UV light. In such treatment formation of water droplet is suppressed and covers the surface with a homogeneous thin wetting layer which penetrates below dirt particles.

Air Purification

Photocatalytic process has the potential to remove the gaseous volatile organic compounds. Applications of photocatalytic components are found in air filters, ventilation and air conditioning system. Photocatalytic oxidation can also be used for the inactivation of infectious microorganisms which can be airborne bioterrorism weapons like *Bacillus anthracis*. The indoor environments such as residences, office buildings, factories, air crafts and space crafts can be cleaned from contaminants by the photocatalytic processes [199].

Water Purification

The process of mineralization of organic compounds can be achieved by photocatalysis to decompose the organic toxic into green components in water reservoirs [201] with the use of concentrating solar type reactors [199]. The suspension of titanium dioxide can be used for disinfection and the inactivation of coli form bacteria and polio virus. Photocatalysis can be used for the removal of sewage containing organisms which are highly resistant to traditional disinfection treatments such as cryptosporidium parvum and noro viruses [202, 203].

Hydrogen Production by Water Cleavage

Previously existed energy sources are constant threat for the environment protection. To replace such energy sources, alternate energies have been developed which are renewable and have lower carbon emissions as compared to conventional energy sources [204].

1. Advantages of hydrogen:
   - Use of renewable source of energy
   - Environmental friendly
   - High energy power [205]
2. Disadvantages of hydrogen.
   - Low hydrogen production efficiencies
   - Current lack of efficient infrastructure to store, transport and distribute hydrogen.
   - Hydrogen production costs [205]

There are two types of configuration that were adapted to a photocatalyst to use in photochemical water splitting: (a) photoelectrochemical cells (b) photocatalytic system. The photoelectrochemical cell used for water decomposition has electrodes immersed in an aqueous electrolyte. Photocatalytic system consists of particles of photocatalysts in aqueous media, in which each particle acts as micro photoelectrode that performs both the oxidation and reduction of water on its surface [205]. Photocatalytic systems have advantages over photo electrochemical cell as these are much simpler and less expensive. The water splitting process for hydrogen generation has two major benefits first, the raw material is abundant and cheap and second the combustion of hydrogen in air produces water. This makes the whole process cycling and non-polluting.

During the process of photocatalytic hydrogen generation organic compounds such as alcohols (methanol, ethanol, isopropanol *etc.*) [206, 207], acids (formic acid, acetic acid *etc.*) and aldehydes (formaldehyde, acetaldehyde *etc.*) are used as electron donor. The mechanism for hydrogen production through organic compounds is discussed as below

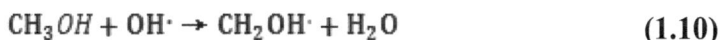

$$h^+ + H_2O \rightarrow OH^. + H^+ \tag{1.9}$$

$$CH_3OH + OH^. \rightarrow CH_2OH^. + H_2O \tag{1.10}$$

$$CH_2OH^. \rightarrow HCHO + H^+ + e^- \tag{1.11}$$

$$2H_2O + e^- \rightarrow H_2 + 2OH^- \tag{1.12}$$

Over all reaction:

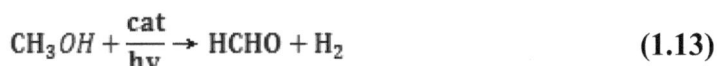

$$CH_3OH + \frac{cat}{hv} \rightarrow HCHO + H_2 \tag{1.13}$$

Formaldehyde could be further oxidized to CO_2 and H_2O [206, 207]

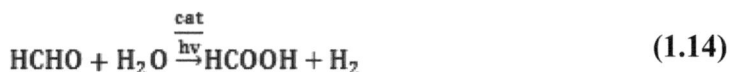

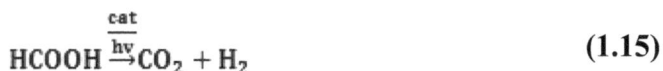

$$HCHO + H_2O \xrightarrow{\overset{cat}{hv}} HCOOH + H_2 \tag{1.14}$$

$$HCOOH \xrightarrow{\overset{cat}{hv}} CO_2 + H_2 \tag{1.15}$$

The reduction and oxidation reactions are the basic mechanism of photocatalytic hydrogen production. For hydrogen production with the use of photocatalyst, the conduction band level must be more negative than the reduction potential of H_2O molecules while the valence band should be more positive than the oxidation potential of water (Fig. **18**) [208].

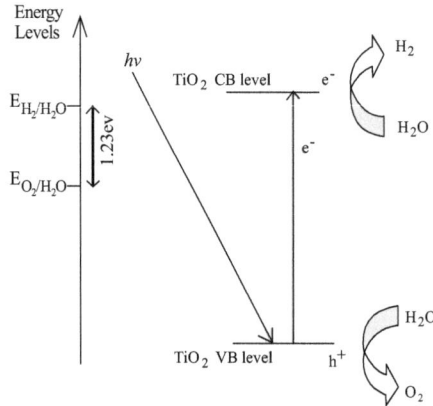

Fig. (18). Mechanism of hydrogen production through water cleavage by using photocatalyst [208].

In photocatalytic phenomena, the photocatalyst can either be supported or suspended in the photocatalytic reactor. So the state of photocatalyst can be defined as:

1. Photocatalytic slurry (suspended) reactors: the photocatalyst particles are freely dispersed in liquid phase. Thus, the photocatalyst is fully incorporated in the liquid phase.

   Advantages of photocatalytic slurry (suspended) reactors;
   - Uniform distribution of catalyst
   - Large photocatalytic surface area to reactor volume ratio
   - Limited mass transfer
   - Well mixed particle suspension
   - Low pressure drop through the reactor

   Disadvantages of photocatalytic slurry (suspended) reactors
   - Post-process filtration is required
   - Suspended medium particle scatter and adsorb the light

2. Photocatalytic reactors with immobilized photocatalyst

Dye-sensitized Solar Cell

Dye sensitized solar cells also known as Gratzel cells are photo electrochemical cells that generate electricity with the aid of semi-conductor nanowires, nanoparticles, dye molecules as a photocatalyst and an iodine solution [49]. These are flexible, inexpensive and easier to manufacture than silicon solar cells. Mechanism of such solar cells lies on the fact that sunlight first enters through the transparent conductive oxide (TCO) top coat glass surface and excites the dye molecules that are attached with the surface of semiconductor nanoparticles. Excited electrons are then transferred to the conduction band of semi- conductor from where they are transferred to TCO and then to the circuit. While during this each dye molecules loses its electrons so another electron must be provided for its survival. Therefore, dye molecule receives electron from iodine in the electrolyte by oxidizing iodide into tri-iodide as shown in Fig. (**19**). The tri-iodide then recovers its missing electron through counter electrode after passing through external circuit [209].

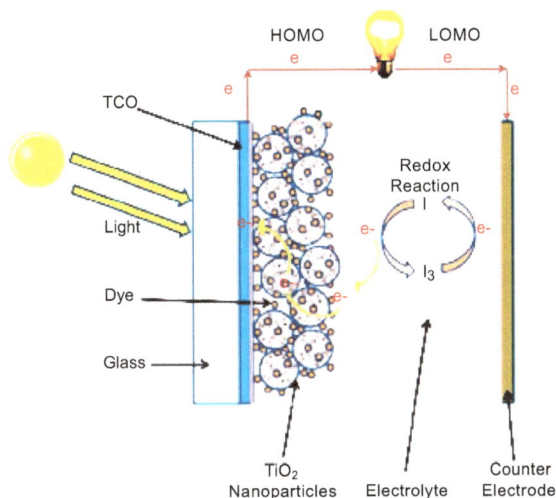

Fig. (19). Working mechanism of dye synthesized solar cell [209].

Zinc oxide based dye synthesized solar cell is most efficient as compared to titania based dye synthesized solar cells due to fast electron transport with low recombination rate and its ease of crystallization and anisotropic growth. It was evaluated that light conversion efficiency of the branched zinc oxide nanowire is much higher than the upstanding ZnO nanowires based dye synthesized solar cell [210]. Since the enhanced surface area for the dye absorption molecules and the low chances of charge recombination due to direct conduction pathways along the crystalline ZnO nano tree having multi generation branches are responsible for superior photocatalytic activity [210].

Water Flow Purification System

High working efficiency of continuous water flow purification system (Fig. **20**) was observed by using ZnO nanowires that were grown on flexible poly-L-lactide nanofibers. There is no need for the separation of photo-chemically active material from the continuous flow water purification system. For the purification water system zinc oxide nanowires were grown on various substrate like woven polyethylene fibers, polyethylene fibers and on silicon substrates [211, 212].

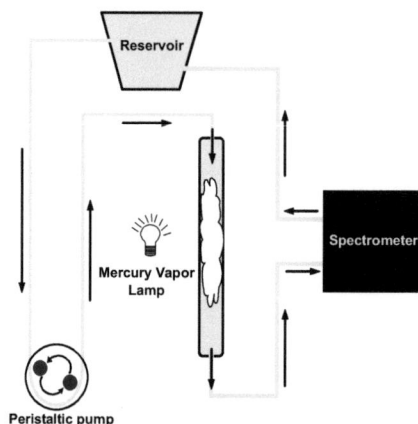

Fig. (20). Photocatalytic water treatment system [212].

CONSENT FOR PUBLICATION

Not applicable.

CONFLICT OF INTEREST

The author (editor) declares no conflict of interest, financial or otherwise.

ACKNOWLEDGEMENTS

This work has been supported by Higher Education Commission of Pakistan (NRPU Grant Number 3660), International Islamic University and Pakistan Institute of Engineering and Applied Sciences, Islamabad Pakistan.

REFERENCES

[1] Li, W. Photocatalysis of oxide semiconductors. *J. Aust. Ceram. Soc.,* **2013**, *49*, 4-6.

[2] Zhao, H.; Tian, F.; Wang, R.; Chen, R. A Review on Bismuth-Related Nanomaterials for Photocatalysis. *Rev. Adv. Sci. Eng.,* **2014**, *3*, 3-27.
[http://dx.doi.org/10.1166/rase.2014.1050]

[3] Kresge, C.; Leonowicz, M.; Roth, W.; Vartuli, J.; Beck, J. Ordered mesoporous molecular sieves synthesized by a liquid-crystal template mechanism. *Nature,* **1992**, *359*, 710-712.
[http://dx.doi.org/10.1038/359710a0]

[4] Koh, C.W.; Lee, U.H.; Song, J.K.; Lee, H.R.; Kim, M.H.; Suh, M.; Kwon, Y.U. Mesoporous titania thin film with highly ordered and fully accessible vertical pores and crystalline walls. *Chem. Asian J.,* **2008**, *3*(5), 862-867.
[http://dx.doi.org/10.1002/asia.200700331] [PMID: 18386267]

[5] Yuan, S.; Sheng, Q.; Zhang, J.; Chen, F.; Anpo, M.; Zhang, Q. Synthesis of La^{3+} doped mesoporous titania with highly crystallized walls. *Microp. Mesop. Mater.,* **2005**, *79*, 93-99.
[http://dx.doi.org/10.1016/j.micromeso.2004.10.028]

[6] Puhlfürß, P.; Voigt, A.; Weber, R.; Morbé, M. Microporous TiO$_2$ membranes with a cut off< 500 Da. *J. Membr. Sci.,* **2000**, *174*, 123-133.

[http://dx.doi.org/10.1016/S0376-7388(00)00380-X]

[7]   Chen, H.; Wang, L. Nanostructure sensitization of transition metal oxides for visible-light photocatalysis. *Beilstein J. Nanotechnol.,* **2014**, *5*, 696-710.
[http://dx.doi.org/10.3762/bjnano.5.82] [PMID: 24991507]

[8]   Chen, X.; Mao, S.S. Titanium dioxide nanomaterials: synthesis, properties, modifications, and applications. *Chem. Rev.,* **2007**, *107*(7), 2891-2959.
[http://dx.doi.org/10.1021/cr0500535] [PMID: 17590053]

[9]   Nozik, A.J.; Beard, M.C.; Luther, J.M.; Law, M.; Ellingson, R.J.; Johnson, J.C. Semiconductor quantum dots and quantum dot arrays and applications of multiple exciton generation to third-generation photovoltaic solar cells. *Chem. Rev.,* **2010**, *110*(11), 6873-6890.
[http://dx.doi.org/10.1021/cr900289f] [PMID: 20945911]

[10]   Eder, D. Carbon nanotube-inorganic hybrids. *Chem. Rev.,* **2010**, *110*(3), 1348-1385.
[http://dx.doi.org/10.1021/cr800433k] [PMID: 20108978]

[11]   Kudo, A.; Kato, H.; Tsuji, I. Strategies for the development of visible-light-driven photocatalysts for water splitting. *Chem. Lett.,* **2004**, *33*, 1534-1539.
[http://dx.doi.org/10.1246/cl.2004.1534]

[12]   Linsebigler, A.L.; Lu, G.; Yates, J.T., Jr Photocatalysis on TiO_2 surfaces: principles, mechanisms, and selected results. *Chem. Rev.,* **1995**, *95*, 735-758.
[http://dx.doi.org/10.1021/cr00035a013]

[13]   Ibhadon, A.O.; Fitzpatrick, P. Heterogeneous photocatalysis: recent advances and applications. *Catalysts,* **2013**, *3*, 189-218.
[http://dx.doi.org/10.3390/catal3010189]

[14]   Herrmann, J-M. Heterogeneous photocatalysis: fundamentals and applications to the removal of various types of aqueous pollutants. *Catal. Today,* **1999**, *53*, 115-129.
[http://dx.doi.org/10.1016/S0920-5861(99)00107-8]

[15]   Colmenares, J.C.; Luque, R.; Campelo, J.M.; Colmenares, F.; Karpiński, Z.; Romero, A.A. Nanostructured photocatalysts and their applications in the photocatalytic transformation of lignocellulosic biomass: an overview. *Materials (Basel),* **2009**, *2*, 2228-2258.
[http://dx.doi.org/10.3390/ma2042228]

[16]   Akpan, U.G.; Hameed, B.H. Parameters affecting the photocatalytic degradation of dyes using TiO_2-based photocatalysts: a review. *J. Hazard. Mater.,* **2009**, *170*(2-3), 520-529.
[http://dx.doi.org/10.1016/j.jhazmat.2009.05.039] [PMID: 19505759]

[17]   Fox, M.A.; Dulay, M.T. Heterogeneous photocatalysis. *Chem. Rev.,* **1993**, *93*, 341-357.
[http://dx.doi.org/10.1021/cr00017a016]

[18]   Sajjad, A.K.L.; Shamaila, S.; Tian, B.; Chen, F.; Zhang, J. Comparative studies of operational parameters of degradation of azo dyes in visible light by highly efficient WO_x/TiO_2 photocatalyst. *J. Hazard. Mater.,* **2010**, *177*(1-3), 781-791.
[http://dx.doi.org/10.1016/j.jhazmat.2009.12.102] [PMID: 20074854]

[19]   Huang, Y.; Zheng, X.; Zhongyi, Y.; Feng, T.; Beibei, F.; Keshan, H. Preparation of nitrogen-doped TiO_2 nanoparticle catalyst and its catalytic activity under visible light. *Chin. J. Chem. Eng.,* **2007**, *15*, 802-807.
[http://dx.doi.org/10.1016/S1004-9541(08)60006-3]

[20]   Reutergådh, L.B.; Iangphasuk, M. Photocatalytic decolourization of reactive azo dye: A comparison between TiO_2 and us photocatalysis. *Chemosphere,* **1997**, *35*, 585-596.
[http://dx.doi.org/10.1016/S0045-6535(97)00122-7]

[21]   Beydoun, D.; Amal, R.; Low, G.; McEvoy, S. Role of nanoparticles in photocatalysis. *J. Nanopart. Res.,* **1999**, *1*, 439-458.
[http://dx.doi.org/10.1023/A:1010044830871]

[22] Kalyanasundaram, K.; Grätzel, M. Applications of functionalized transition metal complexes in photonic and optoelectronic devices. *Coord. Chem. Rev.,* **1998**, *177*, 347-414.
[http://dx.doi.org/10.1016/S0010-8545(98)00189-1]

[23] Jaeger, C.D.; Fan, F-R.F.; Bard, A.J. Semiconductor electrodes. 26. Spectral sensitization of semiconductors with phthalocyanine. *J. Am. Chem. Soc.,* **1980**, *102*, 2592-2598.
[http://dx.doi.org/10.1021/ja00528a012]

[24] Giraudeau, A.; Fan, F-R.F.; Bard, A.J. Semiconductor electrodes. 30. Spectral sensitization of the semiconductors titanium oxide (n-TiO$_2$) and tungsten oxide (n-WO$_3$) with metal phthalocyanines. *J. Am. Chem. Soc.,* **1980**, *102*, 5137-5142.
[http://dx.doi.org/10.1021/ja00536a001]

[25] Stipkala, J.M.; Castellano, F.N.; Heimer, T.A.; Kelly, C.A.; Livi, K.J.; Meyer, G.J. Light-induced charge separation at sensitized sol-gel processed semiconductors. *Chem. Mater.,* **1997**, *9*, 2341-2353.
[http://dx.doi.org/10.1021/cm9703177]

[26] Kalyanasundaram, K.; Grätzel, M.; Pelizzetti, E. Interfacial electron transfer in colloidal metal and semiconductor dispersions and photodecomposition of water. *Coord. Chem. Rev.,* **1986**, *69*, 57-125.
[http://dx.doi.org/10.1016/0010-8545(86)85009-3]

[27] Memming, R. Electron transfer processs with excited molecules at semiconductor electrodes. *Prog. Surf. Sci.,* **1984**, *17*, 7-73.
[http://dx.doi.org/10.1016/0079-6816(84)90012-1]

[28] Howe, R. Recent developments in photocatalysis. *Dev. Chem. Eng. Miner. Process.,* **1998**, *6*, 55-84.
[http://dx.doi.org/10.1002/apj.5500060105]

[29] Fujishima, A.; Rao, T.N.; Tryk, D.A. Titanium dioxide photocatalysis. *J. Photochem. Photobiol. Photochem. Rev.,* **2000**, *1*, 1-21.
[http://dx.doi.org/10.1016/S1389-5567(00)00002-2]

[30] Chen, C.; Ma, W.; Zhao, J. Semiconductor-mediated photodegradation of pollutants under visible-light irradiation. *Chem. Soc. Rev.,* **2010**, *39*(11), 4206-4219.
[http://dx.doi.org/10.1039/b921692h] [PMID: 20852775]

[31] Chen, L-C.; Ting, J-M.; Lee, Y-L.; Hon, M-H. A binder-free process for making all-plastic substrate flexible dye-sensitized solar cells having a gel electrolyte. *J. Mater. Chem.,* **2012**, *22*, 5596-5601.
[http://dx.doi.org/10.1039/c2jm15360b]

[32] Shamaila, S.; Sajjad, A.K.L.; Chen, F.; Zhang, J. Mesoporous titania with high crystallinity during synthesis by dual template system as an efficient photocatalyst. *Catal. Today,* **2011**, *175*, 568-575.
[http://dx.doi.org/10.1016/j.cattod.2011.03.041]

[33] Chen, H-W.; Liang, C-P.; Huang, H-S.; Chen, J-G.; Vittal, R.; Lin, C-Y.; Wu, K.C-W.; Ho, K-C. Electrophoretic deposition of mesoporous TiO$_2$ nanoparticles consisting of primary anatase nanocrystallites on a plastic substrate for flexible dye-sensitized solar cells. *Chem. Commun. (Camb.),* **2011**, *47*(29), 8346-8348.
[http://dx.doi.org/10.1039/c1cc12514a] [PMID: 21691640]

[34] Pelaez, M.; Nolan, N.T.; Pillai, S.C.; Seery, M.K.; Falaras, P.; Kontos, A.G.; Dunlop, P.S.; Hamilton, J.W.; Byrne, J.A.; O'Shea, K. A review on the visible light active titanium dioxide photocatalysts for environmental applications. *Appl. Catal. B,* **2012**, *125*, 331-349.
[http://dx.doi.org/10.1016/j.apcatb.2012.05.036]

[35] Ma, Y.; Wang, X.; Jia, Y.; Chen, X.; Han, H.; Li, C. Titanium dioxide-based nanomaterials for photocatalytic fuel generations. *Chem. Rev.,* **2014**, *114*(19), 9987-10043.
[http://dx.doi.org/10.1021/cr500008u] [PMID: 25098384]

[36] Hamal, D.B.; Klabunde, K.J. Synthesis, characterization, and visible light activity of new nanoparticle photocatalysts based on silver, carbon, and sulfur-doped TiO$_2$. *J. Colloid Interface Sci.,* **2007**, *311*(2), 514-522.

[http://dx.doi.org/10.1016/j.jcis.2007.03.001] [PMID: 17418857]

[37]   Li, X.Z.; Li, F.B. Study of Au/Au($^{3+}$)-TiO$_2$ photocatalysts toward visible photooxidation for water and wastewater treatment. *Environ. Sci. Technol.,* **2001**, *35*(11), 2381-2387.
[http://dx.doi.org/10.1021/es001752w] [PMID: 11414049]

[38]   Behar, D.; Rabani, J. Kinetics of hydrogen production upon reduction of aqueous TiO$_2$ nanoparticles catalyzed by Pd(0), Pt(0), or Au(0) coatings and an unusual hydrogen abstraction; steady state and pulse radiolysis study. *J. Phys. Chem. B,* **2006**, *110*(17), 8750-8755.
[http://dx.doi.org/10.1021/jp060971m] [PMID: 16640431]

[39]   Kovács, G.; Pap, Z.; Coteţ, C.; Coşoveanu, V.; Baia, L.; Danciu, V. Photocatalytic, Morphological and Structural Properties of the TiO$_2$-SiO$_2$-Ag Porous Structures Based System. *Materials (Basel),* **2015**, *8*(3), 1059-1073.
[http://dx.doi.org/10.3390/ma8031059] [PMID: 28787988]

[40]   Chou, C-S.; Yang, R-Y.; Yeh, C-K.; Lin, Y-J. Preparation of TiO$_2$/nano-metal composite particles and their applications in dye-sensitized solar cells. *Powder Technol.,* **2009**, *194*, 95-105.
[http://dx.doi.org/10.1016/j.powtec.2009.03.039]

[41]   Cai, Y.; Zhang, J.; Sun, G.; Wang, Q.; Ye, H.; Qiao, H.; Wei, Q. Fabrication, structural morphology and photocatalytic activity of porous TiO$_2$ nanofibres through combination of sol-gel, electrospinning and doping-removal techniques. *Mater. Technol.,* **2014**, *29*, 40-46.
[http://dx.doi.org/10.1179/1753555713Y.0000000098]

[42]   Nguyen-Phan, T-D.; Shin, E.W. Morphological effect of TiO$_2$ catalysts on photocatalytic degradation of methylene blue. *J. Ind. Eng. Chem.,* **2011**, *17*, 397-400.
[http://dx.doi.org/10.1016/j.jiec.2011.05.013]

[43]   Kandiel, T.A.; Robben, L.; Alkaim, A.; Bahnemann, D. Brookite *versus* anatase TiO$_2$ photocatalysts: phase transformations and photocatalytic activities. *Photochem. Photobiol. Sci.,* **2013**, *12*(4), 602-609.
[http://dx.doi.org/10.1039/C2PP25217A] [PMID: 22945758]

[44]   Hwang, S.H.; Song, J.; Jung, Y.; Kweon, O.Y.; Song, H.; Jang, J. Electrospun ZnO/TiO$_2$ composite nanofibers as a bactericidal agent. *Chem. Commun. (Camb.),* **2011**, *47*(32), 9164-9166.
[http://dx.doi.org/10.1039/c1cc12872h] [PMID: 21761035]

[45]   Sajjad, S.; Leghari, S.A.K.; Zhang, J. Copper impregnated ionic liquid assisted mesoporous titania: visible light photocatalyst. *RSC Advances,* **2013**, *3*, 12678-12687.
[http://dx.doi.org/10.1039/c3ra23347b]

[46]   Amna, T.; Hassan, M.S.; Pandurangan, M.; Khil, M-S.; Lee, H-K.; Hwang, I. Characterization and potent bactericidal effect of Cobalt doped Titanium dioxide nanofibers. *Ceram. Int.,* **2013**, *39*, 3189-3193.
[http://dx.doi.org/10.1016/j.ceramint.2012.10.003]

[47]   Amna, T.; Hassan, M.S.; Yousef, A.; Mishra, A.; Barakat, N.A.; Khil, M-S.; Kim, H.Y. Inactivation of foodborne pathogens by NiO/TiO$_2$ composite nanofibers: a novel biomaterial system. *Food Bioprocess Technol.,* **2013**, *6*, 988-996.
[http://dx.doi.org/10.1007/s11947-011-0741-1]

[48]   Johar, M.A.; Afzal, R.A.; Alazba, A.A.; Manzoor, U. Photocatalysis and Bandgap Engineering Using ZnO Nanocomposites. *Adv. Mater. Sci. Eng.,* **2015**, *2015*, 1, 22.
[http://dx.doi.org/10.1155/2015/934587]

[49]   Özgür, Ü.; Alivov, Y.I.; Liu, C.; Teke, A.; Reshchikov, M.; Doğan, S.; Avrutin, V.; Cho, S-J.; Morkoc, H. A comprehensive review of ZnO materials and devices. *J. Appl. Phys.,* **2005**, *98*, 041301.
[http://dx.doi.org/10.1063/1.1992666]

[50]   Kajbafvala, A.; Ghorbani, H.; Paravar, A.; Samberg, J.P.; Kajbafvala, E.; Sadrnezhaad, S. Effects of morphology on photocatalytic performance of Zinc oxide nanostructures synthesized by rapid microwave irradiation methods. *Superlattices Microstruct.,* **2012**, *51*, 512-522.

[http://dx.doi.org/10.1016/j.spmi.2012.01.015]

[51]   Xie, J.; Hao, Y.J.; Zhou, Z.; Meng, X.C.; Yao, L.; Bian, L.; Wei, Y. Fabrication and characterization of ZnO particles with different morphologies for photocatalytic degradation of pentachlorophenol. *Res. Chem. Intermed., 2014, 40*, 1937-1946.
[http://dx.doi.org/10.1007/s11164-013-1091-6]

[52]   Fouad, O.; Ismail, A.; Zaki, Z.; Mohamed, R. Zinc oxide thin films prepared by thermal evaporation deposition and its photocatalytic activity. *Appl. Catal. B, 2006, 62*, 144-149.
[http://dx.doi.org/10.1016/j.apcatb.2005.07.006]

[53]   Lu, F.; Cai, W.; Zhang, Y. ZnO hierarchical micro/nanoarchitectures: solvothermal synthesis and structurally enhanced photocatalytic performance. *Adv. Funct. Mater., 2008, 18*, 1047-1056.
[http://dx.doi.org/10.1002/adfm.200700973]

[54]   Baruah, S.; Jaisai, M.; Imani, R.; Nazhad, M.M.; Dutta, J. Photocatalytic paper using zinc oxide nanorods. *Sci. Technol. Adv. Mater., 2010, 11*(5), 055002.
[http://dx.doi.org/10.1088/1468-6996/11/5/055002] [PMID: 27877367]

[55]   Sapkota, A.; Anceno, A.J.; Baruah, S.; Shipin, O.V.; Dutta, J. Zinc oxide nanorod mediated visible light photoinactivation of model microbes in water. *Nanotechnology, 2011, 22*(21), 215703.
[http://dx.doi.org/10.1088/0957-4484/22/21/215703] [PMID: 21451231]

[56]   Wang, W.; Zhang, L. Photocatalytic degradation of *E. Coli* membrane cell in the presence of ZnO nanowires. *J. Wuhan Uni. Techno.-. Mater. Sci. Ed., 2011, 26*, 222-225.

[57]   McFarland, E.W.; Tang, J. A photovoltaic device structure based on internal electron emission. *Nature, 2003, 421*(6923), 616-618.
[http://dx.doi.org/10.1038/nature01316] [PMID: 12571591]

[58]   Li, B.; Wang, Y. Synthesis, microstructure, and photocatalysis of ZnO/CdS nano-heterostructure. *J. Phys. Chem. Solids, 2011, 72*, 1165-1169.
[http://dx.doi.org/10.1016/j.jpcs.2011.07.010]

[59]   Cun, W.; Jincai, Z.; Xinming, W.; Bixian, M.; Guoying, S.; Pingan, P.; Jiamo, F. Preparation, characterization and photocatalytic activity of nano-sized ZnO/SnO$_2$ coupled photocatalysts. *Appl. Catal. B, 2002, 39*, 269-279.
[http://dx.doi.org/10.1016/S0926-3373(02)00115-7]

[60]   Zhang, Z.; Shao, C.; Li, X.; Zhang, L.; Xue, H.; Wang, C.; Liu, Y. Electrospun Nanofibers of ZnO–SnO$_2$ Heterojunction with High Photocatalytic Activity. *J. Phys. Chem. C, 2010, 114*, 7920-7925.
[http://dx.doi.org/10.1021/jp100262q]

[61]   Joshi, B.N.; Yoon, H.; Na, S-H.; Choi, J-Y.; Yoon, S.S. Enhanced photocatalytic performance of graphene–ZnO nanoplatelet composite thin films prepared by electrostatic spray deposition. *Ceram. Int., 2014, 40*, 3647-3654.
[http://dx.doi.org/10.1016/j.ceramint.2013.09.060]

[62]   Ahmad, M.; Ahmed, E.; Ahmed, W.; Elhissi, A.; Hong, Z.; Khalid, N. Enhancing visible light responsive photocatalytic activity by decorating Mn-doped ZnO nanoparticles on graphene. *Ceram. Int., 2014, 40*, 10085-10097.
[http://dx.doi.org/10.1016/j.ceramint.2014.03.184]

[63]   Schaaff, T.G.; Rodinone, A.J. Preparation and characterization of silver sulfide nanocrystals generated from silver (i)-thiolate polymers. *J. Phys. Chem. B, 2003, 107*, 10416-10422.
[http://dx.doi.org/10.1021/jp034979x]

[64]   Subash, B.; Krishnakumar, B.; Pandiyan, V.; Swaminathan, M.; Shanthi, M. An efficient nanostructured Ag 2 S–ZnO for degradation of Acid Black 1 dye under day light illumination. *Separ. Purif. Tech., 2012, 96*, 204-213.
[http://dx.doi.org/10.1016/j.seppur.2012.06.002]

[65]   Khanchandani, S.; Srivastava, P.K.; Kumar, S.; Ghosh, S.; Ganguli, A.K. Band gap engineering of

ZnO using core/shell morphology with environmentally benign Ag_2S sensitizer for efficient light harvesting and enhanced visible-light photocatalysis. *Inorg. Chem.,* **2014**, *53*(17), 8902-8912.
[http://dx.doi.org/10.1021/ic500518a] [PMID: 25144692]

[66]   Uddin, M.T.; Nicolas, Y.; Olivier, C.; Toupance, T.; Servant, L.; Müller, M.M.; Kleebe, H-J.; Ziegler, J.; Jaegermann, W. Nanostructured SnO_2-ZnO heterojunction photocatalysts showing enhanced photocatalytic activity for the degradation of organic dyes. *Inorg. Chem.,* **2012**, *51*(14), 7764-7773.
[http://dx.doi.org/10.1021/ic300794j] [PMID: 22734686]

[67]   Zheng, L.; Zheng, Y.; Chen, C.; Zhan, Y.; Lin, X.; Zheng, Q.; Wei, K.; Zhu, J. Network structured SnO_2/ZnO heterojunction nanocatalyst with high photocatalytic activity. *Inorg. Chem.,* **2009**, *48*(5), 1819-1825.
[http://dx.doi.org/10.1021/ic802293p] [PMID: 19235945]

[68]   Sun, S.; Wang, W. Advanced chemical compositions and nanoarchitectures of bismuth based complex oxides for solar photocatalytic application. *RSC Advances,* **2014**, *4*, 47136-47152.
[http://dx.doi.org/10.1039/C4RA06419D]

[69]   Shamaila, S.; Sajjad, A.K.L.; Chen, F.; Zhang, J. WO_3/BiOCl, a novel heterojunction as visible light photocatalyst. *J. Colloid Interface Sci.,* **2011**, *356*(2), 465-472.
[http://dx.doi.org/10.1016/j.jcis.2011.01.015] [PMID: 21320705]

[70]   Sajjad, S.; Leghari, S.A.K.; Zhang, J. Nonstoichiometric Bi_2O_3: efficient visible light photocatalyst. *Rsc. Adv.,* **2013**, *3*, 1363-1367.
[http://dx.doi.org/10.1039/C2RA22239F]

[71]   Oshikiria, M.; Boero, M.; Ye, J.; Zou, Z.; Kido, G. Electronic structures of promising photocatalysts InMO4 „MÄV, Nb, Ta... and $BiVO_4$ for water decomposition in the visible wavelength region. *J. Chem. Phys.,* **2002**, *117*(15), 7313-7318.
[http://dx.doi.org/10.1063/1.1507101]

[72]   Tang, J.; Zou, Z.; Ye, J. Photocatalytic decomposition of organic contaminants by Bi_2WO_6 under visible light irradiation. *Catal. Lett.,* **2004**, *92*, 53-56.
[http://dx.doi.org/10.1023/B:CATL.0000011086.20412.aa]

[73]   Saison, T.; Chemin, N.; Chanéac, C.; Durupthy, O.; Ruaux, V.; Mariey, L.; Maugé, F.o.; Beaunier, P.; Jolivet, J-P. Bi_2O_3, $BiVO_4$, and Bi_2WO_6: impact of surface properties on photocatalytic activity under visible light. *J. Phys. Chem. C,* **2011**, *115*, 5657-5666.
[http://dx.doi.org/10.1021/jp109134z]

[74]   Fu, H.; Zhang, S.; Xu, T.; Zhu, Y.; Chen, J. Photocatalytic degradation of RhB by fluorinated Bi_2WO_6 and distributions of the intermediate products. *Environ. Sci. Technol.,* **2008**, *42*(6), 2085-2091.
[http://dx.doi.org/10.1021/es702495w] [PMID: 18409641]

[75]   Ye, H.; Lee, J.; Jang, J.S.; Bard, A.J. Rapid screening of $BiVO_4$-based photocatalysts by scanning electrochemical microscopy (SECM) and studies of their photoelectrochemical properties. *J. Phys. Chem. C,* **2010**, *114*, 13322-13328.
[http://dx.doi.org/10.1021/jp104343b]

[76]   Spanhel, L.; Weller, H.; Henglein, A. Photochemistry of semiconductor colloids. 22. Electron ejection from illuminated cadmium sulfide into attached titanium and zinc oxide particles. *J. Am. Chem. Soc.,* **1987**, *109*, 6632-6635.
[http://dx.doi.org/10.1021/ja00256a012]

[77]   Wang, X.; Yin, L.; Liu, G.; Wang, L.; Saito, R.; Lu, G.Q.M.; Cheng, H-M. Polar interface-induced improvement in high photocatalytic hydrogen evolution over ZnO–CdS heterostructures. *Energy Environ. Sci.,* **2011**, *4*, 3976-3979.
[http://dx.doi.org/10.1039/c0ee00723d]

[78]   G. Enhanced photocatalytic hydrogen evolution by prolonging the lifetime of carriers in ZnO/CdS heterostructures. *Chem. Commun. (Camb.),* **2009**, *23*, 3452-3454.

[79]   Gao, X-F.; Li, H-B.; Sun, W-T.; Chen, Q.; Tang, F-Q.; Peng, L-M. CdTe quantum dots-sensitized TiO$_2$ nanotube array photoelectrodes. *J. Phys. Chem. C,* **2009**, *113*, 7531-7535.
        [http://dx.doi.org/10.1021/jp810727n]

[80]   Seabold, J.A.; Shankar, K.; Wilke, R.H.; Paulose, M.; Varghese, O.K.; Grimes, C.A.; Choi, K-S. Photoelectrochemical properties of heterojunction CdTe/TiO$_2$ electrodes constructed using highly ordered TiO$_2$ nanotube arrays. *Chem. Mater.,* **2008**, *20*, 5266-5273.
        [http://dx.doi.org/10.1021/cm8010666]

[81]   Yang, H.; Fan, W.; Vaneski, A.; Susha, A.S.; Teoh, W.Y.; Rogach, A.L. Heterojunction Engineering of CdTe and CdSe Quantum Dots on TiO$_2$ Nanotube Arrays: Intricate Effects of Size Dependency and Interfacial Contact on Photoconversion Efficiencies. *Adv. Funct. Mater.,* **2012**, *22*, 2821-2829.
        [http://dx.doi.org/10.1002/adfm.201103074]

[82]   Tian, Y.; Tatsuma, T. Mechanisms and applications of plasmon-induced charge separation at TiO$_2$ films loaded with gold nanoparticles. *J. Am. Chem. Soc.,* **2005**, *127*(20), 7632-7637.
        [http://dx.doi.org/10.1021/ja042192u] [PMID: 15898815]

[83]   Tian, Y.; Tatsuma, T. Plasmon-induced photoelectrochemistry at metal nanoparticles supported on nanoporous TiO$_2$. *Chem. Commun. (Camb.),* **2004**, *16*(16), 1810-1811.
        [http://dx.doi.org/10.1039/b405061d] [PMID: 15306895]

[84]   Yao, Y.; Li, G.; Ciston, S.; Lueptow, R.M.; Gray, K.A. Photoreactive TiO$_2$/carbon nanotube composites: synthesis and reactivity. *Environ. Sci. Technol.,* **2008**, *42*(13), 4952-4957.
        [http://dx.doi.org/10.1021/es800191n] [PMID: 18678032]

[85]   Wang, W.; Serp, P.; Kalck, P.; Faria, J.L. Visible light photodegradation of phenol on MWNT-TiO$_2$ composite catalysts prepared by a modified sol-gel method. *J. Mol. Catal. Chem.,* **2005**, *235*, 194-199.
        [http://dx.doi.org/10.1016/j.molcata.2005.02.027]

[86]   Novoselov, K.S.; Geim, A.K.; Morozov, S.V.; Jiang, D.; Zhang, Y.; Dubonos, S.V.; Grigorieva, I.V.; Firsov, A.A. Electric field effect in atomically thin carbon films. *Science,* **2004**, *306*(5696), 666-669.
        [http://dx.doi.org/10.1126/science.1102896] [PMID: 15499015]

[87]   Bai, X.; Wang, L.; Zhu, Y. Visible photocatalytic activity enhancement of ZnWO$_4$ by graphene hybridization. *ACS Catal.,* **2012**, *2*, 2769-2778.
        [http://dx.doi.org/10.1021/cs3005852]

[88]   Woan, K.; Pyrgiotakis, G.; Sigmund, W. Photocatalytic Carbon Nanotube-TiO$_2$ Composites. *Adv. Mater.,* **2009**, *21*, 2233-2239.
        [http://dx.doi.org/10.1002/adma.200802738]

[89]   Baker, S.N.; Baker, G.A. Luminescent carbon nanodots: emergent nanolights. *Angew. Chem. Int. Ed. Engl.,* **2010**, *49*(38), 6726-6744.
        [http://dx.doi.org/10.1002/anie.200906623] [PMID: 20687055]

[90]   Shen, J.; Zhu, Y.; Yang, X.; Li, C. Graphene quantum dots: emergent nanolights for bioimaging, sensors, catalysis and photovoltaic devices. *Chem. Commun. (Camb.),* **2012**, *48*(31), 3686-3699.
        [http://dx.doi.org/10.1039/c2cc00110a] [PMID: 22410424]

[91]   Li, H.; Kang, Z.; Liu, Y.; Lee, S-T. Carbon nanodots: synthesis, properties and applications. *J. Mater. Chem.,* **2012**, *22*, 24230-24253.
        [http://dx.doi.org/10.1039/c2jm34690g]

[92]   Zhang, Z.; Zhang, J.; Chen, N.; Qu, L. Graphene quantum dots: an emerging material for energy-related applications and beyond. *Energy Environ. Sci.,* **2012**, *5*, 8869-8890.
        [http://dx.doi.org/10.1039/c2ee22982j]

[93]   Yu, H.; Zhang, H.; Huang, H.; Liu, Y.; Li, H.; Ming, H.; Kang, Z. ZnO/carbon quantum dots nanocomposites: one-step fabrication and superior photocatalytic ability for toxic gas degradation under visible light at room temperature. *New J. Chem.,* **2012**, *36*, 1031-1035.
        [http://dx.doi.org/10.1039/c2nj20959d]

[94]   Li, H.; He, X.; Kang, Z.; Huang, H.; Liu, Y.; Liu, J.; Lian, S.; Tsang, C.H.A.; Yang, X.; Lee, S.T. Water-soluble fluorescent carbon quantum dots and photocatalyst design. *Angew. Chem. Int. Ed. Engl.,* **2010**, *49*(26), 4430-4434.
[http://dx.doi.org/10.1002/anie.200906154] [PMID: 20461744]

[95]   Lian, S.; Huang, H.; Zhang, J.; Kang, Z.; Liu, Y. One-step solvothermal synthesis of ZnO-carbon composite spheres containing different amounts of carbon and their use as visible light photocatalysts. *Solid State Commun.,* **2013**, *155*, 53-56.
[http://dx.doi.org/10.1016/j.ssc.2012.11.003]

[96]   Wang, F.; Liu, Y.; Ma, Z.; Li, H.; Kang, Z.; Shen, M. Enhanced photoelectrochemical response in SrTiO$_3$ films decorated with carbon quantum dots. *New J. Chem.,* **2013**, *37*, 290-294.
[http://dx.doi.org/10.1039/C2NJ40988G]

[97]   Guo, C.X.; Dong, Y.; Yang, H.B.; Li, C.M. Graphene Quantum Dots as a Green Sensitizer to Functionalize ZnO Nanowire Arrays on F-Doped SnO$_2$ Glass for Enhanced Photoelectrochemical Water Splitting. *Adv. Energy Mater.,* **2013**, *3*, 997-1003.
[http://dx.doi.org/10.1002/aenm.201300171]

[98]   Tang, D.; Zhang, H.; Huang, H.; Liu, R.; Han, Y.; Liu, Y.; Tong, C.; Kang, Z. Carbon quantum dots enhance the photocatalytic performance of BiVO$_4$ with different exposed facets. *Dalton Trans.,* **2013**, *42*(18), 6285-6289.
[http://dx.doi.org/10.1039/c3dt50567g] [PMID: 23519004]

[99]   Li, H.; Liu, R.; Liu, Y.; Huang, H.; Yu, H.; Ming, H.; Lian, S.; Lee, S-T.; Kang, Z. Carbon quantum dots/Cu$_2$O composites with protruding nanostructures and their highly efficient (near) infrared photocatalytic behavior. *J. Mater. Chem.,* **2012**, *22*, 17470-17475.
[http://dx.doi.org/10.1039/c2jm32827e]

[100] Zhang, H.; Huang, H.; Ming, H.; Li, H.; Zhang, L.; Liu, Y.; Kang, Z. Carbon quantum dots/Ag$_3$PO$_4$ complex photocatalysts with enhanced photocatalytic activity and stability under visible light. *J. Mater. Chem.,* **2012**, *22*, 10501-10506.
[http://dx.doi.org/10.1039/c2jm30703k]

[101] Zhang, H.; Ming, H.; Lian, S.; Huang, H.; Li, H.; Zhang, L.; Liu, Y.; Kang, Z.; Lee, S-T. Fe$_2$O$_3$/carbon quantum dots complex photocatalysts and their enhanced photocatalytic activity under visible light. *Dalton Trans.,* **2011**, *40*(41), 10822-10825.
[http://dx.doi.org/10.1039/c1dt11147g] [PMID: 21935522]

[102] Bamwenda, G.R.; Arakawa, H. The visible light induced photocatalytic activity of tungsten trioxide powders. *Appl. Catal. A Gen.,* **2001**, *210*, 181-191.
[http://dx.doi.org/10.1016/S0926-860X(00)00796-1]

[103] Deb, S.K. Opportunities and challenges in science and technology of WO$_3$ for electrochromic and related applications. *Sol. Energy Mater. Sol. Cells,* **2008**, *92*, 245-258.
[http://dx.doi.org/10.1016/j.solmat.2007.01.026]

[104] Gondal, M.; Sayeed, M.; Alarfaj, A. Activity comparison of Fe$_2$O$_3$, NiO, WO$_3$, TiO$_2$ semiconductor catalysts in phenol degradation by laser enhanced photo-catalytic process. *Chem. Phys. Lett.,* **2007**, *445*, 325-330.
[http://dx.doi.org/10.1016/j.cplett.2007.07.094]

[105] Solarska, R.; Santato, C.; Jorand-Sartoretti, C.; Ulmann, M.; Augustynski, J. Photoelectrolytic oxidation of organic species at mesoporous tungsten trioxide film electrodes under visible light illumination. *J. Appl. Electrochem.,* **2005**, *35*, 715-721.
[http://dx.doi.org/10.1007/s10800-005-1400-x]

[106] Kim, J.K.; Shin, K.; Cho, S.M.; Lee, T-W.; Park, J.H. Synthesis of transparent mesoporous tungsten trioxide films with enhanced photoelectrochemical response: application to unassisted solar water splitting. *Energy Environ. Sci.,* **2011**, *4*, 1465-1470.
[http://dx.doi.org/10.1039/c0ee00469c]

[107]   Santato, C.; Ulmann, M.; Augustynski, J. Enhanced visible light conversion efficiency using nanocrystalline WO_3 films. *Adv. Mater.,* **2001**, *13*, 511-514.
[http://dx.doi.org/10.1002/1521-4095(200104)13:7<511::AID-ADMA511>3.0.CO;2-W]

[108]   Ke, D.; Liu, H.; Peng, T.; Liu, X.; Dai, K. Preparation and photocatalytic activity of WO_3/TiO_2 nanocomposite particles. *Mater. Lett.,* **2008**, *62*, 447-450.
[http://dx.doi.org/10.1016/j.matlet.2007.05.060]

[109]   Cheng, P.; Deng, C.; Liu, D.; Dai, X. Titania surface modification and photovoltaic characteristics with tungsten oxide. *Appl. Surf. Sci.,* **2008**, *254*, 3391-3396.
[http://dx.doi.org/10.1016/j.apsusc.2007.11.018]

[110]   Sajjad, A.K.L.; Shamaila, S.; Tian, B.; Chen, F.; Zhang, J. One step activation of WO x/TiO_2 nanocomposites with enhanced photocatalytic activity. *Appl. Catal. B,* **2009**, *91*, 397-405.
[http://dx.doi.org/10.1016/j.apcatb.2009.06.005]

[111]   Leghari, S.A.K.; Sajjad, S.; Zhang, J. Large mesoporous micro-spheres of WO_3/TiO_2 composite with enhanced visible light photo activity. *RSC Advances,* **2013**, *3*, 15354-15361.
[http://dx.doi.org/10.1039/c3ra41782d]

[112]   Ohno, T.; Tanigawa, F.; Fujihara, K.; Izumi, S.; Matsumura, M. Photocatalytic oxidation of water on TiO_2-coated WO_3 particles by visible light using Iron (III) ions as electron acceptor. *J. Photochem. Photobiol. Chem.,* **1998**, *118*, 41-44.
[http://dx.doi.org/10.1016/S1010-6030(98)00374-8]

[113]   Erbs, W.; Desilvestro, J.; Borgarello, E.; Graetzel, M. Visible-light-induced oxygen generation from aqueous dispersions of tungsten (VI) oxide. *J. Phys. Chem.,* **1984**, *88*, 4001-4006.
[http://dx.doi.org/10.1021/j150662a028]

[114]   Bamwenda, G.R.; Arakawa, H. The visible light induced photocatalytic activity of tungsten trioxide powders. *Appl. Catal. A.,* **2001**, *210*, 181-191.
[http://dx.doi.org/10.1016/S0926-860X(00)00796-1]

[115]   Sayama, K.; Arakawa, H. Photocatalytic decomposition of water and photocatalytic reduction of carbon dioxide over zirconia catalyst. *J. Phys. Chem.,* **1993**, *97*, 531-533.
[http://dx.doi.org/10.1021/j100105a001]

[116]   Wang, H.; Xu, P.; Wang, T. The preparation and properties study of photocatalytic nanocrystalline/nanoporous WO_3 thin films. *Mater. Des.,* **2002**, *23*, 331-336.
[http://dx.doi.org/10.1016/S0261-3069(01)00040-1]

[117]   Pelizzetti, E.; Borgarello, M.; Minero, C.; Pramauro, E.; Borgarello, E.; Serpone, N. Photocatalytic degradation of polychlorinated dioxins and polychlorinated biphenyls in aqueous suspensions of semiconductors irradiated with simulated solar light. *Chemosph,* **1988**, *17*, 499-510.
[http://dx.doi.org/10.1016/0045-6535(88)90025-2]

[118]   Sclafani, A.; Palmisano, L.; Marcı, G.; Venezia, A. Influence of platinum on catalytic activity of polycrystalline WO_3 employed for phenol photodegradation in aqueous suspension. *Sol. Energy Mater. Sol. Cells,* **1998**, *51*, 203-219.
[http://dx.doi.org/10.1016/S0927-0248(97)00215-8]

[119]   Wang, H.; Xu, P.; Wang, T. Doping of Nb_2O_5 in photocatalytic nanocrystalline/nanoporous WO_3 films. *Thin Solid Films,* **2001**, *388*, 68-72.
[http://dx.doi.org/10.1016/S0040-6090(01)00853-7]

[120]   Iliev, V.; Tomova, D.; Bilyarska, L.; Prahov, L.; Petrov, L. Phthalocyanine modified TiO_2 or WO_3-catalysts for photooxidation of sulfide and thiosulfate ions upon irradiation with visible light. *J. Photochem. Photobiol. Chem.,* **2003**, *159*, 281-287.
[http://dx.doi.org/10.1016/S1010-6030(03)00170-9]

[121]   Ma, Z.; Hua, W.; Tang, Y.; Gao, Z. Catalytic decomposition of CFC-12 over solid acids WO_3/M_xO_y (M= Ti, Sn, Fe). *J. Mol. Catal. Chem.,* **2000**, *159*, 335-345.

[http://dx.doi.org/10.1016/S1381-1169(00)00191-6]

[122] Sakthivel, S.; Geissen, S-U.; Bahnemann, D.; Murugesan, V.; Vogelpohl, A. Enhancement of photocatalytic activity by semiconductor heterojunctions: α-Fe$_2$O$_3$, WO$_3$ and CdS deposited on ZnO. *J. Photochem. Photobiol. Chem.,* **2002**, *148*, 283-293.
[http://dx.doi.org/10.1016/S1010-6030(02)00055-2]

[123] Li, D.; Haneda, H. Enhancement of photocatalytic activity of sprayed nitrogen-containing ZnO powders by coupling with metal oxides during the acetaldehyde decomposition. *Chemosphere,* **2004**, *54*(8), 1099-1110.
[http://dx.doi.org/10.1016/j.chemosphere.2003.09.022] [PMID: 14664838]

[124] Pietruszka, B.; Di Gregorio, F.; Keller, N.; Keller, V. High-efficiency WO$_3$/carbon nanotubes for olefin skeletal isomerization. *Catal. Today,* **2005**, *102*, 94-100.
[http://dx.doi.org/10.1016/j.cattod.2005.02.014]

[125] Wang, S.; Shi, X.; Shao, G.; Duan, X.; Yang, H.; Wang, T. Preparation, characterization and photocatalytic activity of multi-walled carbon nanotube-supported tungsten trioxide composites. *J. Phys. Chem. Solids,* **2008**, *69*, 2396-2400.
[http://dx.doi.org/10.1016/j.jpcs.2008.04.029]

[126] Li, B.; Gu, T.; Ming, T.; Wang, J.; Wang, P.; Wang, J.; Yu, J.C. (Gold core)@(ceria shell) nanostructures for plasmon-enhanced catalytic reactions under visible light. *ACS Nano,* **2014**, *8*(8), 8152-8162.
[http://dx.doi.org/10.1021/nn502303h] [PMID: 25029556]

[127] Stetsovych, V.; Pagliuca, F.; Dvořák, F.; Duchoň, T.; Vorokhta, M.; Aulická, M.; Lachnitt, J.; Schernich, S.; Matolínová, I.; Veltruská, K.; Skála, T.; Mazur, D.; Mysliveček, J.; Libuda, J.; Matolín, V. Epitaxial cubic Ce$_2$O$_3$ films *via* Ce–CeO$_2$ interfacial reaction. *J. Phys. Chem. Lett.,* **2013**, *4*(6), 866-871.
[http://dx.doi.org/10.1021/jz400187j] [PMID: 26291348]

[128] Esch, F.; Fabris, S.; Zhou, L.; Montini, T.; Africh, C.; Fornasiero, P.; Comelli, G.; Rosei, R. Electron localization determines defect formation on ceria substrates. *Science,* **2005**, *309*(5735), 752-755.
[http://dx.doi.org/10.1126/science.1111568] [PMID: 16051791]

[129] Yue, L.; Zhang, X-M. Structural characterization and photocatalytic behaviors of doped CeO$_2$ nanoparticles. *J. Alloys Compd.,* **2009**, *475*, 702-705.
[http://dx.doi.org/10.1016/j.jallcom.2008.07.096]

[130] Arul, N.S.; Mangalaraj, D.; Chen, P.C.; Ponpandian, N.; Meena, P.; Masuda, Y. Enhanced photocatalytic activity of cobalt-doped CeO$_2$ nanorods. *J. Sol-Gel Sci. Technol.,* **2012**, *64*, 515-523.
[http://dx.doi.org/10.1007/s10971-012-2883-7]

[131] Jorge, A.B.; Sakatani, Y.; Boissière, C.; Laberty-Roberts, C.; Sauthier, G.; Fraxedas, J.; Sanchez, C.; Fuertes, A. Nanocrystalline N-doped ceria porous thin films as efficient visible-active photocatalysts. *J. Mater. Chem.,* **2012**, *22*, 3220-3226.
[http://dx.doi.org/10.1039/c2jm15230d]

[132] Primo, A.; Marino, T.; Corma, A.; Molinari, R.; García, H. Efficient visible-light photocatalytic water splitting by minute amounts of gold supported on nanoparticulate CeO$_2$ obtained by a biopolymer templating method. *J. Am. Chem. Soc.,* **2011**, *133*(18), 6930-6933.
[http://dx.doi.org/10.1021/ja2011498] [PMID: 21506541]

[133] Liyanage, A.D.; Perera, S.D.; Tan, K.; Chabal, Y.; Balkus, K.J., Jr Synthesis, characterization, and photocatalytic activity of Y-Doped CeO$_2$ nanorods. *ACS Catal.,* **2014**, *4*, 577-584.
[http://dx.doi.org/10.1021/cs400889y]

[134] Yang, Z-M.; Huang, G-F.; Huang, W-Q.; Wei, J-M.; Yan, X-G.; Liu, Y-Y.; Jiao, C.; Wan, Z.; Pan, A. Novel Ag$_3$PO$_4$/CeO$_2$ composite with high efficiency and stability for photocatalytic applications. *J. Mater. Chem.,* **2014**, *2*, 1750-1756.
[http://dx.doi.org/10.1039/C3TA14286H]

[135]   Wetchakun, N.; Chaiwichain, S.; Inceesungvorn, B.; Pingmuang, K.; Phanichphant, S.; Minett, A.I.; Chen, J. $BiVO_4/CeO_2$ nanocomposites with high visible-light-induced photocatalytic activity. *ACS Appl. Mater. Interfaces,* **2012**, *4*(7), 3718-3723.
        [http://dx.doi.org/10.1021/am300812n] [PMID: 22746549]

[136]   Hu, S.; Zhou, F.; Wang, L.; Zhang, J. Preparation of Cu_2O/CeO_2 heterojunction photocatalyst for the degradation of acid orange 7 under visible light irradiation. *Catal. Commun.,* **2011**, *12*, 794-797.
        [http://dx.doi.org/10.1016/j.catcom.2011.01.027]

[137]   Jing, D.; Guo, L. Efficient hydrogen production by a composite CdS/mesoporous zirconium titanium phosphate photocatalyst under visible light. *J. Phys. Chem. C,* **2007**, *111*, 13437-13441.
        [http://dx.doi.org/10.1021/jp071700u]

[138]   Barpuzary, D.; Qureshi, M. Enhanced photovoltaic performance of semiconductor-sensitized ZnO-CdS coupled with graphene oxide as a novel photoactive material. *ACS Appl. Mater. Interfaces,* **2013**, *5*(22), 11673-11682.
        [http://dx.doi.org/10.1021/am403268w] [PMID: 24152060]

[139]   Zhao, W.; Bai, Z.; Ren, A.; Guo, B.; Wu, C. Sunlight photocatalytic activity of CdS modified TiO_2 loaded on activated carbon fibers. *Appl. Surf. Sci.,* **2010**, *256*, 3493-3498.
        [http://dx.doi.org/10.1016/j.apsusc.2009.12.062]

[140]   Nagai, Y.; Yamamoto, T.; Tanaka, T.; Yoshida, S.; Nonaka, T.; Okamoto, T.; Suda, A.; Sugiura, M. X-ray absorption fine structure analysis of local structure of CeO_2–ZrO_2 mixed oxides with the same composition ratio (Ce/Zr= 1). *Catal. Today,* **2002**, *74*, 225-234.
        [http://dx.doi.org/10.1016/S0920-5861(02)00025-1]

[141]   Li, W.; Xie, S.; Li, M.; Ouyang, X.; Cui, G.; Lu, X.; Tong, Y. CdS/CeOx heterostructured nanowires for photocatalytic hydrogen production. *J. Mater. Chem. A Mater. Energy Sustain.,* **2013**, *1*, 4190-4193.
        [http://dx.doi.org/10.1039/c3ta10394c]

[142]   Zhang, X.; Zhang, N.; Xu, Y-J.; Tang, Z-R. One-dimensional CdS nanowires–CeO_2 nanoparticles composites with boosted photocatalytic activity. *New J. Chem.,* **2015**, *39*, 6756-6764.
        [http://dx.doi.org/10.1039/C5NJ00976F]

[143]   Han, W-Q.; Wu, L.; Zhu, Y. Formation and oxidation state of $CeO_{(2-x)}$ nanotubes. *J. Am. Chem. Soc.,* **2005**, *127*(37), 12814-12815.
        [http://dx.doi.org/10.1021/ja054533p] [PMID: 16159271]

[144]   Sun, C.; Li, H.; Wang, Z.; Chen, L.; Huang, X. Synthesis and characterization of polycrystalline CeO_2 nanowires. *Chem. Lett.,* **2004**, *33*, 662-663.
        [http://dx.doi.org/10.1246/cl.2004.662]

[145]   Wang, Z.L.; Feng, X. Polyhedral shapes of CeO_2 nanoparticles. *J. Phys. Chem. B,* **2003**, *107*, 13563-13566.
        [http://dx.doi.org/10.1021/jp036815m]

[146]   Mai, H-X.; Sun, L-D.; Zhang, Y-W.; Si, R.; Feng, W.; Zhang, H-P.; Liu, H-C.; Yan, C-H. Shape-selective synthesis and oxygen storage behavior of ceria nanopolyhedra, nanorods, and nanocubes. *J. Phys. Chem. B,* **2005**, *109*(51), 24380-24385.
        [http://dx.doi.org/10.1021/jp055584b] [PMID: 16375438]

[147]   Wu, Q.; Zhang, F.; Xiao, P.; Tao, H.; Wang, X.; Hu, Z.; Lu, Y. Great influence of anions for controllable synthesis of CeO_2 nanostructures: from nanorods to nanocubes. *J. Phys. Chem. C,* **2008**, *112*, 17076-17080.
        [http://dx.doi.org/10.1021/jp804140e]

[148]   Higashine, Y.; Fujihara, S. Facile synthesis of single-crystalline CeO_2 nanorods from aqueous $CeCl_3$ solutions. *J. Ceram. Soc. Jpn.,* **2007**, *115*, 916-919.
        [http://dx.doi.org/10.2109/jcersj2.115.916]

[149] Li, Y.; Sun, Q.; Kong, M.; Shi, W.; Huang, J.; Tang, J.; Zhao, X. Coupling oxygen ion conduction to photocatalysis in mesoporous nanorod-like ceria significantly improves photocatalytic efficiency. *J. Phys. Chem. C,* **2011**, *115*, 14050-14057.
[http://dx.doi.org/10.1021/jp202720g]

[150] Arul, N.S.; Mangalaraj, D.; Kim, T.W.; Chen, P.C.; Ponpandian, N.; Meena, P.; Masuda, Y. Synthesis of CeO_2 nanorods with improved photocatalytic activity: comparison between precipitation and hydrothermal process. *J. Mater. Sci. Mater. Electron.,* **2013**, *24*, 1644-1650.
[http://dx.doi.org/10.1007/s10854-012-0989-x]

[151] Kominami, H.; Tanaka, A.; Hashimoto, K. Gold nanoparticles supported on cerium (IV) oxide powder for mineralization of organic acids in aqueous suspensions under irradiation of visible light of λ= 530nm. *Appl. Catal. A Gen.,* **2011**, *397*, 121-126.
[http://dx.doi.org/10.1016/j.apcata.2011.02.029]

[152] Mitzi, D.B.; Chondroudis, K.; Kagan, C.R. Organic-inorganic electronics. *IBM J. Res. Develop.,* **2001**, *45*, 29-45.
[http://dx.doi.org/10.1147/rd.451.0029]

[153] Mitzi, D.B. Templating and structural engineering in organic–inorganic perovskites. *Dalton Trans.,* **2001**, *1*, 1-12.

[154] Eperon, G.E.; Stranks, S.D.; Menelaou, C.; Johnston, M.B.; Herz, L.M.; Snaith, H.J. Formamidinium lead trihalide: a broadly tunable perovskite for efficient planar heterojunction solar cells. *Energy Environ. Sci.,* **2014**, *7*, 982-988.
[http://dx.doi.org/10.1039/c3ee43822h]

[155] Frost, J.M.; Butler, K.T.; Brivio, F.; Hendon, C.H.; van Schilfgaarde, M.; Walsh, A. Atomistic origins of high-performance in hybrid halide perovskite solar cells. *Nano Lett.,* **2014**, *14*(5), 2584-2590.
[http://dx.doi.org/10.1021/nl500390f] [PMID: 24684284]

[156] Dammak, T.; Elleuch, S.; Bougzhala, H.; Mlayah, A.; Chtourou, R.; Abid, Y. Synthesis, vibrational and optical properties of a new three-layered organic–inorganic perovskite $(C_4H_9NH_3)_4Pb_3I_4Br_6$. *J. Lumin.,* **2009**, *129*, 893-897.
[http://dx.doi.org/10.1016/j.jlumin.2009.04.020]

[157] Mitzi, D.B.; Feild, C.; Harrison, W.; Guloy, A. Conducting tin halides with a layered organic-based perovskite structure. *Nature,* **1994**, *369*, 467-469.
[http://dx.doi.org/10.1038/369467a0]

[158] Hong, X.; Ishihara, T.; Nurmikko, A. Photoconductivity and electroluminescence in lead iodide based natural quantum well structures. *Solid State Commun.,* **1992**, *84*, 657-661.
[http://dx.doi.org/10.1016/0038-1098(92)90210-Z]

[159] Cheng, Z.; Wang, Z.; Xing, R.; Han, Y.; Lin, J. Patterning and photoluminescent properties of perovskite-type organic/inorganic hybrid luminescent films by soft lithography. *Chem. Phys. Lett.,* **2003**, *376*, 481-486.
[http://dx.doi.org/10.1016/S0009-2614(03)01017-0]

[160] Tabuchi, Y.; Asai, K.; Rikukawa, M.; Sanui, K.; Ishigure, K. Preparation and characterization of natural lower dimensional layered perovskite-type compounds. *J. Phys. Chem. Solids,* **2000**, *61*, 837-845.
[http://dx.doi.org/10.1016/S0022-3697(99)00402-3]

[161] Kitazawa, N.; Watanabe, Y. Optical properties of natural quantum-well compounds $(C_6H_5$-C_nH_{2n}-$NH_3)_2PbBr_4$(n= 1-4). *J. Phys. Chem. Solids,* **2010**, *71*, 797-802.
[http://dx.doi.org/10.1016/j.jpcs.2010.02.006]

[162] Kitazawa, N.; Yaemponga, D.; Aono, M.; Watanabe, Y. Optical properties of organic–inorganic hybrid films prepared by the two-step growth process. *J. Lumin.,* **2009**, *129*, 1036-1041.
[http://dx.doi.org/10.1016/j.jlumin.2009.04.023]

[163] Ema, K.; Ishi, J.; Kunugita, H.; Ban, T.; Kondo, T. All-optical serial-to-parallel conversion of T-bits/s signals using a four-wave-mixing process. *Opt. Quant. Elec.,* **2001**, *33*, 1077-1087.
[http://dx.doi.org/10.1023/A:1017560609085]

[164] Brehier, A.; Parashkov, R.; Lauret, J-S.; Deleporte, E. Strong exciton-photon coupling in a microcavity containing layered perovskite semiconductors. *Appl. Phys. Lett.,* **2006**, *89*, 1110.
[http://dx.doi.org/10.1063/1.2369533]

[165] Era, M.; Morimoto, S.; Tsutsui, T.; Saito, S. Organic-inorganic heterostructure electroluminescent device using a layered perovskite semiconductor $(C_6H_5C_2H_4NH_3)_2PbI_4$. *Appl. Phys. Lett.,* **1994**, *65*, 676-678.
[http://dx.doi.org/10.1063/1.112265]

[166] Shibuya, K.; Koshimizu, M.; Takeoka, Y.; Asai, K. Scintillation properties of $(C_6H_{13} NH_3)_2 PbI_4$: exciton luminescence of an organic/inorganic multiple quantum well structure compound induced by 2.0 MeV protons. *Nuclear Instr. Meth. Phys. Res. Sec. B: Beam Inter. Mater. Ato.,* **2002**, *194*, 207-212.
[http://dx.doi.org/10.1016/S0168-583X(02)00671-7]

[167] Liu, M.; Johnston, M.B.; Snaith, H.J. Efficient planar heterojunction perovskite solar cells by vapour deposition. *Nature,* **2013**, *501*(7467), 395-398.
[http://dx.doi.org/10.1038/nature12509] [PMID: 24025775]

[168] Snaith, H.J. Perovskites: the emergence of a new era for low-cost, high-efficiency solar cells. *J. Phys. Chem. Lett.,* **2013**, *4*, 3623-3630.
[http://dx.doi.org/10.1021/jz4020162]

[169] Edri, E.; Kirmayer, S.; Cahen, D.; Hodes, G. High open-circuit voltage solar cells based on organic-inorganic lead bromide perovskite. *J. Phys. Chem. Lett.,* **2013**, *4*(6), 897-902.
[http://dx.doi.org/10.1021/jz400348q] [PMID: 26291353]

[170] Heo, J.H.; Im, S.H.; Noh, J.H.; Mandal, T.N.; Lim, C.-S.; Chang, J.A.; Lee, Y.H.; Kim, H.-j.; Sarkar, A.; Nazeeruddin, M.K.. Efficient inorganic-organic hybrid heterojunction solar cells containing perovskite compound and polymeric hole conductors. *Nature. Photo.,* **2013**, *7*, 486-491.

[171] Burschka, J.; Pellet, N.; Moon, S-J.; Humphry-Baker, R.; Gao, P.; Nazeeruddin, M.K.; Grätzel, M. Sequential deposition as a route to high-performance perovskite-sensitized solar cells. *Nature,* **2013**, *499*(7458), 316-319.
[http://dx.doi.org/10.1038/nature12340] [PMID: 23842493]

[172] Kandjani, S. A.; Mirershadi, S.; Nikniaz, A. *Inorganic-Organic Perovskite Solar Cells.,* **2015**.
[http://dx.doi.org/10.5772/58970]

[173] Leijtens, T.; Eperon, G.E.; Pathak, S.; Abate, A.; Lee, M.M.; Snaith, H.J. Overcoming ultraviolet light instability of sensitized TiO_2 with meso-superstructured organometal tri-halide perovskite solar cells. *Nat. Commun.,* **2013**, *4*, 2885.
[http://dx.doi.org/10.1038/ncomms3885] [PMID: 24301460]

[174] Noh, J.H.; Im, S.H.; Heo, J.H.; Mandal, T.N.; Seok, S.I. Chemical management for colorful, efficient, and stable inorganic-organic hybrid nanostructured solar cells. *Nano Lett.,* **2013**, *13*(4), 1764-1769.
[http://dx.doi.org/10.1021/nl400349b] [PMID: 23517331]

[175] Christians, J.A.; Miranda Herrera, P.A.; Kamat, P.V. Transformation of the excited state and photovoltaic efficiency of $CH_3NH_3PbI_3$ perovskite upon controlled exposure to humidified air. *J. Am. Chem. Soc.,* **2015**, *137*(4), 1530-1538.
[http://dx.doi.org/10.1021/ja511132a] [PMID: 25590693]

[176] Niu, G.; Li, W.; Meng, F.; Wang, L.; Dong, H.; Qiu, Y. Study on the stability of $CH_3NH_3PbI_3$ films and the effect of post-modification by aluminum oxide in all-solid-state hybrid solar cells. *J. Mater. Chem. A Mater. Energy Sustain.,* **2014**, *2*, 705-710.
[http://dx.doi.org/10.1039/C3TA13606J]

[177]   Tang, L-d.; Mei, H.; Wang, B.; Peng, S. Study on structure, thermal stabilization and light absorption of lead-bromide perovskite light harvesters. *J. Mater. Sci. Mater. Electron.,* **2015**, *26*, 8726-8731.
[http://dx.doi.org/10.1007/s10854-015-3549-3]

[178]   Wang, C-C.; Li, J-R.; Lv, X-L.; Zhang, Y-Q.; Guo, G. Photocatalytic organic pollutants degradation in metal–organic frameworks. *Energy Environ. Sci.,* **2014**, *7*, 2831-2867.
[http://dx.doi.org/10.1039/C4EE01299B]

[179]   Du, J-J.; Yuan, Y-P.; Sun, J-X.; Peng, F-M.; Jiang, X.; Qiu, L-G.; Xie, A-J.; Shen, Y-H.; Zhu, J-F. New photocatalysts based on MIL-53 metal-organic frameworks for the decolorization of methylene blue dye. *J. Hazard. Mater.,* **2011**, *190*(1-3), 945-951.
[http://dx.doi.org/10.1016/j.jhazmat.2011.04.029] [PMID: 21531507]

[180]   Cavka, J.H.; Jakobsen, S.; Olsbye, U.; Guillou, N.; Lamberti, C.; Bordiga, S.; Lillerud, K.P. A new zirconium inorganic building brick forming metal organic frameworks with exceptional stability. *J. Am. Chem. Soc.,* **2008**, *130*(42), 13850-13851.
[http://dx.doi.org/10.1021/ja8057953] [PMID: 18817383]

[181]   Gomes Silva, C.; Luz, I.; Llabrés i Xamena, F.X.; Corma, A.; García, H. Water stable Zr-benzenedicarboxylate metal-organic frameworks as photocatalysts for hydrogen generation. *Chemistry,* **2010**, *16*(36), 11133-11138.
[http://dx.doi.org/10.1002/chem.200903526] [PMID: 20687143]

[182]   Dan-Hardi, M.; Serre, C.; Frot, T.; Rozes, L.; Maurin, G.; Sanchez, C.; Férey, G. A new photoactive crystalline highly porous titanium(IV) dicarboxylate. *J. Am. Chem. Soc.,* **2009**, *131*(31), 10857-10859.
[http://dx.doi.org/10.1021/ja903726m] [PMID: 19621926]

[183]   Laurier, K.G.; Vermoortele, F.; Ameloot, R.; De Vos, D.E.; Hofkens, J.; Roeffaers, M.B. Iron(III)-based metal-organic frameworks as visible light photocatalysts. *J. Am. Chem. Soc.,* **2013**, *135*(39), 14488-14491.
[http://dx.doi.org/10.1021/ja405086e] [PMID: 24015906]

[184]   Lee, J.; Farha, O.K.; Roberts, J.; Scheidt, K.A.; Nguyen, S.T.; Hupp, J.T. Metal-organic framework materials as catalysts. *Chem. Soc. Rev.,* **2009**, *38*(5), 1450-1459.
[http://dx.doi.org/10.1039/b807080f] [PMID: 19384447]

[185]   Wang, C-C.; Zhang, Y-Q.; Li, J.; Wang, P. Photocatalytic CO_2 reduction in metal–organic frameworks: a mini review. *J. Mol. Struct.,* **2015**, *1083*, 127-136.
[http://dx.doi.org/10.1016/j.molstruc.2014.11.036]

[186]   Li, J-R.; Kuppler, R.J.; Zhou, H-C. Selective gas adsorption and separation in metal-organic frameworks. *Chem. Soc. Rev.,* **2009**, *38*(5), 1477-1504.
[http://dx.doi.org/10.1039/b802426j] [PMID: 19384449]

[187]   Morris, R.E.; Wheatley, P.S. Gas storage in nanoporous materials. *Angew. Chem. Int. Ed. Engl.,* **2008**, *47*(27), 4966-4981.
[http://dx.doi.org/10.1002/anie.200703934] [PMID: 18459091]

[188]   Suh, M.P.; Park, H.J.; Prasad, T.K.; Lim, D-W. Hydrogen storage in metal-organic frameworks. *Chem. Rev.,* **2012**, *112*(2), 782-835.
[http://dx.doi.org/10.1021/cr200274s] [PMID: 22191516]

[189]   Sumida, K.; Rogow, D.L.; Mason, J.A.; McDonald, T.M.; Bloch, E.D.; Herm, Z.R.; Bae, T-H.; Long, J.R. Carbon dioxide capture in metal-organic frameworks. *Chem. Rev.,* **2012**, *112*(2), 724-781.
[http://dx.doi.org/10.1021/cr2003272] [PMID: 22204561]

[190]   Llabrés i Xamena, F.X.; Corma, A.; Garcia, H. Applications for metal-organic frameworks (MOFs) as quantum dot semiconductors. *J. Phys. Chem. C,* **2007**, *111*, 80-85.
[http://dx.doi.org/10.1021/jp063600e]

[191]   Mahata, P.; Madras, G.; Natarajan, S. Novel photocatalysts for the decomposition of organic dyes based on metal-organic framework compounds. *J. Phys. Chem. B,* **2006**, *110*(28), 13759-13768.

[http://dx.doi.org/10.1021/jp0622381] [PMID: 16836321]

[192] Ai, L.; Zhang, C.; Li, L.; Jiang, J. Iron terephthalate metal–organic framework: Revealing the effective activation of hydrogen peroxide for the degradation of organic dye under visible light irradiation. *Appl. Catal. B,* **2014**, *148*, 191-200.
[http://dx.doi.org/10.1016/j.apcatb.2013.10.056]

[193] Zhang, C-F.; Qiu, L-G.; Ke, F.; Zhu, Y-J.; Yuan, Y-P.; Xu, G-S.; Jiang, X. A novel magnetic recyclable photocatalyst based on a core–shell metal–organic framework $Fe_3O_4@$ MIL-100 (Fe) for the decolorization of methylene blue dye. *J. Mater. Chem. A Mater. Energy Sustain.,* **2013**, *1*, 14329-14334.
[http://dx.doi.org/10.1039/c3ta13030d]

[194] Wen, T.; Zhang, D-X.; Liu, J.; Lin, R.; Zhang, J. A multifunctional helical Cu(I) coordination polymer with mechanochromic, sensing and photocatalytic properties. *Chem. Commun. (Camb.),* **2013**, *49*(50), 5660-5662.
[http://dx.doi.org/10.1039/c3cc42241k] [PMID: 23677170]

[195] Tang, L-d.; Mei, H.; Wang, B.; Peng, S. Study on structure, thermal stabilization and light absorption of lead-bromide perovskite light harvesters. *J. Mater. Sci. Mater. Electron.,* **2015**, *26*, 8726-8731.
[http://dx.doi.org/10.1007/s10854-015-3549-3]

[196] Wang, J-L.; Wang, C.; Lin, W. Metal–organic frameworks for light harvesting and photocatalysis. *ACS Catal.,* **2012**, *2*, 2630-2640.
[http://dx.doi.org/10.1021/cs3005874]

[197] Wen, T.; Zhang, D-X.; Zhang, J. Two-dimensional copper(I) coordination polymer materials as photocatalysts for the degradation of organic dyes. *Inorg. Chem.,* **2013**, *52*(1), 12-14.
[http://dx.doi.org/10.1021/ic302273h] [PMID: 23244571]

[198] Hashimoto, K.; Irie, H.; Fujishima, A. TiO_2 photocatalysis: a historical overview and future prospects. *Jpn. J. Appl. Phys.,* **2005**, *44*, 8269.
[http://dx.doi.org/10.1143/JJAP.44.8269]

[199] Chaturvedi, S.; Dave, P.N.; Shah, N. Applications of nano-catalyst in new era. *J. Saudi Chem. Soc.,* **2012**, *16*, 307-325.
[http://dx.doi.org/10.1016/j.jscs.2011.01.015]

[200] Hoffmann, M.R.; Martin, S.T.; Choi, W.; Bahnemann, D.W. Environmental applications of semiconductor photocatalysis. *Chem. Rev.,* **1995**, *95*, 69-96.
[http://dx.doi.org/10.1021/cr00033a004]

[201] Chong, M.N.; Jin, B.; Chow, C.W.; Saint, C. Recent developments in photocatalytic water treatment technology: a review. *Water Res.,* **2010**, *44*(10), 2997-3027.
[http://dx.doi.org/10.1016/j.watres.2010.02.039] [PMID: 20378145]

[202] Watts, R.J.; Kong, S.; Orr, M.P.; Miller, G.C.; Henry, B.E. Photocatalytic inactivation of coliform bacteria and viruses in secondary wastewater effluent. *Water Res.,* **1995**, *29*, 95-100.
[http://dx.doi.org/10.1016/0043-1354(94)E0122-M]

[203] Otaki, M.; Hirata, T.; Ohgaki, S. Aqueous microorganisms inactivation by photocatalytic reaction. *Water Sci. Technol.,* **2000**, *42*, 103-108.

[204] Dincer, I. Renewable energy and sustainable development: a crucial review. *Renew. Sustain. Energy Rev.,* **2000**, *4*, 157-175.
[http://dx.doi.org/10.1016/S1364-0321(99)00011-8]

[205] Liao, C-H.; Huang, C-W.; Wu, J. Hydrogen production from semiconductor-based photocatalysis *via* water splitting. *Catalysts,* **2012**, *2*, 490-516.
[http://dx.doi.org/10.3390/catal2040490]

[206] Maeda, K.; Domen, K. Photocatalytic water splitting: recent progress and future challenges. *J. Phys. Chem. Lett.,* **2010**, *1*, 2655-2661.

[http://dx.doi.org/10.1021/jz1007966]

[207] Chen, X.; Shen, S.; Guo, L.; Mao, S.S. Semiconductor-based photocatalytic hydrogen generation. *Chem. Rev.,* **2010**, *110*(11), 6503-6570.
[http://dx.doi.org/10.1021/cr1001645] [PMID: 21062099]

[208] Ishii, T.; Kato, H.; Kudo, A. H_2 evolution from an aqueous methanol solution on $SrTiO_3$ photocatalysts codoped with chromium and tantalum ions under visible light irradiation. *J. Photochem. Photobiol.,* **2004**, *163*, 181-186.
[http://dx.doi.org/10.1016/S1010-6030(03)00442-8]

[209] Nuraje, N.; Asmatulu, R.; Kudaibergenov, S. Metal oxide-based functional materials for solar energy conversion: a review. *Curr. Inorg. Chem.,* **2012**, *2*, 124-146.
[http://dx.doi.org/10.2174/1877944111202020124]

[210] Suh, D-I.; Lee, S-Y.; Kim, T-H.; Chun, J-M.; Suh, E-K.; Yang, O-B.; Lee, S-K. The fabrication and characterization of dye-sensitized solar cells with a branched structure of ZnO nanowires. *Chem. Phys. Lett.,* **2007**, *442*, 348-353.
[http://dx.doi.org/10.1016/j.cplett.2007.05.093]

[211] Baruah, S.; Thanachayanont, C.; Dutta, J. Growth of ZnO nanowires on nonwoven polyethylene fibers. *Sci. Technol. Adv. Mater.,* **2008**, *9*(2), 025009.
[http://dx.doi.org/10.1088/1468-6996/9/2/025009] [PMID: 27877984]

[212] Sugunan, A.; Guduru, V.K.; Uheida, A.; Toprak, M.S.; Muhammed, M. Radially Oriented ZnO Nanowires on Flexible Poly☐l☐Lactide Nanofibers for Continuous☐Flow Photocatalytic Water Purification. *J. Am. Ceram. Soc.,* **2010**, *93*, 3740-3744.
[http://dx.doi.org/10.1111/j.1551-2916.2010.03986.x]

CHAPTER 4

Clay Based Nanocomposites and Their Environmental Applications

Iftikhar Ahmad[1,3], **Farman Ali**[2,*] and **Fazal Rahim**[1]

[1] *Pakistan Institute of Engineering and Applied Science (PIEAS), Nilore 45650, Islamabad, Pakistan*

[2] *Department of Chemistry, Hazara University, Mansehra- 21300, Pakistan*

[3] *Institute of Radiotherapy and Nuclear Medicine (IRNUM), University Campus, Peshawar, Pakistan*

Abstract: In the modern world, the quality of the environment (water, soil, and air) is greatly compromised by numerous pollutants and contaminations from diverse sources. The situation is even exacerbating on daily basis. Tracking and treating the existing contaminants while preventing further pollution may maintain and improve the quality of environment. Clays and clay based nanocomposites provide an efficient and cost effective solution to this end. In this chapter, we begin by introducing clays, composition of clay, clay minerals and various tools used for characterization of clays and clay based nanocomposites. The role of clay based nanocomposites for environmental protection is presented. In particular, the removal of heavy metal ions, toxic organic compounds, hazardous dyes and antibiotics from aqueous environment has been discussed and recent studies are summarized. Purification and remediation of contaminated soil and air with the help of clay based nanocomposites are also discussed.

Keywords: Clays, Clay-nanocomposites, Environmental Applications, Properties, Synthesis.

INTRODUCTION

Clean environment (*i.e.* soil, water, and air) is essentially one of the basic needs of healthy human life. However, meeting and maintaining the rising requirement of clean environment is one of the most formidable global challenges. A wide range of contaminants such as inorganic gases (carbon dioxide, carbon monoxide, nitric oxide, sulfur dioxide, *etc.*), organic hydrocarbons, ketones and aldehydes, heavy metals (Hg(II), Cd(II), Cr (VI), As(III), As(V), Pb(II), *etc.*) and other miscell-aneous toxins like endrin, aldrin, dieldvin, toxaphene, heptachlor, chlordane,

---

[*] **Corresponding author Farman Ali**: Department of Chemistry, Hazara University, Mansehra- 21300, Pakistan; Tel: +92 997414136, Fax: +92-997-530045; Email: farmanqau@gmail.com

hexachlorobenzene, mirex, dioxins, polychlorinated biphenyls, furans, *etc.* severely threaten the environmental quality [1]. Further, pollutants from diverse sources such as fertilizer runoff, abandoned mining sites, chemical spills, and airborne particulate and gaseous contamination from automobiles even worsen the environmental quality on daily basis. Detecting and treating prevailing pollutants and preventing or minimizing new contaminations are among the challenges. Nanotechnology and nanomaterials provide one prospective solution towards maintaining and improving air, water, and soil quality.

Nanotechnology is an emerging field of applied sciences that is focused on the engineering of materials at the atomic and molecular scale. Specifically, it encompasses the design, synthesis and characterization of nano-scale materials and their applications. Nano-systems and devices frequently exhibit novel physical, chemical and biological properties. Among others, a unique characteristic of nanomaterials is the significantly enhanced surface area to volume ratio, which potentially opens new research avenues in surface-based sciences such as adsorption [2].

Nanotechnology has been progressively evolved in a variety of consumer markets, such as computer electronics, energy production, communication, medicine and the food industry. Likewise, nanotechnology has the capability to improve the environment. Specifically, nanotechnology can be exploited to design *green* industrial processes that result in environmental friendly products. Further, nano-scale materials have the potential to detect, prevent, and segregate environmental pollutants. For instance, various nano-scale materials are used as adsorbents for a wide range of decontamination applications. These nano-adsorbents exhibit nano-scale pores resulting in their large specific surface area, increased mechanical and thermal stability, more active sites, enhanced permeability, low intra-particle resistances, *etc.* [3]. More importantly, the mechanical, electrical and optical properties of nano-adsorbents can be tuned towards the desired application by manipulating the size, interfacial phenomena and quantum effects occurring at the nano-scale [4].

More recently, nanomaterials and associated techniques have been extensively interrogated for efficient removal of environmental pollutants and remediation. Starting with the early application of iron nanoparticles for removal of chlorinated compounds from contaminated soil and water, nanomaterials has been applied to various industrial and domestic fields [1, 5 - 7]. With the advancements in characterization techniques such as X-ray diffraction, electron microscopy, spectroscopy and other related instrumental analytic procedures, the field has witnessed significant improvements towards understanding the structure and chemistry of such nanomaterials, resulting in several successful demonstrations.

With the progressively expanding spectrum of nanotechnology applications in the environmental treatment and remediation, several important issues need to be properly addressed. These might include: the cost and availability of advanced nanotechnology materials/ devices; to determine the potential benefits of nanotechnology for prevention or reduction of environmental pollutants; explore the possible environmental degradation from nanotechnology; modeling and simulation strategies for deeper understanding and optimization of nanomaterials for environmental remediation; potential emerging directions in environmental decontamination due to novel sensors; impact of nanotechnology rapid expansion on health care as related to the environment; and effects of employed nanoparticles in the atmosphere.

In this review chapter, we focus on the clay based nanomaterials for environmental applications. Specifically, we start with a general classification of nanomaterials for environmental remediation with particular emphasis on clay based nanomaterials. Afterwards, the underlying processes of clay based nanotechnology for environmental treatment has been reviewed and categorized. Removal of pollutants from ground water, soil and wastewater has been discussed in detail. Finally, future perspectives are presented.

Classification of Nanomaterials for Environmental Treatment

Nanomaterials for environmental treatment could be categorized in four groups.

- Carbon based nanomaterials: the primary composition of these materials is based on carbon. The tunable properties of these nanomaterials enable a wide range of applications in addressing environmental challenges such as sorbents, high-flux membranes, sensors, antimicrobial agents, depth filters, and pollution prevention strategy. These nanomaterials are available in diverse shapes such as hollow spheres, ellipsoids (fullerenes), or cylinders (nanotubes). The large specific surface area (about 3000 m^2 /g) and layered structures enable carbon nanotubes promising adsorbents. Further, carbon nanotubes have the capability to establish π-π interactions which can be exploited for removal of organic pollutants from waste water. Interestingly, carbon nanotubes can be cleaned and reused with the help of ultrasonification and autoclaving.

- Metal based nanomaterials: the primary composition of these materials is based on metals. Typical examples include nano gold, nano silver, quantum dots and metal oxides (*e.g.* titanium dioxide). The ability of adaptable surface tailoring and functionalization of these nanomaterials enables targeting of diverse pollutants, thereby allowing for designing multifunctional nano devices towards the specific application. For instance, surface modified iron oxide nanoparticles have been successfully demonstrated for the extraction of heavy metals (*i.e.*

copper and chromium) and arsenic salts from water.

- Dendrimers: this group contains nano-scale polymers arranged in repetitive and tree-like branched architecture. The expression of large surface area due to numerous chain ends in dendrimers qualifies for better adsorption capabilities and contamination removal from the environment. Indeed, dendrimers have been shown to effectively adsorb and eliminate various organic pollutants (aromatics, methyl isobutyl ether, trihalogen methane, *etc.*) and heavy metals from water. Further, dendrimers have also been established in vapor phase decontaminations. Specifically, surface engineered dendrimers have illustrated the capability to sponge hydrogen cyanide (HCN) from air streams. Importantly, regeneration of dendrimers for repeated use can be achieved with the help of simple washing with solvent (*e.g.* acetonitrile).

- Nano-composites: this group contains multiphase nanomaterials. In particular, nanoparticles are admixed in the nano-scale clays matrix to enhance the mechanical, thermal, and flame-retardant properties. In this chapter, we focus on this group of nanomaterials towards environmental decontamination.

Clay and Clay Composition

Diverse definitions of clay exist in different scientific communities. For instance, the term "clay" is used by geologists to define geological materials of size < 4 μm, by soil scientists to describe the fraction of soil with particles size < 2 μm, by colloidal scientists as threshold particle size < 1 μm, by mineralogists as a material comprised of fine grained minerals, which have nominal plasticity due to water contents and harden when fired or dried [8].

It is noteworthy that the terms clay and clay minerals are frequently used interchangeably nevertheless have slightly different meanings [8]. Contrary to 'clay', the term 'clay mineral' is not sensitive to particle size. Specifically, 'clay mineral' refers to: phyllosilicates, a significantly large family of minerals with characteristic layered structures.

The accurate knowledge of clay composition and minerals is essential for efficient management of the environment and its remediation. To this end, advance characterization tools such as X-ray diffraction (XRD), spectroscopy and electron microscopy in combination with various analytic techniques have revealed significant insights into clay composition and structure. Typically, clay is a collection of naturally occurring minerals, with diverse variation in composition. Specifically, clays are aluminosilicates arranged in sheets (called phyllosillicate) and then organized in structural layers. Individual sheets of clays are composed of tetrahedral silica $[SiO_4]^{4-}$ (abbreviated as "T") and octahedral alumina $[AlO_3(OH)_3]^{6-}$ (denoted by "O"). Two, three or four sheets of silica / alumina are firmly combined to form each clay layer. The apices of the tetrahedrons and octahedrons are occupied by oxygen while their interior contains metal cations. These structural units are arranged in such a way that a hexagonal network is constructed with each sheet [9, 10]. Further, the layers have affinity to organize themselves in the form of stacks with the van der Waals interaction defining a regular gap between them, which is called an 'interlayer'. The total number of sheets and their ratio in the basic structural units and the central metal cation in the tetrahedrons and octahedrons and define the net charge of the layers. Typically, the characteristic charge of the clay leads to its hydrophilic and consequently incompatible nature with a significant number of polymers. This important issue is discussed below in detail.

The ionic character (and charge) of the layers is influenced by many factors such as electrochemical environment at the time of clay sedimentation, reaction kinetics occurring during the genesis, evolution and formation of the clay minerals, *etc*. The natural tendency to preserve electrical neutrality is promoted by cations (*e.g.* $K+$, Ca^{2+}, Na^+, Mg^{2+} and H^+ adsorbed) in the interlayer space. Further, a relatively simple treatment of the clay can result in the replacement of the given host cations by another desirable cation towards a particular application.

Based on natural locations, clay can be divided in two classes: first, residual clay typically found in the place of origin; second, transported or sedimentary clay, displaced from their origin deposit through an agent of erosion and deposited at a new and presumably distant position. Residual clays are usually produced by chemical surface weathering in three ways: by the chemical decay of rocks, just as granite, containing silica and alumina; by the solution of rocks, such as limestone, containing clay impurities, which, being insoluble, are deposited as clay; and by the disintegration and solution of shale. It may be noted that shale is a dark fine-grained sedimentary rock primarily composed of layers of compressed silt, clay or mud [11].

Type of Clay Minerals

The various clay minerals mainly differ in terms of their layered structure and may be classified into four major groups, including the montmorillonite/ smectite group, kaolinite group, illite group, and chlorite group [11]. We will briefly discuss each of these groups individually.

Smectite group of clay minerals refer to the family of non-metallic clays predominantly composed of hydrated calcium aluminum sodium silicate and strong tendency of expansion. The general chemical formula for this group is (Ca, Na, H) (Al, Mg, Fe, Zn)$_2$ (Si, Al)$_4$O$_{10}$(OH)$_2$xH$_2$O, where the variable amount of water that the particular group members could contain is represented by x. For example, an important member of this group is Talc with chemical formula of $Mg_3Si_4O_{10}(OH)_2$. Few other members of this clay group include montmorillonite, saponite, pyrophyllite, nontronite, *etc*. The layer structure of this group contains three-sheet phyllosilicates; two silicate sheets sandwiching an aluminum oxide/hydroxide sheet $(Al_2(OH)_4)$ in between, resulting in a T-O-T stacking sequence. The charge of the unit cell (three-sheet layer) is 0.5-1.2 e [10]. An important feature of smectite clays is the absence of hydroxyl group in their interlayer space, enabling the reaction of saline coupling agents at the edges of the plates only. Smectite are frequently used to slow down the progress of water through rocks or soil. The smaller particle size (~ 1 micron) and larger aspect ratio allow them to significantly influence the properties of rocks and soils. Further, they are used as absorbent for removal of pollutants to purify (and de-color) liquids. Talc, a member of smectites, has been traditionally used in facial powder and medicine [12]. Smectites are also used as fillers in rubbers and paints.

The kaolinite group, the most commonly found clay minerals, has closely related three members; kaolinite, dickite, and nacrite. The members of this group share the same chemical formula of $Al_2Si_2O_5(OH)_4$, indicating that the group members are polymorphs *i.e.* they have same chemical formula but different structures [13]. As can be observed from the formula, kaolinites are composed of two groups; silicate (Si_2O_5) and aluminum oxide/hydroxide $(Al_2 (OH)_4)$; the tetrahedral silicate sheets are tightly bonded to octahedral sheets of alumina, resulting in a two sheet phyllosilicates structure, where the O:T ratio is 1:1 [14]. Further, the net charge of a single two-sheet layer (called unit cell, uc) is 0 [10]. Kaolinites have been mostly used in paper industry. Additional uses include as fillers in ceramics, rubber, paint, and plastics.

Illite group represent the non-expanding micaceous minerals. Structurally, illite is similar to smectites in the sense that it is also composed of silicate-alumin--silicate layer sequence. However, illite typically has more Mg, Si, Fe, and water. The general chemical formula for this group is $(K, H)Al_2 (Si, Al)_4O_{10} (OH)_2 \cdot xH_2O$ where the variable amount of water molecules that different group members could contain is represented by x. It is noteworthy that the variable amounts of water would rest between the sandwiched layers [13]. Illite is an important component of rock minerals and a common constituent of shales. This mineral is also used as filler in drilling mud.

The chlorite group is not necessarily considered as part of clays and is sometimes placed as a separate group in phyllosillicate. It is a relatively large group with common members of amesite, cookeite, chamosite and daphnite. The general chemical formula is $X_{4-6}Y_4O_{10}(OH, O)_8$, where the X describe one or more of iron, aluminum, lithium, magnesium, nickel, manganese, zinc or occasionally chromium. The Y typically represents either aluminum or silicon. The chlorites are usually four-sheet silicates, with the ratio of T:O:O (2:1:1) and charge of the unit cell (four-sheet layer) is 1.1-3.3 e [11, 15]. The four types of clay minerals discussed are summarized in Table **1**.

Table 1. Types of Clay Minerals

Group	Layer Type	Group Members	Layer Charge	General Chemical Formula	Remarks
Smectite	2:1	montmorillonite, saponite, pyrophyllite, nontronite, talc, vermiculite	0.5-1.2	(Ca, Na, H) (Al, Mg, Fe, Zn)$_2$ (Si,Al)$_4$O$_{10}$(OH)$_2$ xH$_2$O	x indicate the variable amount of water molecules
Kaolinite	1:1	kaolinite, dickite, nacrite	~ 0	Al$_2$Si$_2$O$_5$(OH)$_4$	group members are polymorphs
Illite	2:1	illite	1.4-2.0	(K,H)Al$_2$ (Si,Al)$_4$O$_{10}$ (OH)$_2$ xH$_2$O	x indicate the variable amount of water molecules
Chlorite	2:1:1	amesite, cookeite, chamosite, daphnite	Variable	(Fe, Al, Li, Mg, Zn, Cr)$_{4-6}$(Al, Si)$_4$ O$_{10}$(OH, O)$_8$	sometimes placed as a separate group in phyllosillicate

Clay Based Synthetic Nanocomposites

Nanocomposite materials are formed by the nano-scale dispersion of nanoparticles such as nano clays in a matrix or substrate towards superior mechanical, electrical, and thermal properties. The low cost, abundant availability and nontoxicity of the clay and the technological advancements in the processing of clay based nanocomposites have been opening new and potential application, especially for remediation of several contaminants in water and soil. In particular, clay based polymeric nanocomposites are the most widely investigated research arena.

It is noteworthy that typically hydrophilic clay and hydrophobic polymer are not compatible for nanocomposite fabrication and the clay mineral often causes agglomeration in the organic polymer matrix. Consequently, surface tailoring of clay minerals is essentially required to achieve polymer nanocomposites. Specifically, the clays are treated organically so that it becomes hydrophobic and compatible with the particular polymers for clay/polymer fabrication. Such surface modified clays are usually referred as organo clays. The most frequently exploited engineering for clays is to the replacement of interlayer inorganic cations with organic ones. Specifically, Na$^+$, Ca^{2+} *etc.* are substituted with ammonium cations. Also, the interlayer space is caused to swell resulting in reduced intra-layer attraction. Consequently, diffusion and accommodation of polymer into the interlayer space is favored.

Understanding and optimizing the processing conditions for the fabrication and development of clay-polymer nanocomposites is of prime interest in designing materials with desired set of properties. The most common approaches for the fabrication of clay-polymer nanocomposite are:

1. Intercalated Nanocomposites; the monolayer chains of extended polymer are introduced into the layered structure of clay minerals enabling a well ordered multilayer stacking that have alternative polymer layers and clay platelets (Fig. **1**).
2. Exfoliated Nanocomposites; the clay platelets are uniformly and completely dispersed in a continuous organic polymer matrix (Fig. **2**).

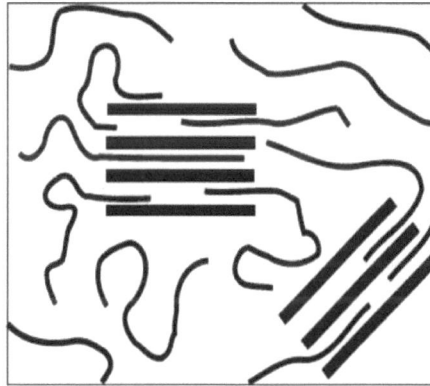

Fig. (1). The idealized representations of intercalated clay nanocomposites [16].

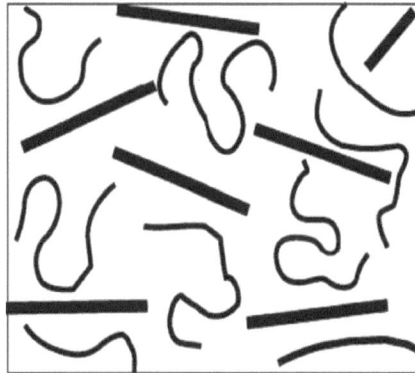

Fig. (2). The idealized representations of exfoliated clay nanocomposites [16].

However, it may be noted that the partial exfoliated nanocomposites are most commonly found in polymer nanocomposites. Further details about these methods for the synthesis of clay- polymer nanocomposites can be found elsewhere [16, 17].

Characterization of Nanocomposites

The properties of Nanocomposites are strongly influenced by the nature of components and other important factors such as the dimensions, microstructure and the uniformity of the dispersed phase. A variety of characterization techniques are being developed and used for interrogating the microstructure and properties of nanocomposites including x-ray diffraction (XRD), scanning electron microscopy (SEM), transmission electron microscopy (TEM), atomic force microscopy (AFM), fourier transform infrared (FTIR) spectroscopy, nuclear magnetic resonance (NMR) spectroscopy, Thermogravimetric analysis (TGA), Differential scanning calorimetry (DSC), *etc*. A more comprehensive list of such characterization tools along with their particular applications are summarized in Table **2**.

Table 2. Typical characterization techniques for clay-based nanocomposites.

Sr. No.	Characterization Technique	Characteristic Property of Clay-based Nanocomposite
1	XRD	Interlayer distance of clays Degree of clay platelets dispersion Nanocomposite morphology (*e.g.* intercalated *vs.* exfoliated)
2	TEM	Microstructure and spatial distribution Structural defects
3	SEM	Surface morphology Degree of dispersion of clay particles
4	FTIR	Composition analysis
5	AFM	Crystallization analysis of polymer Surface morphology Particle size and distribution
6	TGA	Thermal stability
7	NMR	Local dynamics of organic polymer chains Surface chemistry
8	DSC	Melting and crystallization analysis Local dynamics of polymer chains
9	Rheometry Mechanical test	Young's modulus Tensile strength Viscoelastic properties

(Table 2) contd.....

Sr. No.	Characterization Technique	Characteristic Property of Clay-based Nanocomposite
10	Cone calorimetry	Flame retardancy Thermal stability

Clay Based Nanocomposites for Environmental Cleaning

Environmental degradation from all diverse sources (direct or indirect) affects the human health and quality of life in an adverse manner. The air in the modern world has been significantly polluted with numerous contaminants such as carbon monoxide, carbon dioxide, chlorofluorocarbons, nitrogen oxides, sulphur dioxide, heavy metals (chromium, arsenic, lead, mercury, cadmium, zinc, *etc.*), organic chemicals and many others. Water degradation and shortage are caused by many factors like sewage, leaking fertilizers, oil spills, industrial by-products, fossil fuels, and others. Soil, being more complex matrix of chemicals and organisms, presents a challenging target for decontamination and remediation. Specifically, each fraction of the soil matrix including the colloidal, the clay, the slit and the sand fraction have its specific set of properties that can be exploited for the matrix decontamination. For each of these integral parts of the environment, a variety of treatment procedures have been investigated to maintain and improve the environmental quality. In the following, we only focus on the clay based nanotechnology for the improvement of the environment.

It is noteworthy that the environmental contaminants are typically measured in parts per billion (ppb) or parts per million (ppm). Furthermore, the contaminant's toxicity is expressed by a threshold 'toxic level' defined for each contaminant by environmental regulatory bodies. For instance, the toxic level for mercury is 0.002 ppm in water whereas for arsenic is 10 ppm in soil. In addition, contaminants are generally found as mixtures, demanding for extreme care in determining the overall toxicity level.

Clays and Clay Based Nanocomposites for Water Cleaning

Water is the most essential and important component of life on earth. Contamination and degradation of aquatic environment by many compounds is one of the major global concerns of our society. In particular, industrial effluents such as organic and inorganic wastes, heavy metal ions, dyes, aromatic compounds, *etc.* pose considerable risk to drinking water sources. Clay based nanotechnology has the potential for decontamination and remediation of aqueous system, as discussed below.

Removal of Heavy Metals from Water

The contamination of aqueous system by heavy metals even at trace levels is considered to be significantly harmful for all living species [6, 18]. It may be noted that heavy metals are typically defined as elements with atomic weights in the range of 63.5- 200.6, and having specific gravity greater than 5.0 [19]. The most frequent heavy metal contaminants of water include; arsenic (As), lead (Pb), mercury (Hg), cadmium (Cd), copper (Cu), chromium (Cr), cobalt (Co), nickel (Ni), manganese (Mn), tin (Sn), *etc.* Each of these heavy metals has its own routes for entering the aqueous system and causing the adverse effects on the human health. For instance, Hg enter the water system through various routes such as industrial activities, household, acid rain causing leaching of soil and has been found to damage to the kidneys, nervous system and vision. Mining and industrial waste, automobile exhaust and incinerator ash are considered as the prime sources of Pb contamination that eventually lead to anemia, kidneys damage, nervous system deterioration, impairment of protein syntheses *etc.* In addition, the sources of Cd include electroplating, plastic industries, mining, and sewage. Human exposure to Cd appears to have lethal health impacts in terms of provoking cancer, kidney and bone damage, mucous membrane destruction, and even impairment of progesterone and testosterone production.

Currently, a variety of methods and techniques has been used for the removal of heavy metals from aqueous environment. These approaches include adsorption, ion-exchange, chemical precipitation, membrane filtration, flotation, coagulation–flocculation, electrochemical methods, *etc.* While each of these techniques have its own set of merits and demerits; a specific techniques is usually the interplay between the efficiency and expense. For example, commercially activated carbons (adsorbents) or synthetic ion exchange resins have been investigated for the removal of heavy metals from aquatic environments; the results indicate that while these absorbents are more costly, the relatively poor decontamination efficiency limits the use of resins.

Clay and clay based nanocomposites provide a potent alternative choice for the removal of heavy metals from aqueous environment as it offer many promising advantages including high capacity of ion sorption, large specific surface area, inexpensive, chemical and mechanical stability and large swelling ability over the conventional methods. Surface engineering of clay materials and clay-polymer nanocomposites has been shown to significantly improve the capability of clay to remove heavy metals from aqueous media [17, 18, 20]. Further, recently a novel water remediation approach that combines adsorption and biodegradation has been evolved. Specifically, the contaminants are adsorbed on the clay nanocomposites and then degraded into less toxic forms with the help of specific microorganisms [21]. A brief summary of different types of clays used for removal of various heavy metal ions from water has been provided in Table **3**.

Table 3. Summary of studies for the removal of heavy metals using clay and clay based nanocomposites.

Clay Type	Decontaminated Heavy Metal	Remarks	Ref.
Montmorillonite	Cr (VI)	Effect of heat activation, surface modification and heat treatment on the adsorption efficiency was studied	[24]
	Cd (II)	Fe-montmorillonite was modified by poly-hydroxyl ferric	[25]
	As (III) & As (V)	Ti-pillared montmorillonite was Prepared by hydrolysis	[26]
	Cu (II)	Acid (H_2SO_4) modified clay minerals were used Adsorption kinetics were studied	[27]
	Hg (II)	Cations (sodium, potassium and calcium) treated clay and influence of pH, ionic strength were studied	[22]
	Cu (II)	Chitosan-montmorillonite biocomposite beads were prepared by crosslinking with pentasodium tripolyphosphate.	[28]
	Co (II)	Chitosan-montmorillonite composites were analyzed	[29]
Bentonite	Zn (II)	Acid (HCl) modified bentonite	[30]
	Pb (II)	Raw, acid activated and manganese oxide-coated bentonite was explored	[31]
	Pb (II)	Native and acid (HCl) modified bentonite	[32]
	Ni(II) & Cd(II)	Epichlorohydrin cross-linked chitosan–clay composite beads	[33]
Kaolin	Pb (II)	The influences of different (initial metal ion concentration, pH, contact time, dosage and temperature) on the adsorption were studied	[34]
	Cu (II)	Natural and acid (H_2SO_4) modified clay minerals were	[27]
	Cr (III)	kaolin was used as support material for biofilm formation of *Bacillus sp*	[35]
	Pb (II)	Polymer–clay (Kaolinite-polyvinyl alcohol) based composite	[36]
	Cd (II)	Raw, acid activated and manganese oxide-coated bentonite was explored	[27]
	Pb (II) & Cd (II)	Electro-kinetic remediation with Cd(II) and Pb(II) removal	[37]
Smectite, Illite, Calcite	Cu(II), Zn(II)	To compare the adsorptive capacity of different natural clays	[38]

The adsorption efficiency of clay and clay based nanocomposites for heavy metals from aqueous solution has been explored by many researches and found that it is influenced by several important factors such as absorbent dose, pH, temperature, initial concentration, the presence of other compounds, *etc.* Each of these parameters plays crucial role in the optimization of heavy metals removal from the aqueous environment. The adsorption process and capacity is significantly influenced by the pH of a given solution as it controls many important aspects such as the surface layer charge of the clay, the exchange capacity, the solution chemistry of heavy metals including precipitation, complexation, hydrolysis, redox reaction, *etc.* Generally, the uptake/removal of heavy metals by the clay and clay based nanocomposites initially increases with pH followed by a reduction beyond a certain cutoff value of pH [22]. In addition, temperature also plays crucial role in defining the adsorption capacity of heavy metals onto clays and its nanocomposites. Typically, physical adsorption is known to have inverse relation while chemical adsorption has direct relation with the amount of heavy metals extracted by the adsorbent [23].

Removal of Hazardous Dyes from Water

Color and dyes are an integral part of human life. Global production of thousands of different types of dyes is about one million tons annually. A significant part of these synthetic dyes and their by-products are eventually released in to the environment imposing a risk to human health. To reduce and eliminate hazard to living organisms, it is essential to detect and remove these chemical contamination from the environment.

Synthetic dyes can be categorized into about 30 groups based on their structure. Some major dye groups are azo, anthraquinone, di- and tri-arylmethane, phthalocyanine, indigoid, oxazine, azine, thiazine, *etc.* However, the largest synthetic dye group is the acid dyes; anionic compounds, with reactive groups forming covalent bonds. Other common dyes classes are metal complex dyes, basic dyes, direct dyes, disperse dyes, mordant dyes, pigment dyes, anionic dyes, solvent dyes, ingrain dyes, sulfur dyes, *etc.*

Typically dyes are designed and fabricated to be highly stable both chemically and photolytically. Consequently, dyes are extremely persistent in natural environments. Some dyes are very resistant to degradation due to its stable chemical structure. Depending on their chemical composition and structures, synthetic dyes may be highly toxic including carcinogenic, mutagenic, teratogenic, *etc.* to human. Furthermore, long term exposure of synthetic dyes has been observed to have severe damage to liver, kidney, central nervous system, and reproductive system. Consequently, the release and accumulation of synthetic dyes provoke an ecotoxic hazard that may ultimately affect human health by many channels, such as transport through the food chain.

A variety of different techniques including physical, chemical, and biological has been employed to remove dyes from contaminated water. These technologies include adsorption, chemical oxidation, coagulation, membrane separation, aerobic and anaerobic microbial degradation, *etc*. Certain hazardous dyes in water are not acceptable even at very low concentrations (~1 ppm); adsorption using various adsorbents is supposed as the more appropriate choice of treatment of such water.

A considerable amount of research has been devoted for finding efficient and cheap adsorbents for the removal of dyes from water [39 - 42]. However, some essential issues in this regard still need to be explored. More importantly, the adsorption capacity strongly depends on the class of dye; some adsorbent show excellent performance for certain class of dyes but can hardly extract dye from other classes. Further, industrial scale application of adsorption also appears to be limited due to cost-effectiveness.

Clay and clay based nanocomposites offer a much better dye adsorbents from aqueous environment. Due to large surface area, dye adsorption capabilities of clays are comparable to activated carbons, a much expensive choice. Interestingly, studies have also revealed that some clays possess high adsorption performance towards several classes of dyes. Also, pre-treatment and surface modification of clay nanocomposites provide significant enhancement in adsorption efficiency and capacity. A summary of some recent studies pertaining to the adsorption of dyes using clay and its modified forms has been given in Table **4**. It is evident that natural clay and its various modified form show significant dye removal capacities which appears to depend on the class of dye.

Table 4. Summary of some recent studies regarding removal of dyes from water using clay nanocomposites.

Clay Type	Dye	Max. Adsorption	Remarks	Ref.
Bentonite	Methylene blue	2.22 mmol/g	Natural and acid (HCl) modified bentonite	[30]
	Acid green 25	3.723 mmol/g	Cetyl trimethylammonium bromide modified bentonite	[46]
	Reactive red 120	81.97 mg/g	Cetylpyridinium modified resadiye-bentonite	[47]
	Acid blue 129	2.76 mmol/g	Cetyltrimethylammonium bromide modified bentonite	[48]
	Methylene blue	256 mg/g	Rarasaponin modified bentonite	[49]

(Table 4) contd.....

Clay Type	Dye	Max. Adsorption	Remarks	Ref.
Bentonite	Cango red	Natural bentonite 19.5 mg/g, thermal activation 54.64 mg/g, acid activation 69.44 mg/g, combination acid and thermal 75.75 mg/g	Natural, thermal activated, acid activated and combined acid and thermal activated bentonite	[50]
	Evans blue	0.516 mmol/g	Rarasaponin modified bentonite	[51]
	Methylene blue	142.86 mg/g	Chitosan cross-linked bentonite composite	[52]
	Orange II	53.78 mg/g	Alkyltriphenyl phosphonium modified bentonite	[53]
	Amido black 10B	323.6 mg/g	Chitosan cross-linked bentonite composite	[54]
Montmorillonite	Methyl orange	Na-montmorillonite 22.83 mg/g, Cethyltrimethyl ammonium bromide modified 42.04 mg/g, Anionic surfactant sodium stearate 121.97 mg/g	Anionic and cationic surfactants modified montmorillonite	[55]
Sepiolite	Everzol Black B	120.5 mg/g	Mayas Sepiolite	[56]
Kaolin	Malachite green	0.919 mmol/g	Rarasaponin modified kaolin	[57]
	Coomassie Brilliant Blue R 250	30.08 mg/g	Acid treated kaolinite	[58]
Attapulgite	Congo red	189.39 mg/g	Hexadecyl trimethyl ammonium bromide modified attapulgite	[59]
Rectorite	Methylene blue	37 mg/g	Acid (HCl) modified rectorite	[60]
Alunite	Acid red 88	832.31 mg/g	Calcined alunite	[61]

As mentioned earlier, the adsorption process and capacity are strongly influenced by factors such as pH, temperature, contact time, equilibrium dye concentration, *etc*. However, pH [43, 44] and temperature [45] seems to have more strong influence on the dye adsorption, as indicated by many studies.

Removal of Antibiotics from Water

Presently available antibiotics are mostly semi synthetic modifications of different natural compounds isolated from living organisms and have been proven to be vital drugs to treat various bacterial infections. In addition to the use of antibiotics for human therapy, they are also extensively used for animal farming and agricultural purposes. It has been reported that about 30–90% of the given antibiotic dose remain undegradable in both human and animal body systems;

presumably excreted as active compound in to the environment afterwards [46]. The active and undegradable antibiotic release into the environment has affected the structure and activity of microorganism and developed the so-called resistant bacteria [47]. Compromise to the environmental microbial populations is one of the greatest challenges that will adversely impact the environment.

To properly remove the antibiotic contamination from water, pre-treatment of water and sewage should be implemented which can potentially eliminate (by up to 80%) certain antibiotics such as fluoroquinolones or tetracyclines [48]. Water chlorination also helps to stimulate degradation of antibiotics [49, 50]. Several other techniques including activated carbon filtration, coagulation, ionic treatment or micelle-clay systems are also capable for the removal of various antibiotics from water [51]. However, all these techniques being promising for specific antibiotics, other antibiotics usually remain unaffected after water treatment [52].

Removal of antibiotic contamination from environment with the help of adsorption can be effectively and efficiently performed by utilizing clay and clay based nanocomposites as adsorbents as revealed by a large number of studies [46, 53 - 56]. For instance, montmorillonite have been employed for adsorption and intercalation of the commonly used antibiotic (ciprofloxacin) from model aquatic environments [53]. The results indicated that the most important mechanism for adsorption of ciprofloxacin on montmorillonite is the cation exchange. Similar conclusion of cation exchange has been also reported for the adsorption of ciprofloxacin onto other clay matrices such as rectorite and illite [57]. Further, the observed increase in the basal spacing of montmorillonite was correlated to the intercalation of ciprofloxacin. Considerable adsorption contribution has been provided by hydrogen bonding between the basal oxygen atoms of carboxylic groups of ciprofloxacin present on the external surface of clay minerals. This finding has been supported by other similar studies. Specifically, the intercalation phenomenon was observed in the study of tetracycline adsorption onto Na-montmorillonite [58].

The type of adsorption mechanism of various antibiotics on the clay and clay based nanocomposites has been attributed to different factors. For instance, pH has been found to play crucial role in defining the adsorption mechanism. A recent study of quinolone antibiotic nalidixic acid adsorption onto montmorillonite and kaolinite at different pH of the synthetic aqueous model solution revealed interesting findings [59]. For pH below pKa, neutral form of the antibiotic dominates resulting primarily in hydrophobic interaction for the antibiotic uptake. For pH above pKa, the electrostatic repulsion between the negatively charged clay surface and anionic form of antibiotic potentially reduce the nalidixic acid uptake. Further, strong indication of intercalation of the antibiotic in the interlayer space was suggested by the expansion of basal spacing [58].

Bentonite has been also explored as an adsorbent for the extraction of antibiotics from aqueous models. Excellent removal efficiency (99%) of bentonite for the adsorption of the antibiotic (ciprofloxacin) from aqueous solution has been reported [60]. In addition, bentonite achieved the highest ciprofloxacin removal capacity among bentonite, activated carbon, zeolite and pumice [61]. These studies were conducted on synthetic aqueous solution as the representative model for real aquatic environments. However, the amoxicillin removal efficiency of bentonite from real wastewater was found to be 88%. These studies indicate that the performance of clay such as bentonite for the removal of antibiotics such as ciprofloxacin and amoxicillin from wastewater is higher or almost comparable to activated carbon [46].

Removal of Organic Pollutants from Water

Organic pollution is a general term that includes many organic compounds that may originates from industrial effluents, domestic sewage, agriculture and urban run-off. Specifically, organic pollutants include fertilizers, pesticides, phenols, hydrocarbons, detergents, plasticizers, oils, biphenyls, pharmaceuticals, proteins, greases and carbohydrates [62]. Organic pollutants are potentially hazardous to our environment. For example, the organic pollutants present in water contains significant amount of suspended solids enabling reduction of available light to photosynthetic organisms. Consequently, the natural characteristics of the river bed are altered; posing serious problems to the life of many habitats. Further, the decomposition of organic pollutants also cause imbalance in the consumption and replenishment of the dissolved oxygen.

Persistent organic pollutants (POPs), the most common organic pollutants are of great environmental concern due to their long term persistence in the environment (resistant to degradation), toxicity and bioaccumulation in the food chain. POPs are carbon-based compounds that include polychlorinated dibenzo-pdioxins and dibenzofurans (PCDD/Fs), polychlorinated biphenyls (PCBs), organochlorine pesticides (OCPs) and furans [63].

Efficient techniques for the remediation and removal of organic pollutants from water are essentially required. Different methods such as adsorption, coagulation, ion exchange, filtration with coagulation, precipitation, ozonation, reverse osmosis and many other have been used for the extraction of organic pollutants from water. Most of these techniques are often highly expensive. That said, adsorption appears a relatively suitable methods due to its simple design and low cost involvement.

Clays and clay based nanocomposites offer an attractive adsorbent for the removal of organic contaminants from water. A summary of organic pollutants removal using natural clay and its nanocomposites from water has been given in Table **5**. Many organic contaminants such as dichloroacetic acid, carbon tetrachloride, phenol, humid acid, O-dichlorobenzene, blue green algae (cyanobacterial microcystis aeruginosa), atrazine, sulfentrazone, imazaquin, alachlor, naphthalene, phenolic derivative, salicylic acid, nitro benzene, carbamazepine naproxen, salicylic acid, clofibric acid, and carbamazepine has been removed from aqueous environment using different clays and clay based nanocomposites [75].

Table 5. Summary of organic decontamination applications of natural clay and its nanocomposites for water.

Clay Type	Organic Pollutant	Efficiency (%)	Remarks	Ref.
Bentonite	Dichloroacetic acid	92 (%)	Bentonite-based absorptive ozonation	[64]
	Carbon tetrachloride	70 (%)	Quaternary ammonium modified bentonite	[65]
	Naproxen, salicylic acid, clofibric acid, carbamazepine	2.69 μmol/g 5.55 μmol/	Inorganic-organic-intercalated bentonites	[66]
	Phenol	333 mg /g	Bentonite modified with acetyl trimethyl ammonium bromide	[67]
	Humid acid	95 (%)	Bentonite coagulation	[68]
	O-Dichlorobenzene	74 (%)	Bentonite coagulation	[68]
Montmorillonite	Cyanobacterial microcystis aeruginosa	92 (%)	Metal oxides modified-montmorillonite	[69]
	Atrazine	90-99 (%)	Organically modified montmorillonite	[70]
	Sulfentrazone, imazaquin, alachlor	100 (%)	Vesicle-clay complex	[71]
	Naphthalene, phenolic derivative	99 (%)	Crystal violet- montmorillonite composite	[72]
Kaolin	Salicylic acid	-	Kaolin	[73]
Smectite	Carbamazepine	-	Modified smectite clays	[74]

Clay Based Nanocomposites for Soil Cleaning

Soil degradation can be classified into: physical, chemical and biological degradation. Physical degradation affects the air and water-holding capacity, permeability, root development and biological activity of soil [93]. The two most significant activities responsible for physical degradation of soil are agriculture and forestry. The chemical degradation has been related to contamination, salinization and acidification, and nutrient depletion. Oil and heavy metals are the primary soil contaminants, while gasoline, metal industries and vehicle service stations are considered as the typical sources of local soil contamination. Biological degradation is concerned with the soil organic matter (SOM) degradation which is in term correlated to the conversion of grassland, forests and natural vegetation to arable land, deep ploughing of arable soils. Further, accumulation of soluble salts in the soils (salinization) also falls in this category.

Whereas most research has been devoted to decontamination and remediation of aqueous environment, much less has been contributed towards soil remediation. Nevertheless, some studies have reported on different aspects of soil decontamination. For instance, montmorillonite-TiO_2 composite has been designed and used as photocatalyst for the degradation of γ-hexachlorocyclohexane (γ -HCH) in soils. Specifically, soil samples were loaded with the said composite photocatalysts and then exposed to ultraviolet (UV) light irradiation. The study revealed that the photocatalytic activities of the composite depend on the content of TiO_2. The degradation of γ –HCH was confirmed by the intermediates products such as penta and trichlorocyclohexene, and dichlorobenzene which were progressively degraded with the photo degradation evolution [76]. Further, TiO_2 pillared montmorillonite clay was used for the adsorption and photocatalytic degradation of various organic compounds hydrophobicity's (di-n-butyl phthalate, dimethyl phthalate, diethyl phthalate and bisphenol-A). It was found that the hydrothermal treatment of TiO_2 pillared montmorillonite composite enhanced the photocatalytic degradation efficiency [77].

Clay and Clay Based Nanocomposites for Air Cleaning

Clean air is essential for healthy life. However, air often contains significantly high levels of contamination and pollutants that are extremely harmful to human health. The commonly found pollutants in air include particulate pollution, carbon monoxide, ground level ozone, nitrogen oxides, sulfur oxides, and lead. Increasing air pollution has imposed serious risks to human health, plants and wildlife. The health problems associated with air contamination include increased frequency in respiratory symptoms, heart or lung related diseases, and even premature death. Therefore, it is necessary to put significant efforts towards air pollution reduction and remediation.

Mostly, the modern purification systems for removing volatile organic compounds from air are based on photocatalysts, adsorbents. Clay and clay based nanocomposite offers attractive and efficient adsorption and photocatalysts tools.

For instance, the phocatalytic activity of Halloysite -TiO$_2$ clay composite for purification of air from toxic gases such as NO$_x$ and volatile substances like toluene in gas phase has been assessed. In the fabrication, TiO$_2$ nanoparticles were homogeneously distributed on the halloysite clay surface. The prepared nanocomposite showed significantly higher photo-decomposition activity for NO$_x$ under both visible light irradiation (>510nm), under UV-visible light irradiation as compared to the commercial TiO$_2$, called Degusaa-P25. The probed halloysite-TiO$_2$ nanocomposite was also exceptionally better than P25, another commercially available TiO$_2$. Further, the dynamic adsorption capacity of activated carbon–clay composites towards toluene contaminated air has been explored [78].

CONSENT FOR PUBLICATION

Not applicable.

CONFLICT OF INTEREST

The author (editor) declares no conflict of interest, financial or otherwise.

ACKNOWLEDGEMENTS

Declared none.

REFERENCES

[1]   Zhang, L.; Fang, M. Nanomaterials in pollution trace detection and environmental improvement. *Nano Today,* **2010**, *5*(2), 128-142.
       [http://dx.doi.org/10.1016/j.nantod.2010.03.002]

[2]   Masciangioli, T.; Zhang, W.X. Environmental technologies at the nanoscale. *Environ. Sci. Technol.,* **2003**, *37*(5), 102A-108A.
       [http://dx.doi.org/10.1021/es0323998] [PMID: 12666906]

[3]   Gangadhar, G.; Maheshwari, U.; Gupta, S. Application of nanomaterials for the removal of pollutants from effluent streams. *Nanosci. Nanotechnol. Asia,* **2012**, *2*(2), 140-150.
       [http://dx.doi.org/10.2174/2210681211202020140]

[4]   Grassian, V.H. when size really matters : Size-dependent properties and surface chemistry of metal and metal oxide nanoparticles in gas and liquid phase environments. *J. Phys. Chem. C,* **2008**, *112*(47), 18303-18313.
       [http://dx.doi.org/10.1021/jp806073t]

[5]   Feng, Z.; Zhu, S.; Martins de Godoi, D.R.; Samia, A.C.S.; Scherson, D. Adsorption of Cd2+ on carboxyl-terminated superparamagnetic iron oxide nanoparticles. *Anal. Chem.,* **2012**, *84*(8), 3764-3770.
       [http://dx.doi.org/10.1021/ac300392k] [PMID: 22428526]

[6]   Hua, M.; Zhang, S.; Pan, B.; Zhang, W.; Lv, L.; Zhang, Q. Heavy metal removal from water/wastewater by nanosized metal oxides: a review. *J. Hazard. Mater.,* **2012**, *211-212*, 317-331.
       [http://dx.doi.org/10.1016/j.jhazmat.2011.10.016] [PMID: 22018872]

[7]   Kharisov, B.I.; Dias, H.V.R.; Kharissova, O.V.; Jiménez-Pérez, V.M.; Péreza, B.O.; Flores, B.M.

Iron-Containing Nanomaterials: Synthesis, Properties, and Environmental Applications. *RSC Advances,* **2012**, *2*(25), 9325-9358.
[http://dx.doi.org/10.1039/c2ra20812a]

[8]   Zhou, C.H.; Keeling, J. Fundamental and applied research on clay minerals : From climate and environment to nanotechnology. *Appl. Clay Sci.,* **2013**, *74*, 3-9.
[http://dx.doi.org/10.1016/j.clay.2013.02.013]

[9]   Konta, J. Clay and man : Clay raw materials in the service of man. *Appl. Clay Sci.,* **1995**, *10*, 275-335.
[http://dx.doi.org/10.1016/0169-1317(95)00029-4]

[10] Lee, M.S.; Tiwari, D. Organo and inorgano-organo-modified clays in the remediation of aqueous solutions : An overview. *Appl. Clay Sci.,* **2012**, *59-60*, 84-102.
[http://dx.doi.org/10.1016/j.clay.2012.02.006]

[11] Uddin, F. Clays, nanoclays, and montmorillonite minerals. *Metall. Mater. Trans., A Phys. Metall. Mater. Sci.,* **2008**, *39*(12), 2004-2814.
[http://dx.doi.org/10.1007/s11661-008-9603-5]

[12] Tzonou, A.; Polychronopoulou, A.; Hsieh, C.C.; Rebelakos, A.; Karakatsani, A.; Trichopoulos, D. Hair dyes, analgesics, tranquilizers and perineal talc application as risk factors for ovarian cancer. *Int. J. Cancer,* **1993**, *55*(3), 408-410.
[http://dx.doi.org/10.1002/ijc.2910550313] [PMID: 8375924]

[13] Nadeau, P.H.; Bain, D.C. Composition of some smectites and diagenetic illitic clays and Implications for their origin. *Clays Clay Miner.,* **1986**, *34*(4), 455-464.
[http://dx.doi.org/10.1346/CCMN.1986.0340412]

[14] Wang, Y.H.; Siu, W.K. Structure characteristics and mechanical properties of kaolinite soils; Surface charges and structural characterizations. *Can. Geotech. J.,* **2006**, *43*(6), 587-600.
[http://dx.doi.org/10.1139/t06-026]

[15] De Caritat, P.; Hutcheon, I.; Walshe, J.L. Chlorite geothermometry: A review. *Clays Clay Miner.,* **1993**, *41*(2), 219-239.
[http://dx.doi.org/10.1346/CCMN.1993.0410210]

[16] Choudalakis, G.; Gotsis, A.D. Permeability of polymer/clay nanocomposites: A review. *Eur. Polym. J.,* **2009**, *45*(4), 967-984.
[http://dx.doi.org/10.1016/j.eurpolymj.2009.01.027]

[17] Zeng, Q.H.; Yu, A.B.; Lu, G.Q.; Paul, D.R. Clay-based polymer nanocomposites: research and commercial development. *J. Nanosci. Nanotechnol.,* **2005**, *5*(10), 1574-1592.
[http://dx.doi.org/10.1166/jnn.2005.411] [PMID: 16245517]

[18] Fu, F.; Wang, Q. Removal of heavy metal ions from wastewaters: a review. *J. Environ. Manage.,* **2011**, *92*(3), 407-418.
[http://dx.doi.org/10.1016/j.jenvman.2010.11.011] [PMID: 21138785]

[19] Srivastava, N.K.; Majumder, C.B. Novel biofiltration methods for the treatment of heavy metals from industrial wastewater. *J. Hazard. Mater.,* **2008**, *151*(1), 1-8.
[http://dx.doi.org/10.1016/j.jhazmat.2007.09.101] [PMID: 17997034]

[20] Barakat, M.A. New trends in removing heavy metals from industrial wastewater. *Arab. J. Chem.,* **2011**, *4*(4), 361-377.
[http://dx.doi.org/10.1016/j.arabjc.2010.07.019]

[21] Sarkar, B.; Xi, Y.; Megharaj, M.; Krishnamurti, G.S.R.; Bowman, M.; Rose, H. Bioreactive organoclay: a new technology for environmental remediation. *Crit. Rev. Environ. Sci. Technol.,* **2012**, *42*(5), 435-488.
[http://dx.doi.org/10.1080/10643389.2010.518524]

[22] Dos Santos, V.C.G.; Grassi, M.T.; Abate, G. Sorption of Hg(II) by modified K10 montmorillonite: Influence of pH, ionic strength and the treatment with different cations. *Geoderma,* **2015**, *237-238*,

129-136.
[http://dx.doi.org/10.1016/j.geoderma.2014.08.018]

[23]   Rivera-Hernández, J.R.; Green-Ruiz, C. Geosorption of As(III) from aqueous solution by red clays: kinetic studies. *Bull. Environ. Contam. Toxicol.,* **2014**, *92*(5), 596-601.
[http://dx.doi.org/10.1007/s00128-014-1233-6] [PMID: 24549918]

[24]   Akar, S.T.; Yetimoglu, Y.; Gedikbey, T. Removal of chromium (VI) ions from aqueous solutions by using Turkish montmorillonite clay: effect of activation and modification. *Desalination,* **2009**, *244*(1-3), 97-108.
[http://dx.doi.org/10.1016/j.desal.2008.04.040]

[25]   Wu, P.; Wu, W.; Li, S.; Xing, N.; Zhu, N.; Li, P.; Wu, J.; Yang, C.; Dang, Z. Removal of Cd^{2+} from aqueous solution by adsorption using Fe-montmorillonite. *J. Hazard. Mater.,* **2009**, *169*(1-3), 824-830.
[http://dx.doi.org/10.1016/j.jhazmat.2009.04.022] [PMID: 19443105]

[26]   Na, P.; Jia, X.; Yuan, B.; Li, Y.; Na, J.; Chen, Y. Arsenic adsorption on Ti-pillared montmorillonite. *J. Chem. Technol. Biotechnol.,* **2010**, *85*(5), 708-714.
[http://dx.doi.org/10.1002/jctb.2360]

[27]   Bhattacharyya, K.G.; Sen Gupta, S. Removal of Cu(II) by natural and acid-activated clays: An insight of adsorption isotherm, kinetic and thermodynamics. *Desalination,* **2011**, *272*(1-3), 66-75.
[http://dx.doi.org/10.1016/j.desal.2011.01.001]

[28]   Pereira, F.A.R.; Sousa, K.S.; Cavalcanti, G.R.S.; Fonseca, M.G.; de Souza, A.G.; Alves, A.P.M. Chitosan-montmorillonite biocomposite as an adsorbent for copper (II) cations from aqueous solutions. *Int. J. Biol. Macromol.,* **2013**, *61*(8), 471-478.
[http://dx.doi.org/10.1016/j.ijbiomac.2013.08.017] [PMID: 23973496]

[29]   Wang, H.; Tang, H.; Liu, Z.; Zhang, X.; Hao, Z.; Liu, Z. Removal of cobalt(II) ion from aqueous solution by chitosan-montmorillonite. *J. Environ. Sci. (China),* **2014**, *26*(9), 1879-1884.
[http://dx.doi.org/10.1016/j.jes.2014.06.021] [PMID: 25193838]

[30]   Hajjaji, M.; El Arfaoui, H. Adsorption of methylene blue and zinc ions on raw and acid-activated bentonite from Morocco. *Appl. Clay Sci.,* **2009**, *46*(4), 418-421.
[http://dx.doi.org/10.1016/j.clay.2009.09.010]

[31]   Eren, E.; Afsin, B.; Onal, Y. Removal of lead ions by acid activated and manganese oxide-coated bentonite. *J. Hazard. Mater.,* **2009**, *161*(2-3), 677-685.
[http://dx.doi.org/10.1016/j.jhazmat.2008.04.020] [PMID: 18501507]

[32]   Kul, A.R.; Koyuncu, H. Adsorption of Pb(II) ions from aqueous solution by native and activated bentonite: kinetic, equilibrium and thermodynamic study. *J. Hazard. Mater.,* **2010**, *179*(1-3), 332-339.
[http://dx.doi.org/10.1016/j.jhazmat.2010.03.009] [PMID: 20356674]

[33]   Tirtom, V.N.; Dinçer, A.; Becerik, S.; Aydemir, T.; Çelik, A. Comparative adsorption of Ni(II) and Cd(II) ions on epichlorohydrin crosslinked chitosan-clay composite beads in aqueous solution. *Chem. Eng. J.,* **2012**, *197*, 379-386.
[http://dx.doi.org/10.1016/j.cej.2012.05.059]

[34]   Tang, Q.; Tang, X.W.; Li, Z.Z.; Chen, Y.M.; Kou, N.Y.; Sun, Z.F. Adsorption and desorption behaviour of Pb(II) on a natural kaolin: equilibrium, kinetic and thermodynamic studies. *J. Chem. Technol. Biotechnol.,* **2009**, *84*(9), 1371-1380.
[http://dx.doi.org/10.1002/jctb.2192]

[35]   Fathima, A.; Rao, J.R.; Unni Nair, B. Trivalent chromium removal from tannery effluent using kaolin-supported bacterial biofilm of Bacillus sp isolated from chromium polluted soil. *J. Chem. Technol. Biotechnol.,* **2012**, *87*(2), 271-279.
[http://dx.doi.org/10.1002/jctb.2710]

[36]   Unuabonah, E.I.; El-Khaiary, M.I.; Olu-Owolabi, B.I.; Adebowale, K.O. Predicting the dynamics and performance of a polymer-clay based composite in a fixed bed system for the removal of lead (II) ion.

Chem. Eng. Res. Des., **2012**, *90*(8), 1105-1115.
[http://dx.doi.org/10.1016/j.cherd.2011.11.009]

[37]   Mascia, M.; Vacca, A.; Palmas, S. Effect of surface equilibria on the electrokinetic behaviour of Pb and Cd ions in kaolinite. *J. Chem. Technol. Biotechnol.,* **2014**, *90*, 1290-1298.
[http://dx.doi.org/10.1002/jctb.4435]

[38]   Musso, T.B.; Parolo, M.E.; Pettinari, G.; Francisca, F.M. Cu(II) and Zn(II) adsorption capacity of three different clay liner materials. *J. Environ. Manage.,* **2014**, *146*, 50-58.
[http://dx.doi.org/10.1016/j.jenvman.2014.07.026] [PMID: 25156265]

[39]   Vakili, M.; Rafatullah, M.; Ibrahim, M.H.; Abdullah, A.Z.; Salamatinia, B.; Gholami, Z. Oil palm biomass as an adsorbent for heavy metals. *Rev. Environ. Contam. Toxicol.,* **2014**, *232*, 61-88.
[http://dx.doi.org/10.1007/978-3-319-06746-9_3] [PMID: 24984835]

[40]   Sharma, P.; Kaur, H.; Sharma, M.; Sahore, V. A review on applicability of naturally available adsorbents for the removal of hazardous dyes from aqueous waste. *Environ. Monit. Assess.,* **2011**, *183*(1-4), 151-195.
[http://dx.doi.org/10.1007/s10661-011-1914-0] [PMID: 21387170]

[41]   Demirbas, A. Agricultural based activated carbons for the removal of dyes from aqueous solutions: a review. *J. Hazard. Mater.,* **2009**, *167*(1-3), 1-9.
[http://dx.doi.org/10.1016/j.jhazmat.2008.12.114] [PMID: 19181447]

[42]   Gupta, V.K.; Suhas, Application of low-cost adsorbents for dye removal--a review. *J. Environ. Manage.,* **2009**, *90*(8), 2313-2342.
[http://dx.doi.org/10.1016/j.jenvman.2008.11.017] [PMID: 19264388]

[43]   Yagub, M.T.; Sen, T.K.; Afroze, S.; Ang, H.M. Dye and its removal from aqueous solution by adsorption: a review. *Adv. Colloid Interface Sci.,* **2014**, *209*, 172-184.
[http://dx.doi.org/10.1016/j.cis.2014.04.002] [PMID: 24780401]

[44]   Bhattacharyya, R.; Ray, S.K. Micro- and nano-sized bentonite filled composite superabsorbents of chitosan and acrylic copolymer for removal of synthetic dyes from water. *Appl. Clay Sci.,* **2014**, *101*, 510-520.
[http://dx.doi.org/10.1016/j.clay.2014.09.015]

[45]   Bhattacharyya, K.G.; SenGupta, S.; Sarma, G.K. Interactions of the dye, Rhodamine B with kaolinite and montmorillonite in water. *Appl. Clay Sci.,* **2014**, *99*, 7-17.
[http://dx.doi.org/10.1016/j.clay.2014.07.012]

[46]   Putra, E.K.; Pranowo, R.; Sunarso, J.; Indraswati, N.; Ismadji, S. Performance of activated carbon and bentonite for adsorption of amoxicillin from wastewater: mechanisms, isotherms and kinetics. *Water Res.,* **2009**, *43*(9), 2419-2430.
[http://dx.doi.org/10.1016/j.watres.2009.02.039] [PMID: 19327813]

[47]   Martinez, J.L. Environmental pollution by antibiotics and by antibiotic resistance determinants. *Environ. Pollut.,* **2009**, *157*(11), 2893-2902.
[http://dx.doi.org/10.1016/j.envpol.2009.05.051] [PMID: 19560847]

[48]   Dolliver, H.; Gupta, S. Antibiotic losses in leaching and surface runoff from manure-amended agricultural land. *J. Environ. Qual.,* **2008**, *37*(3), 1227-1237.
[http://dx.doi.org/10.2134/jeq2007.0392] [PMID: 18453442]

[49]   Dodd, M.C.; Huang, C.H. Aqueous chlorination of the antibacterial agent trimethoprim: reaction kinetics and pathways. *Water Res.,* **2007**, *41*(3), 647-655.
[http://dx.doi.org/10.1016/j.watres.2006.10.029] [PMID: 17173950]

[50]   Li, D.; Yang, M.; Hu, J.; Zhang, Y.; Chang, H.; Jin, F. Determination of penicillin G and its degradation products in a penicillin production wastewater treatment plant and the receiving river. *Water Res.,* **2008**, *42*(1-2), 307-317.
[http://dx.doi.org/10.1016/j.watres.2007.07.016] [PMID: 17675133]

[51]　Choi, K.J.; Kim, S.G.; Kim, S.H. Removal of antibiotics by coagulation and granular activated carbon filtration. *J. Hazard. Mater.,* **2008**, *151*(1), 38-43.
[http://dx.doi.org/10.1016/j.jhazmat.2007.05.059] [PMID: 17628341]

[52]　Brown, K.D.; Kulis, J.; Thomson, B.; Chapman, T.H.; Mawhinney, D.B. Occurrence of antibiotics in hospital, residential, and dairy effluent, municipal wastewater, and the Rio Grande in New Mexico. *Sci. Total Environ.,* **2006**, *366*(2-3), 772-783.
[http://dx.doi.org/10.1016/j.scitotenv.2005.10.007] [PMID: 16313947]

[53]　Wu, Q.; Li, Z.; Hong, H.; Yin, K.; Tie, L. Adsorption and intercalation of ciprofloxacin on montmorillonite. *Appl. Clay Sci.,* **2010**, *50*(2), 204-211.
[http://dx.doi.org/10.1016/j.clay.2010.08.001]

[54]　Li, Z.; Chang, P.H.; Jean, J.S.; Jiang, W.T.; Wang, C.J. Interaction between tetracycline and smectite in aqueous solution. *J. Colloid Interface Sci.,* **2010**, *341*(2), 311-319.
[http://dx.doi.org/10.1016/j.jcis.2009.09.054] [PMID: 19883920]

[55]　Li, Z.; Hong, H.; Liao, L.; Ackley, C.J.; Schulz, L.A.; MacDonald, R.A.; Mihelich, A.L.; Emard, S.M. A mechanistic study of ciprofloxacin removal by kaolinite. *Colloids Surf. B Biointerfaces,* **2011**, *88*(1), 339-344.
[http://dx.doi.org/10.1016/j.colsurfb.2011.07.011] [PMID: 21802909]

[56]　Sturini, M.; Speltini, A.; Maraschi, F.; Rivagli, E.; Pretali, L.; Malavasi, L. Sunlight photodegradation of marbofloxacin and enrofloxacin adsorbed on clay minerals. *J. Photochem. Photobiol. Chem.,* **2015**, *299*, 103-109.
[http://dx.doi.org/10.1016/j.jphotochem.2014.11.015]

[57]　Wu, Q.; Li, Z.; Hong, H.; Li, R.; Jiang, W.T. Desorption of ciprofloxacin from clay mineral surfaces. *Water Res.,* **2013**, *47*(1), 259-268.
[http://dx.doi.org/10.1016/j.watres.2012.10.010] [PMID: 23123088]

[58]　Chang, P.H.; Li, Z.; Jiang, W.T.; Jean, J.S. Adsorption and intercalation of tetracycline by swelling clay minerals. *Appl. Clay Sci.,* **2009**, *46*(1), 27-36.
[http://dx.doi.org/10.1016/j.clay.2009.07.002]

[59]　Robberson, K.A.; Waghe, A.B.; Sabatini, D.A.; Butler, E.C. Adsorption of the quinolone antibiotic nalidixic acid onto anion-exchange and neutral polymers. *Chemosphere,* **2006**, *63*(6), 934-941.
[http://dx.doi.org/10.1016/j.chemosphere.2005.09.047] [PMID: 16307776]

[60]　Ahmed, M.B.; Zhou, J.L.; Ngo, H.H.; Guo, W. Adsorptive removal of antibiotics from water and wastewater: Progress and challenges. *Sci. Total Environ.,* **2015**, *532*, 112-126.
[http://dx.doi.org/10.1016/j.scitotenv.2015.05.130] [PMID: 26057999]

[61]　Genç, N.; Dogan, E.C. Adsorption kinetics of the antibiotic ciprofloxacin on bentonite, activated carbon, zeolite, and pumice. *Desalination Water Treat.,* **2015**, *53*(3), 785-793.
[http://dx.doi.org/10.1080/19443994.2013.842504]

[62]　Ali, I.; Asim, M.; Khan, T.A. Low cost adsorbents for the removal of organic pollutants from wastewater. *J. Environ. Manage.,* **2012**, *113*, 170-183.
[http://dx.doi.org/10.1016/j.jenvman.2012.08.028] [PMID: 23023039]

[63]　Burkhard, L.P.; Lukasewycz, M.T. Toxicity equivalency values for polychlorinated biphenyl mixtures. *Environ. Toxicol. Chem.,* **2008**, *27*(3), 529-534.
[http://dx.doi.org/10.1897/07-349.1] [PMID: 17967071]

[64]　Gu, L.; Yu, X.; Xu, J.; Lv, L.; Wang, Q. Removal of dichloroacetic acid from drinking water by using adsorptive ozonation. *Ecotoxicology,* **2011**, *20*(5), 1160-1166.
[http://dx.doi.org/10.1007/s10646-011-0680-7] [PMID: 21499868]

[65]　Jie, L.; Jiafen, P. Removal of Carbon Tetrachloride from Contaminated Groundwater Environment by Adsorption Method. *4th International Conference on Bioinformatics and Biomedical Engineering (iCBBE '10), Chengdu, China,* **2010**, pp. 1-4.

[66] Rivera-Jimenez, S.M.; Lehner, M.M.; Cabrera-Lafaurie, W.A.; Hernández-Maldonado, A.J. Removal of naproxen, salicylic acid, clofibric acid, and carbamazepine by water phase adsorption onto inorganic–organic-intercalated bentonites modified with transition metal cations. *Environ. Eng. Sci.*, **2011**, *28*(3), 171-182.
[http://dx.doi.org/10.1089/ees.2010.0213]

[67] Senturk, H.B.; Ozdes, D.; Gundogdu, A.; Duran, C.; Soylak, M. Removal of phenol from aqueous solutions by adsorption onto organomodified Tirebolu bentonite: equilibrium, kinetic and thermodynamic study. *J. Hazard. Mater.*, **2009**, *172*(1), 353-362.
[http://dx.doi.org/10.1016/j.jhazmat.2009.07.019] [PMID: 19656623]

[68] Gu, L.; Zhang, X.; Lei, L.; Liu, X. Concurrent removal of humic acid and o-dichlorobenzene in drinking water by combined ozonation and bentonite coagulation process. *Water Sci. Technol.*, **2009**, *60*(12), 3061-3068.
[http://dx.doi.org/10.2166/wst.2009.678] [PMID: 19955629]

[69] Gao, Z.; Peng, X.; Zhang, H.; Luan, Z.; Fan, B. Montmorillonite – Cu (II)/ Fe (III) oxides magnetic material for removal of cyanobacterial microcystis aeruginosa and its regeneration. *Desalination*, **2009**, *247*(1-3), 337-345.
[http://dx.doi.org/10.1016/j.desal.2008.10.006]

[70] Zadaka, D.; Nir, S.; Radian, A.; Mishael, Y.G. Atrazine removal from water by polycation-clay composites: effect of dissolved organic matter and comparison to activated carbon. *Water Res.*, **2009**, *43*(3), 677-683.
[http://dx.doi.org/10.1016/j.watres.2008.10.050] [PMID: 19038414]

[71] Undabeytia, T.; Nir, S.; Sánchez-Verdejo, T.; Villaverde, J.; Maqueda, C.; Morillo, E. A clay-vesicle system for water purification from organic pollutants. *Water Res.*, **2008**, *42*(4-5), 1211-1219.
[http://dx.doi.org/10.1016/j.watres.2007.09.004] [PMID: 17915281]

[72] Rytwo, G.; Kohavi, Y.; Botnick, I.; Gonen, Y. Use of CV- and TPP-montmorillonite for the removal of priority pollutants from water. *Appl. Clay Sci.*, **2007**, *36*(1-3), 182-190.
[http://dx.doi.org/10.1016/j.clay.2006.04.016]

[73] Bonina, F.P.; Giannossi, M.L.; Medici, L.; Puglia, C.; Summa, V.; Tateo, F. Adsorption of salicylic acid on bentonite and kaolin and release experiments. *Appl. Clay Sci.*, **2007**, *36*(1-3), 77-85.
[http://dx.doi.org/10.1016/j.clay.2006.07.008]

[74] Zhang, W.; Ding, Y.; Boyd, S.A.; Teppen, B.J.; Li, H. Sorption and desorption of carbamazepine from water by smectite clays. *Chemosphere*, **2010**, *81*(7), 954-960.
[http://dx.doi.org/10.1016/j.chemosphere.2010.07.053] [PMID: 20797761]

[75] Srinivasan, R. Advances in application of aatural clay and its composites in removal of biological, organic, and inorganic contaminants from drinking water. *Adv. Mater. Sci. Eng.*, **2011**, *2011*, 1-17.
[http://dx.doi.org/10.1155/2011/872531]

[76] Zhao, X.; Quan, X.; Chen, S.; Zhao, H.M.; Liu, Y. Photocatalytic remediation of γ-hexachlorocyclohexane contaminated soils using TiO_2 and montmorillonite composite photocatalyst. *J. Environ. Sci. (China)*, **2007**, *19*(3), 358-361.
[http://dx.doi.org/10.1016/S1001-0742(07)60059-X] [PMID: 17918601]

[77] Ooka, C.; Yoshida, H.; Horio, M.; Suzuki, K.; Hattori, T. Adsorptive and photocatalytic performance of TiO_2 pillared montmorillonite in degradation of endocrine disruptors having different hydrophobicity. *Appl. Catal. B*, **2003**, *41*, 313-321.
[http://dx.doi.org/10.1016/S0926-3373(02)00169-8]

[78] Yates, M.; Martín-Luengo, M.A.; Argomaniz, L.V.; Velasco, S.N. Design of activated carbon–clay composites for effluent decontamination. *Microporous Mesoporous Mater.*, **2012**, *154*, 87-92.
[http://dx.doi.org/10.1016/j.micromeso.2011.07.006]

CHAPTER 5

Ion Exchange Materials and Their Applications

Anish Khan[1,*], **Fayaz Ali**[1], **Aftab Aslam Parwaz Khan**[1], **Aleksandr Evhenovych Kolosov**[2] and **Abdullah M. Asiri**[1]

[1] *Center of Excellence for Advanced Materials Research, Department of Chemistry, Faculty of Science, King Abdulaziz University, Jeddah 21589, Saudi Arabia*

[2] *Chemical, Polymeric and Silicate Machine Building Department of Chemical Engineering Faculty National Technical University of Ukraine, "Igor Sikorsky Kyiv Polytechnic Institute", Kyiv, Ukraine19 Build, 37 Prospect Peremohy, 03056, Kyiv, Ukraine*

Abstract: The material with small size and high efficiency is the topic of the future. It is valuable and important in the field of separation and purification science. This chapter discusses ion-exchange materials, natural, synthetic, organic, inorganic and organic-inorganic nanocomposite. In this chapter, we discussed about the chromatography of ion-exchange particularly cation-exchanger and its uses as desalination and ion-selective electrode. The Estimation of the impact on the manufacturing procedure on the environment requires a systemic approach and suitable metrics for the quantitative valuation of environmental threats. Thus, this chapter starts with an overview on cation exchange materials for green technology as well as the metrics starting the procedure. Therefore, there are many applications of cation-exchange materials and their derivatives have been explained in this chapter. In addition, the technological advancement in inorganic nanocomposite, cation-exchange materials from old era to modern age of nano are also explained as green chemistry and can be implemented to actual developments. Specifically, two elements are highlighted: (*a*) the usage of new resources for the facilitation of selective and active chemistry and the implementation of the said materials for the removal of hazardous materials for environment.

Keywords: Cation-exchanger, Heavy metal, Ion selective electrode, Life spam, Membrane, Nanocomposite, Polymer, Selectivity.

ION-EXCHANGE PHENOMENON & ITS HISTORICAL BACKGROUND

The ion exchange phenomenon is not of a current origin. Actually, million years before, it was observed in various parts of the globe. For example, some ions like potassium and lithium of petalite of pegmatite veins had been replaced with

* **Corresponding author Anish Khan**: Center of Excellence for Advanced Materials Research (CEAMR), Department of Chemistry, Faculty of Science, King Abdulaziz University, Jeddah 21589, Saudi Arabia; Tel\Fax: +966-59-3541984; Email: akrkhan@kau.edu.sa

Sher Bahadar Khan (Ed.)

rubidium and cesium ions of step wisely fluid from the mega. This is nothing but ion-exchange phenomenon between minerals like petalite (solid phase) and fused salt fluid (liquid phase) [1]. It is well known that ion exchange has been playing a very important role during the course of weathering; aqueous rocks, clay rocks and soils being very effective ion-exchangers. Since life had been created in the sea, ion-exchange through bio-membranes between living organs and outside matters has been giving the essential motive force to life and its evolution. Earlier references can be found in the Holy Bible about Moses' priority, whose were able by an ion-exchange method to convert brackish water into drinking water [2]. Afterward, Aristotle found that some part of salt content from seawater were remove during percolation over certain sand [3]. In Greece and Egypt as well as in China, ancient people were clever enough to use some sands, soils, plants and natural zeolites for enhancing the quality of drinking water by way of softening or desalting. However, they were not aware of the actual phenomenon occurring in the process. Basically, ion exchange is a natural process which occurring from the age, before the birth of civilization, and has been embraced by analytical chemists to convert difficult separation methods into easier and possible methods.

Francis Bacon in 1623 brought the intended use of ion exchange, without knowledge of its theoretical nature, based purely on empirical experiences and he described a method for removing salts from seawater. The first half of the 19th century was characterized by the appearance of the first information leading to the discovery of the ion-exchange principle, based primarily on the work of soil chemists. Thompson, Spence and Way in 1850 described independently that calcium and magnesium ions of certain types of soils could be exchanged for potassium and ammonium ions [4, 5]. They defined the special properties of soil as 'base exchange'. In the second half of the 19th century, agro chemists published a great number of papers dealing with ion exchange in soils. Eichhorn (in 1858) demonstrated exchange processes are reversible in soils [6]. In 1859, Boedecker proposed an empirical equation describing the establishment of equilibrium on inorganic ion-exchange sorbents. In the 20th century, the majority chemists believed that the 'base exchange' in soils is nothing but a sort of absorption. Strong supports to ion-exchange come out with the synthesis of materials from clay, sand and sodium carbonate by Gans [7].

The theory of the discovery and development of ion exchange was reflected in practical applications. Gans [7], developed the basis for the synthesis and technical application of inorganic cation-exchangers at the beginning of the 20th century. He termed the amorphous cation-exchangers based on aluminosilicate gels "permutated", having broad application, were actually the first commercially available ion-exchangers. In 1917, Folin and Bell developed an analytical method, which is based on the above materials and were used for the separation and collection of ammonia in urine. However, the usefulness of these synthetic zeolites was limited because of their low chemical and mechanical stability, ion-exchange capacity that led the chemists to seek alternatives. During the period between the 1930s and 1940s, inorganic ion-exchange sorbents were replaced in almost all fields by the new organic ion-exchangers. The observation of Adam and Holms show ion-exchange properties in the crushed phonograph records, eventually resulted in the more significant development of synthetic ion-exchange resins (high molecular weight organic polymers containing a large number of ionic functional groups) in 1935. Therefore, the phenomenon of ion-exchange could not be neglected by any scientist. However, this phenomenon took nearly 85 years to be fully recognized in chemistry, since the scientific understandings and findings of Thompson and Way.

Just as applications of the organic resins are limited by breakdown in aqueous systems at high temperatures and in existence of high ionizing radiation amounts; for these reasons there had been a resurgence of concern in inorganic exchangers in the 1950s. One of the possible ways of solving these problems involved replacing the organic skeleton of the ion-exchanger by an inorganic skeleton. Pioneering work was carried out in this field by the research team at the Oak Ridge National University led by Kraus, and by the English team led by Amphlett.

Further extensive research and study of inorganic ion-exchange sorbents were carried out in the 1960s and 1980s. Research led from the original amorphous type of ion-exchange sorbents to the study of crystalline ion-exchange materials. Clearfield and co-workers made great contributions in this area. Since last two decades, intense research has continued on the synthesis of a number of new 'organic-inorganic' composite materials having excellent properties that not only led to the determination of many previously insolvable problems as well as met the necessities of modern laboratories. In industries, an interest of inorganic as well as composite ion-exchange materials is enhancing day by day in ion-exchange processes due to their enormous field of applications.

Ion-exchange Chromatography

Ion-exchange chromatography is a powerful tool for chemical separations and was the first of the various liquid chromatography (LC) methods to be used under modern LC conditions. It has grown in response to practical needs. Most environmental samples have complex composition, thus trace, separation and pre-concentration of analyses are essential for accurate as well as precise determination of elements in environmental samples. Ion-exchange chromatography is an excellent technique that permits selective separation by the appropriate combination of ion-exchanger and eluent. Columns of ion-exchange constituents have been extensively applied for the separation of amino acids, inorganic ions (especially rare earths), multi-components of alloys, heavy metals in industrial effluents and fission products of radioactive elements [8]; as well as organic ions and organic compounds that are not ionized at all.

Ion-exchange chromatography uses an ion-exchange material as stationary phase and in size exclusion chromatography separation that takes place as a function of size of the porous media. These are actually arbitrary classification of chromatography and some types of chromatography are considered together as a separate technique.

Ion-exchange Process and its Mechanism

The ion-exchange method became recognized as an analytical means in industries and in laboratories, as primarily practical chemists concerned in performance and effects *etc*. Stoichiometry is the primary situation for an ion-exchange process. It is a recognized fact in organic resins [9]. Stoichiometrically, the exchange of ions took place between two immiscible phases, mobile and stationary. The ion-exchange reaction may be symbolized as follows:

$$\overline{YX} + Z \text{ (aq)} \rightleftharpoons \overline{ZX} + Y \text{ (aq)} \tag{1}$$

Where X is the structural unit (matrix) and Y and Z (taking part in ion exchange) are the replaceable ions of the ion-exchanger. (aq) indicates the aqueous phase and Bar represents the exchanger phase.

The study of thermodynamics and kinetics of ion exchange process is important to understand the mechanism of the process, which occurring on the exchanger surface and explain equilibrium as well as to estimate its theoretical behavior. Such studies are simpler and easier to perform on inorganic ion-exchangers as compare to the organic resins because they possess a rigid matrix and do not swell appreciably. Ion-exchange equilibrium may be explained with the help of two theoretical approaches *viz* (i) Based on Donnan theory, and (ii) Based on law of mass action.

The Donnan theory has an advantage from the theoretical point of view by permitting a more sophisticated explanation of thermodynamic performance in an ion-exchanger. Probably, first time Gane use the law of mass action in its simplest form without including the activity coefficients for the numerical formation of ion-exchange equilibrium. Kielland further accounted this concept and finally, Gaines and Thomas give an appropriate choice of universal treatment [10]. The thermodynamics of cation-exchange on zirconium (IV) phosphate have been evaluated by many workers [11, 12]. In previous studies, the effect of crystallinity on the samples of α -zirconium phosphate by ion-exchange of alkali metal ions/H^+ ions were examined. Calorimetric heats and ion-exchange isotherms were studied on amorphous to highly crystalline samples [13 - 16].

However, the approach of mass action is simpler from the practical point of view. The thermodynamical functions between ion exchange matrix and alkali metals in term of binding nature have interpreted by Nancollas and coworkers [17, 18]. Larsen and Vissers studied the ion-exchange equilibria of Na(I), K(I) and Li(I), on zirconium (IV) phosphate [19], who calculated the thermo dynamical parameters *viz.* $\Delta H°$, $\Delta G°$ and $\Delta S°$ and equilibrium constants. On anion-exchanger, similar studies have also been made by Nancollas [20]. Alkali earth metal ions exchange equilibria on different inorganic ion-exchangers such as iron (III) antimonite [21], tantalum arsenate [22], zirconium (IV) phosphosilicate [23, 24], antimony (V) silicate [25], and alkali metal ions on α -cerium phosphate [26] and iron (III) antimonate [27]. Other interesting studies of thermodynamics have been studied in the laboratories related to the adsorption of pesticides on composite and inorganic ion-exchangers [28, 29]. The study exposed that at low temperature the adsorption is higher and the adsorption capacity is greatly increase in the presence of an ion-exchange material in the soil. The first and detailed attempt was made by Nachod and Wood on kinetic studies of ion-exchange [30]. The reaction rate has been studied by them in which the exchangeable ions are released from the exchanger or in which ions from solutions are removed by solid ion-exchangers. The kinetics of metal ions upon the resin beads have been studied by Boyd *et al.* [31] and clearly explain the phenomenon that govern the ion-exchange processes about the film and particle diffusion. At higher concentration the former is valid while the later at lower concentrations. The kinetic of metal ions on sulphonated polystyrene had studied by Reichenberg, who again confirmed that the rate is independent of the ingoing ion (particle diffusion) at high concentrations; while the reverse is true (film diffusion) at low concentrations.

Ion-exchange Materials: An Introduction and Literature Review

Inorganic Ion-exchange Materials

The synthetic ion-exchanger compounds can produce the desire chemical and physical properties. The synthetic inorganic ion-exchangers are classified on the basis of chemical characteristics as follows:

- Synthetic zeolites (aluminosilicates)
- Acidic salts of polyvalent metals
- Hydrous oxides of metals

- Insoluble hydrated metal hexacyanoferrate (II) and (III) (Ferro cyanides)
- Insoluble salts of heteropolyacids
- Other substances with weak exchange properties

The first inorganic materials used for the large-scale removal of waste effluents was **zeolites**. Zeolites (crystalline alumino-silicate based materials) can be prepared as microcrystalline pellets, powder or beads. In comparison with naturally occurring zeolites, the synthetic zeolites are stable at high temperature and can be engineered with large diversity of chemical properties and pore sizes.

The main limitations of synthetic zeolites are that:

- They have a limited chemical stability at extreme pH ranges (either high or low);
- They have a relatively high cost compared with natural zeolites;
- The materials tend to be brittle, which limits their mechanical stability.
- Their ion specificity is susceptible to interference from similar sized ions;

The zeolites selectivity and capacity can provide an appropriate processing of low strength salt solutions. The actual obtained processing capacities of zeolites are lower than their maximum capacities. Hence, at the early stage of breakthrough the bed is changed and because other ions are present in the waste streams that will occupy some of the exchange sites and as a result the processing capacity is reduced. To examine the performance of locally available zeolites in India, a systematic investigation has been carried out for the removal of thorium, strontium and cesium from solution [32, 33]. The zeolites, after exchange with thorium, strontium or cesium, were treated thermally to fix the ions effectively in the same matrix. Recently, a locally available synthetic mordenite was used to reduce activity in spent fuel storage pool water.

Hydrous oxides are of particular interest since most of them can function as both anion and cation-exchangers. Schematically, their dissociation may be represented as follows:

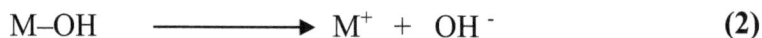

$$M–OH \longrightarrow M^+ + OH^- \tag{2}$$

$$M–OH \longrightarrow M–O^- + H^+ \tag{3}$$

Where "M" represents the central atom. Equation '2' is applied, when the substance is favored by acid conditions and can function as anion-exchanger and equation '3' when the substance by alkaline conditions, can function as cation-exchanger. Dissociation according to both schemes can take place near the isoelectric point, and both type of exchanges may occur simultaneously.

Framework hydrates and particle hydrates are the two main types of hydrous oxides. Framework hydrates are generally consisting of metals in group 5 and 15 in their higher oxidation states. Particle hydrates are both anion and cation-exchangers. The hydrous oxides of the group 3, 4, 13 and 14 metals belong to this group. The powder X-ray diffraction revealed the pyrochlore nature of crystalline hydrous antimony (V) oxide exchanger'. DeRoy *et al.*; published an excellent review on LDHs.

Acidic salts of multivalent metals were synthesized by mixing the salt solutions of the III and IV group elements of the periodic table with the more acidic salts. Generally, these salts are microcrystalline materials or gel like and act as cation exchangers, possessing mostly high thermal, chemical and radiation stability.

Salts of heteropolyacids have a general formula $H_m X.Y_{12} O_{40} . nH_2 O$, where m = 3, 4 or 5, X can be phosphorous, silicon, arsenic, boron or germanium and Y, one of the elements such as tungsten, molybdenum or vanadium. The salts of heteropolyacids with small cations are more soluble, in comparison to the salts with large cations. Their hydrolytic degradation occurs in strongly alkaline solutions.

Insoluble Ferro cyanides can be precipitated by mixing $Na_4[Fe(CN)_6]$, $K_4[Fe(CN)_6]$ or $H_4[Fe(CN)_6]$ solution with the metal salt solutions. The composition of such precipitates may depend on the acidity, order of mixing and the initial ratio of the reacting components. They are chemically stable in acid solutions up to a concentration of 2 M. Cu and Co ferrocyanides have been found to be radiation resistant. They have found various applications in analytical chemistry and in technological practice, because of their highly selective ion-exchange behavior and chemical and mechanical stability.

Nowadays, enormous literature is available on ion-exchanger, particularly on inorganic ion-exchangers. The literature review shows that the materials used as inorganic ion-exchangers, become an established class of materials of great analytical importance. The first monograph that is also of historical importance, which was written by one of the first research workers devoted to the development of modern inorganic sorbents, [34], who describes the beginning of the rapid development of this subject. Barrer, wrote an excellent monograph on contemporary zeolite and clay minerals. In the 1980s, the monograph of the [13, 15, 16]. made a great contribution to understanding the structure and mechanism of sorption processes on the acidic salts of multivalent metals and hydrous oxides.

Insoluble polybasic acid salts of polyvalent metals have shown a great promise in preparative reproducibility, ion-exchange behavior, and both chemical and thermal behavior. Many metals such as aluminum, antimony, cerium, bismuth, iron, cobalt, lead, tin, tantalum, niobium, titanium, tungsten, thorium, zirconium and uranium have been used for the preparation of ion-exchangers. Also, a large number of anionic species such as phosphates, molybdate, tungstate, antimonate, arsenate, telluride, silicate, ferrocyanide, arsenophosphate, vanadate, arsenomolybdate, arsenosilicate, arsenotungstate, arsenovanadate, phosphosilicate, phosphovanadate, phosphomolybdate, phosphotungstate, vanadosilicate and molybdosilicate *etc.* have been used to prepare inorganic ion-exchangers. The majority of works carried out on zirconium, titanium, tin, niobium and tantalum. The literature survey reveals that a good volume of work has been carried out on single as well as three components (salts of heteropolyacids) inorganic ion-exchangers of both amorphous and crystalline nature. As compared with single salts, inorganic ion-exchange materials of double salts such as based on tetravalent metal acid (TMA) salts often exhibit much better ion-exchange behavior. Khan *et al.* [29] have published their findings on arsenophosphate, arsenosilicate and hexacyanoferrate (II) of tin(IV), and amine and silica based tin(IV) hexacyanoferrate(II); arsenophosphate, silicate and phosphate of antimony(V) cation-exchangers. Different phases of these ion-exchangers have been found selective for K^+, Cd^{2+}, Zr^{4+} and Th^{4+} and some kinetic and thermodynamic parameters for M^{2+}-H^+ exchangers have also been investigated on these cation-exchangers. Numerous inorganic ion-exchangers based on Tin(IV) and Th(V) are listed in Table **1**.

Table 1. Preparation and salient features of numerous inorganic ion-exchangers based on Tin(IV) and Th(V).

S. No.	Material	Nature	Composition	Empirical formula	Selectivity	Ref.
(I)	Exchangers based on Tin (IV)					
1.	Stannic phosphate	Amorphous	P: Sn = 1.25 – 1.50	SnO_2. $0.62P_2O_5.nH_2O$	Cu(II), Zn(II), Ni(II), Co(II), Na(I), Li(I), K(I), Rb(I), Cs(I),	
		Crystalline	---	$SnO_2.P_2O_5.2H_2O$	Zr(IV)	
2.	Stannic EDTA	Amorphous	---	---	---	
		Semi-crystalline	Sn:Mo:P= 40:26.5:0.7	---	Cs^+, Sr^{2+}	
4.	Stannic molybdo-phosphate	---	Sn:Mo:P 1:0.33:2.0	---	---	
	Stannic antimonate	Amorphous	Sb/Sn = 1.0	$SnO_2.Sb_2O_5.nH_2O$	Cu(II), Ni(II), Co(II)	

(Table 1) contd.....

S. No.	Material	Nature	Composition	Empirical formula	Selectivity	Ref.
5.	Tin oxide (hydrated)	Amorphous	---	---	$[Fe(CN)_6]^{4+}$, $[Fe(CN)_6]^{3+}$- SCN^-	
6.	Stannic arsenate	Amorphous	Sn/As = 1.84	---	Pb(II), Fe(III), Al(III), Ga(III), In(III)	
		Crystalline	---	$SnO_2.As_2O_5.2H_2O$	Li(I), Na(I), K(I)	
7.	Stannic selenite	Amorphous	Sn/Se = 1.33 Sn/Se = 1.0	$[(SnO_4)(OH)_2 (SeO_3)_3.6H_2O]$	Li(I), Na(I), K(I), Cu(II), Fe(III), Sc(III), La(III)	
8.	Stannic molybdate	Amorphous	Sn/Mo = 1.0	---	Pb(II)	
9.	Stannic vanado pyrophosphate	Micro-crystalline	---	---	Ag(I), Cu(II), Pb(II), Bi(III), Zr(IV)	
10.	Stannic tungstate	Amorphous	Sn/W = 1.33	---	Co(II), Ba(II), Ni(II), Pb(II), Mn(II), Cu(II), Sr(II)	
11.	Stannic vanadate	Amorphous	Sn/V = 1.0	$[(Sn(OH)_3 V_3O_9.4H_2O]_n$	K(I), Na(I), Li(I)	
12.	Stannic silicate	---	---	---	---	
13.	Stannic ferro-cyanide	Amorphous	Sn/Fe = 3.0	$[(SnO)_3(OH)_3 .HFe (CN).3H_2O]_n$	K(I), Ba(II), Na(I)	
14.	Stannous ferrocyanide	Amorphous	Sn/Fe = 1.0	$[SnO.H_4Fe (CN)_6.2.5H_2O]_n$	Cu(II), Ni(II), Mg(II), Mn(II), Y(III)	
15.	Stannic hexametaphosphae	---	---	---	Ag(I), Pb(II)	
16.	Stannic molybdo arsenate	Amorphous	Sn:Mo:As 2:1:1	---	---	
17.	Stannic arsenophosphate	Amorphous	Sn:As:P 1:1:1	$(SnO_2)_5.(H_3PO_4)_3(H_3AsO_4).nH_2O$	Th(IV), Zr(IV), K(I)	
		Crystalline	---	$Sn(HAsO_4) (HPO_4).H_2O$	---	

(Table 1) contd.....

S. No.	Material	Nature	Composition	Empirical formula	Selectivity	Ref.
18.	Stannic sulfide	Amorphous	---	---	Cu(II)	
19.	Stannic pyrophosphate	Amorphous	Sn:PO_4^{3-} 1:2	---	Zr(IV), Th(IV), Y(III), Bi(III)	
20.	Stannic pyroantimonate	---	---	---	---	
21.	Stannic phosphosilicate	Amorphous	Sn:Si:P 2:2:3	$(SnO_2)_2$ $(SiO_2)_2$ (H_3PO_4). nH_2O	Hg(II)	
22.	Stannic selenopyrophosphate	Amorphous	Sn:Se:PO_4^{3-} 1:1:1	$(15SnO.8OH)$ $(10H_2P_2O_7.O_2$ $HSeO_3).5nH_2O$	Ag(I), Pb(II), Sr(II), Zr(IV)	
23.	Stannic selenophosphate	Amorphous	Sn:Se:P 1:1:1	---	---	
		Crystalline	Sn:Se:P 4:1:6	$[(SnO)_4 (OH)$ $(HSeO_3)$ $(H_2PO_4)_6]_n.4H_2O$	---	
24.	Stannic vanadoarsenate	Amorphous	Sn:V:As 1.94:1.14:1	---	Ba(II)	
25.	Stannic tungstoarsenate	Amorphous	Sn:W:As 12:5:2	---	Ba(II), Cu(II)	
26.	Stannic antimonophosphate	Amorphous	---	---	Pb(II), Ce(III), Sm(III)	
		Crystalline	---	---	Pb(II), Sm(III), La(III)	
27.	Stannic vanadophosphate	Crystalline	---	---	Ba(II), Cu(II)	
28.	Stannic fungstoselenate	Crystalline	Sn:Se:W 7:1:18	$[(SnO_2)_7.HSeO_3$ $(HWO_4)_{18}.45H_2O]$	Th(IV), Ce(IV)	
29.	Stannic vanadotungstate	Amorphous	Sn:V:W 2:1:1	---	Al(III)	[35]
30.	Stannic tungstovanadophosphate	Amorphous	Sn:W:V:P 1:1:1:1	---	---	
31.	Tin(IV) antimonite	Amorphous	Sn:Sb = 2:11	$Sn_2[Sb(OH)_6]_{11}$.$8H_2O$	Pb(II)	
32.	Stannic arsenosilicate	---	---	---	---	[36]
33.	Tin(IV) vanadopyrophosphate	Amorphous	---	---	---	[37]

(Table 1) contd.....

S. No.	Material	Nature	Composition	Empirical formula	Selectivity	Ref.
34.	Stannic hexacyano Ferrate (II)	---	---	---	Tl(I), Ba(II), Pb(II), Ce(IV), Th(IV)	
35.	Tin(IV) sulphosilicate	Crystalline	---	---	---	[38]
36.	Stannic hexacyano ferrate(III)	Amourphous	---	---	---	[39]
37.	Stannic setenoarsenate	Amourphous	Sn:Se:As 1:1:1.02	---	Hg(II)	[40]
38.	Stannic iodophosphate	---	---	---	Hg(II)	[41]
39.	Stannic bortophosphate	Amourphous	---	---	---	[42]
40.	Amine bored stannic hexacyano-ferrate (II)	---	---	---	Cd(II), Cu(II), Hg(II)	[43]
41.	Stannic silicomolybedate					[45]
42.	Silica based stannic hexacyano-ferrate (II)	Amorphous	Sn:Fe:Si 5:4:2	---	---	[44]
43.	Sodium stannosilicate	-----	Sn:Si = 1:1	$Na_2(SnO_2)x$ $(SiO_2)y.zH_2O$	Ag(I)	[46]
44.	Stannic vanadophosphate	Crystalline	Sn:V:P = 3:3:10	-------	------	[47]
(II) Exchangers based on Thorium						
1.	Thorium arsenate	Crystalline	As/Th = 1.53	$Th(HAsO_4)_2.$ H_2O	Li(I)	
2.	Thorium phospahte	Amorphous	P/Th = 1.9-2.1	-------	Pb(II), Fe(III), Bi(III)	[48]
		Crystalline	Th/PO₄ = 0.50	$Th(HPO_4)_2.2H_2$ O	Ca(II), Sr(II), Ba(II)	
		Fibrous	------	$ThO_2.P_2O_5.2H_2$ O	------	
3.	Thorium antimonate	Amorphous	Sb:Th = 3.67: 4.27	------	------	
4.	Thorium tungstate	Crystalline	Th/W = 2.0	$Th(OH)_2$ $(HO_{42}.nH_2O$	Cs(I), K(I), Na(I)	[49]
5.	Thorium molebedate	Crystalline	------	------	Fe(III), Zr(IV)	
		Amorphous	Th/Mo = 0.50	-------	Fe(III), Zr(IV), Pb(II)	

(Table 1) contd.....

S. No.	Material	Nature	Composition	Empirical formula	Selectivity	Ref.
6.	Thorium tellurite	-------	Th:Te = 1:2	-------	Pb(II), Co(II), Cu(II)	
7.	Thorium tungstate	Amorphous	------	-------	Bi(III), Hg(II)	[50]
8.	Thorium oxide	Amorphous	-------	$Th(OH)n.nH_2O$	Na(I), Rb(I), Ca(II), Sr(II)	
9.	Thorium phosphosilicate	Amorphous	Th:P:Si = 1:1.2: 8:1.12	$(ThO_2.H_3PO_4H_4 SiO_4).6H_2O$	Hg^{2+}	
10.	Thorium iodate	-------	Th:I = 1:1	$ThO_2.I_2O_5.nH_2O$	-------	

Organic Ion-exchange Materials

Nowadays, synthetic organic resins are the largest groups of ion-exchangers available in bead (0.5–2 mm diameter) or powdered (5–150 μm) form. The resin matrix or framework, is a random network of hydrocarbon chains. This framework of cation-exchangers contains the ionic groups like: $-COO^-$, $-SO_3^-$, $-AsO_3^{2-}$, $-PO_3^{2-}$, *etc.*, and in anion-exchangers: $-NH_2^+$, $-NH_3^+$, $-S^+$, $-N^+$, *etc.* Thus, ion-exchange resins are cross-linked polyelectrolytes. By cross-linking various hydrocarbon chains, these resins are made insoluble. Mesh width of the matrix, movement of mobile ions, swelling ability, hardness and mechanical durability were determines by the degree of cross-linking. Highly cross-linked resins are less porous, harder, swell less in solvents and more resistant to mechanical degradation. An organic-ion exchange material will expand or swell, when it is placed in a solvent or solution. The degree of swelling depends on both the exchanger and solution/solvent, and is effected by a number of conditions, such as:

- The degree of cross-linking,
- The solvent's polarity,
- A strong or weak salvation tendency of the fixed ion groups,
- The exchange capacity,
- The concentration of the external solution,
- The extent of the ionic dissociation of functional groups.
- The size and extent of the salvation of counter ions,

Low cost, high capacity, wide versatility and wide applicability are the main advantages of synthetic organic ion-exchange resins relative to some synthetic inorganic media. Thermal stabilities and limited radiation are their main limitations. Generally, anion-exchange resins are usually limited to less than 70 ^{0}C temperature, while cation-exchange resins are limited to operational temperatures below 150 °C. Most organic resins will exhibit a severe reduction at a total absorbed radiation dose of 10^9 to 10^{10} rads in their ion exchange capacity (10 - 100% capacity loss).

Chelating Ion-exchange Materials

The use of complexing agent or ligand in the solution to increase the efficiency of separation of cation mixtures using conventional anion or cation-exchange resins is well recognized. However, the use of chelating resins is an alternative mode of application for complex formation. Recently, chelating ion-exchange materials have been developed and their applications are explored [51]. New chelating resins have been prepared from complexions, which on the basis of complex formation were applied for separation of metal ions [52]. The incorporation of ligands on resins prepared number of ion-exchangers [53]. The separation metal ions at trace level was carried by 8-hydroxy quinoline sobbed on porasil [54]. The recovery of precious metal ions Au^{3+} and Pt^{4+}, and ammonium-molybdophosphate was possible by using PAN [1-(2-pyridylazo-2-napthol)] sobbed zinc silicate [56]. The greater selectivity of the chelating ion-exchanger as compared with the conventional type of ion-exchanger is their important feature. Particular metal ion affinity depends mainly on the nature of the chelating group for a certain chelating resin. And the selectivity and stability of the metal complexes formed on the resin largely based on pH conditions.

Intercalation Ion-exchangers

Recently, much interest has been put for the development of intercalation compounds and pillared inorganic materials that can be prepared in the matrix of layered inorganic ion-exchangers by introducing some organic molecules. The ready access to the interior of complexes and large ions due to enlargement in pore sizes and inter layer distance are main advantage of a pillared structure. In radioactive waste cleanup, these are very useful. Moreover, intercalation compounds are very important, amongst the new develop ion exchangers, in the field of separation science and technology. These compounds can be simply prepared by introducing organic molecules or ions in the inorganic ion exchangers matrix. Alumina, clays, kaolin, bentonite, alginic acid, pectin *etc.* have been used in medical science throughout the world as adsorbent. Hence, the neutral polar molecules are put in the sheets of layered insoluble compounds in intercalation process.

A new type of zirconium phosphate was synthesized and characterized by Alberti *et al.* [61] and reported with the name of zirconium phosphate hemihydrate [α - Zr $(HPO_4)_{0.5}$. H_2O]. By pillaring methods, a large number of other new materials of zirconium phosphate have also been prepared. α-Zr(IV)(RPO_3)_2.H_2O$ by phenyl containing $-SO_3H$ groups and γ - $Zr(IV)(PO_4)(H_2PO_4).2H_2O$ intercalated by Alberti and coworkers [62] with crown ether. A detailed description of the intercalation of glycols and alkanols into α -$Zr(HPO_4).H_2O$ has given by U. Costantino [63] and also developed zirconium phosphate-phosphite. Monophenyl; diphenyl and triphenyl bridging pillared zirconium phosphates have prepared by Dines *et al.* [65], by using phenyl disulphonic acids to bridge across the layers. Moreover, they showed that pillared analogous or three dimensional phosphonates can be formed by utilizing α, ω -diphosphonic acids. These studies on amine tin(II) hexacyanoferrate(II) by Varshney *et al.* [66], on tin(IV) diethanol amine by Rawat *et al.* [67], on iron(III) diethanol amine by Singh *et al.* [68], and on zirconium(IV) ethylene diamine by Qureshi *et al.* [69], have also been reported.

In literature, recently developed some new intercalation ion-exchangers have been reported. Hudson *et al.* [70] have reported monoamine intercalation into α-$Sn(HPO_4)_2$. H_2O and investigated the ion exchange behavior in the presence of transition metal ions. The selective separation of Cs^+ have reported by Wang *et al.* on zirconium phenyl diphosphonate phosphate [71]. A new inorganic-organic ion exchanger was synthesized by Chudasama *et al.* [72] by anchoring *p*-chlorophenol to $Zn(WO_4)$. pyridinium tungstoarsenate have reported by Malik *et al.* [73], which was selective for Cs^+ and Rb^+.

'Organic-Inorganic' Composite Ion-Exchanger

Composite materials prepared by the combination of inorganic materials and organic polymers are attractive for the purpose of creating high functional polymeric material. Micromolecular composite are those composites having molecular size bigger than the micro scale. In this discipline, the latest development is the conversion of organic ion exchange materials into hybrid ion exchangers. The hybrid ion exchangers preparation is carried out by organic polymers bindings *i.e.* polyacrylonitrile, polyaniline, polystyrene *etc.* An improvement in a number of properties was shown by these polymer based hybrid ion exchangers *viz* radiation stability, improvement in ion-exchange properties, chemical and mechanical properties, and selective nature for the heavy toxic metal ions. Granulometric nature of them is one of the important properties that makes it more suitable for column applications. Hybrid ion exchangers as three-dimensional porous materials can be prepared, in which layers are as layered compounds or cross linked containing carboxylic acid, sulphonic acid or amino groups. So far, the hybrid ion-exchangers prepared are styrene supported zirconium phosphate, pyridinium-tungstoarsenate [78], zirconium (IV) sulphosalisylo phosphate [79]. Recently, in some of these laboratories organic-inorganic composite ion-exchange materials have been synthesized, *i.e* polypyrrole Th(IV) phosphate [80], polyaniline Sn(IV) phosphate [81], have reported by Khan *et al.* Similarly, for the selective separation of Pb^{2+}, Hg^{2+}, Cd^{2+} polyaniline Sn(IV) arsenophosphate [82], Hg^{2+}, polystyrene Zr(IV) tungstophosphate [83], and Hg^{2+} polypyrrole/polyantiminc acid were used and ion- exchange and adsorption of pesticide [84], have also investigated on these materials. Such type of ion-exchange materials has been synthesized by Beena Pandit *et al.*, *i.e.* p-chlorophenol Zr(IV) and tungstate o-chlorophenol Zr(IV) tungstate [85]. Varshney *et al.* [86], investigated pectin based Th(IV) phosphate with great analytical applications. These materials can be used as ion selective electrodes and ion-exchange membranes [87, 88]. Gupta *et al.* [89], synthesize Polyaniline Zr(IV) tungstophosphate, which was applied for the separation of La^{3+} and UO_2^{2+}.

Table 2. General comparison of organic and inorganic ion exchangers

Properties	Inorganic Ion exchanger	Organic ion exchanger	Comments
Selectivity	Available	Available	For some applications, inorganics can be much better than organics such as cesium removal, owing to their greater selectivity.
Chemical stability	Fair to good	Good	for any given pH range the specific organics and inorganics are available
Thermal Stability	Good	Fair to poor	For long term stability inorganics are especially good.
Radiation stability	Good	Fair to poor	In combination, organics are very poor with high temperatures and oxygen.
Regeneration	Good	Good	Inorganics are mostly sorption based, which limits regeneration.
Exchange capacity	Low to high	Low to high	The exchange capacity, the experimental conditions and its chemical environment of the ion being removed will be a function of its nature.
Cost	Medium to High	Low to high	Organics are costlier than most of the common inorganic.
Mechanical strength	Variable	Good	Outside a limited pH range, inorganics may be soft or brittle or may break down.
Availability	Good	Good	From a number of commercial source, both types are available.
Handling	Fair	Good	Inorganic may be brittle, which are more friable angular particles, however organics are generally tough spheres.
Immobilization	Good	Good	Organics in a variety of matrices can be immobilized or can be incinerated; Inorganic to equivalent mineral structures can be converted.

(Table 2) contd.....

Properties	Inorganic Ion exchanger	Organic ion exchanger	Comments
Ease of use	Good	Good	Both types are easy to use in column or batch applications, if available in a granulated form.

Applications of Ion-exchange Materials

Ion-exchangers have a wide variety of domestic, industrial, laboratory and governmental applications. Some better granulometric properties have shown by the composite ion-exchangers that facilitates its stability in column operations especially for filtration, separation and preconcentrating of ionic species. In regeneration of exhausted beds also make the column operation suitable and more convenient. These hybrid ion-exchangers have higher stabilities, good ion-exchange capacity, selectivity for specific heavy metal ions, and reproducibility which indicates its useful environmental applications. As in general in following disciplines, these materials have their applications:

• Separation and preconcentration of metal ions [90]
• Water softening [89]
• Nuclear separations [91]
• Synthesis of organic pharmaceutical compounds [92]
• Nuclear medicine [93]
• Catalysis [94]
• Electrodialysis [95]
• Redox systems
• Hydrometallurgy [96]
• Ion-exchange membranes
• Effluent treatment
• Chemical and biosensors
• Ion memory effect [97]
• Ion-selective electrodes
• Ion-exchange fibers [98]
• Proton conductors [99]

Ion-exchange Membrane

"Membrane" precise and complete definition is difficult, because any definition which cover all the facets of membrane behavior is unavailable. A membrane is a structure or phase interposed between two compartments or phases which completely prevents or obstructs grass mass moment, but permits passage between the two adjacent compartments or phases with various degree of restriction and acting as a physico-chemical machine as explained by Sollner. It is described usually as heterogeneous phase in simple terms, which acting as barrier for the flow of ionic and molecular species. To indicate the external physico-chemical performance and internal physical structure, the term heterogeneous has been used. Therefore, in general most of the membranes are considered as heterogeneous, despite the fact that membranes are conventionally prepared from coherent gels which is called homogeneous.

In a mass separation process the usefulness of a membrane is determined by its selectivity, by its mechanical, chemical and thermal stability and as well as its overall mass transport rate. The membrane chemical nature is of prime importance when components with less or more identical in molecular dimensions and similar electrical or chemical properties have to be separated. To a large extent its useful lifetime determines the thermal, chemical and mechanical stability of the membrane. In pressure driven process the mechanical properties of a membrane are of special significance such as ultra-filtration, reverse osmosis, *etc*. Ideally, a membrane should not vary its valuable properties when the composition of feed solution is altered or when it is derived out drastically. Generally, for economic reasons, the flow rate of the permeable components must be as high as possible through a given membrane area under a given driving force. The use of membranes in mass separation processes significantly expand their present applications. Membranes having specific transport properties, higher flux rates and longer lifetimes are required.

Ion-selective Electrodes (ISE)

These types of electrodes are mainly membrane-based devices; which consist of ion-conducting materials to separate the sample from the inside of the electrode. The membrane is usually non-porous, water insoluble and mechanically stable. The membrane composition is designed to yield the potential of the ion of interest (*via* selective binding processes). The purpose is to find membranes that leaves co-ions behind and will selectively bind the analyte ions. To impart high selectivity, the membrane materials which possessing different ion-recognition properties have been developed.

The Ion-Selective Electrodes (ISEs) are commonly known as "Electrochemical Sensors" or "Ion Sensors". In the past decades, the ion-selective electrodes exhibited the typical behavior of expansion, which is followed by consolidation. The new electrodes have been rapidly grown for ion activity measurement, new materials and new formats of construction.

Physico-chemical Properties of Ion-selective Electrodes

To study the electrode characteristics, the following parameters were investigated:

membrane potential or electrode response, slope response curve and lower detection limit, working pH range, response time, *etc.*

Electrode Response or Membrane Potential

The determination of potentials depends on the use of ion-selective electrodes. Directly, the potentials cannot be determined however from *e.m.f.* values, it can be easily derived for the complete electrochemical cells. Electrochemical cell comprises of two reference electrodes and the membrane separating solutions 1 and 2. Equation (4) expressed the membrane potential:

$$E_m = \frac{RT}{Z_A F}\left[\ln\frac{[a_A]_2}{[a_A]_1} - (Z_y - Z_A)\int_1^2 t_y d\ln a\pm\right] \tag{4}$$

Where $[a_A]_1$ and $[a_A]_2$ is the activities of the counter ions in the solution 1 and 2, A = counter ion, Y = co-ion, Z = charge on ions, $a_\pm$ = mean ionic activity of the electrolyte, t_Y = transference number of co-ions in the membrane phase. From the equation, it is quite evident that 'E_m' is the sum of Donnan Potential and diffusion. The right-hand side consists of two terms in equation 4, the second term denotes the diffusion potential and first term represents the thermodynamic limiting value.

In general, one solution concentration (say 1) is kept constant (usually 0.1 M) known as reference/internal solution and an SCE is dipped in this reference solution which is termed as an internal reference electrode. The one compact unit of the membrane with internal reference electrode and internal solution as a whole is called as membrane electrode. This membrane electrode is then dipped in another solution known as 2, usually referred as test solution or external solution, having an external reference electrode. This potentiometric cell *e.m.f.* is given by the following expression:

$$E_{cell} = E_{SCE} + E_{L(2)} + E_m + E_{L(1)} - E_{SCE} \tag{5}$$

As the internal solution activity is kept constant and the $E_{L(1)}$ and $E_{L(2)}$ values are also almost constant, therefore the term in parenthesis may be taken as a constant, ($E°$). In addition, the $E_{L(1)}$ and $E_{L(2)}$ values are negligible (due to salt bridge). The equation (5) reduces to-

$$E_{cell} = E° + \frac{RT}{Z_A F}\ln[a_A]_2 \tag{6}$$

From equation (6), it is quite clear that the cell potential would change by changing the activity or concentration of the cation in test or external solution 2. The value of $RT/Z_A F$ comes out to be $0.059/Z_A$ volts at 25 °C. If the slope of a plot between log activity and cell potential comes out to be $0.059/Z_A$ volts, the membrane is said to give Nernstian response, and these slopes are called as Nernstian slope and plots are called Nernst plots.

Selectivity Coefficients

One of the most important factors of ion-selective electrodes (ISEs) is selectivity coefficient, on the basis of which electrode potential applications in a given system can be predicted. Generally, ISEs are membrane-based devices, consisting of ion-conductive materials. Inside the electrode, a filling solution is taken into consideration from inside the electrode, which contain the ion of interest at a constant activity. The membrane is usually water insoluble, non-porous and mechanically stable.

For a particular ion, no electrode is absolutely selective. Thus, the selectivity depends on selectivity coefficients. So, for the analytical chemist it is important to realize the significance of selectivity coefficient of a specific electrode. Problems might be created from inconsistent in the values of selectivity coefficient. For the determination of selectivity coefficient various methods have been suggested, however, in two main groups it falls, namely- (1) Separate-solution method and (2) Mixed-solution method.

Separate-solution Methods

In this method, the electrode potential E_i and E_j are measured separately in solutions containing 'i' and 'j', whose activities are a_i and a_j, respectively and are given by the following equations:

$$E_j = E^o + (2.303 \, RT/Z_i f) \log K_{ij}^{pot} \, a_j \tag{7}$$

$$E_i = E^o + (2.303 \, RT/Z_i f) \log a_i \tag{8}$$

K_{ij}^{pot} can be calculated either with equal potential or equal activity method. In both cases, the electrode standard potentials are tactually assumed equal in the presence of ion 'i' and in the presence of ion 'j'. According to the equal activities method, the solution of ion 'i' and 'j' at same concentration are prepared and the potentials are measured. From the equations (7) and (8), we get

$$logK_{ij}^{pot} = \frac{E_j - E_i}{2.303RT/Z_iF} \ log \frac{a_i}{(a_j)^{Z_i/Z_j}} \tag{9}$$

The term $2.303RT/Z_iF$ is the Nernst plot slope. From Nernstian behavior, most of the solid membranes show deviation, therefore from the theoretical slope $2.303RT/Z_iF$, the experimental slope (S) usually differs. Thus, instead of Nernstian slope it is a practice to use 'S' for the calculation of K_{ij}^{pot}. Equation (9) takes the form.

$$logK_{ij}^{pot} = \frac{E_j - E_i}{S} \ log \frac{a_i}{(a_j)^{Z_i/Z_j}} \tag{10}$$

Thus, selectivity coefficient K_{ij}^{pot} can be calculated by using equation (10). To determine the selectivity coefficients, the separate solution technique is simple and a number of K_{ij}^{pot} values on the basis of different potentials and activities to be measured.

Mixed Solution Methods

In this mixed, the electrode potentials of the solutions comprising of both the primary ion 'i' and the interfering ion 'j' are measured. The procedure for selectivity coefficients determination by Mixed Solution Method is given as.

Procedure

In this method, firstly in solution of only primary ion 'i', the potentials of the electrode E_i and E_{ij} are measured and then measured for a mixture of primary and interfering ion 'j', respectively.

$$K_{ij}^{pot} = log \ [10^{(E_{i+j} - E_{i/m})} - 1] + loga_i - Z_i/Z_j \ log \ a_j \tag{11}$$

Response Time

The promptness of the response of the electrode is an important factor that recommends the use of ISEs or membrane electrode. The time needed to attain equilibrium value is the response time of an ion-selective electrode. However, the response time interpretation varies from a one group to others.

Effect of pH

With polymer binders like PVC, the membrane electrode response is changing in the pH value of the solutions. So, the effect of pH study is necessary and for accurate measurements, the favorable working range of pH has to be estimated.

Since, in membrane electrodes for the construction of the membrane one or other polymeric binder was used, therefore it is necessary to finds out the effect of pH on the response of electrode. For measurements of the ions provided by the solution, the electrode could be safely used in the range, where the response of electrode does not change with the pH, and is used as working pH range of the electrode.

Life Span of Membrane Electrode

In continuous service, Ion-exchanger membrane electrodes can be utilized for one to three months. This short lifetime of the ion- exchanger may be related to the gradual loss through the porous membrane. The ion-exchanger and the membrane internal filling solution are replaced when the response of the electrode becomes drifts or noisy.

So, to find out the life time of the electrode, the response of the electrode was noted every week and the response curve was drawn. Usually at initial period of the response data, some changes in the response are noted but after the week or so, the electrode response over a period of time remains constant. After this period, the electrode starts behaving unpredictable, therefore this electrode cannot be used for further any measurements. The period in which the electrode rcsponse is almost equal to constant can be called as life of electrode. The life of studied membranes ranges from 45 to 120 days.

CONSENT FOR PUBLICATION

Not applicable.

CONFLICT OF INTEREST

The author (editor) declares no conflict of interest, financial or otherwise.

ACKNOWLEDGEMENT

We are thankful to the Center of Excellence for Advanced Materials Research (CEAMR), Department of Chemistry, Faculty of Science, King Abdulaziz University, Jeddah, Saudi Arabia for this study.

REFERENCES

[1]   Sharma, S.C. *Composite Materials*; Narosa Pub. House: New Delhi, India, **2000**.

[2]   The Second Book of Moses. Exodus, Ch. 15, P. 25.

[3]   Aristotle, Works, about 330 BC. *Clarenndon Press London,* **1927**, *7*, 933b.

[4]   Thompson, H.S. On the absorbent power of soils. *J. R. Agric. Soc. Engl.,* **1650**, *11*, 68-74.

[5]   Way, J.T. On the power of soils to absorb manure. *J. R. Agric. Soc. Engl.,* **1850**, *11*, 313-379.

[6]   Eichorn, E. On the action of dilute salt solutions on silicates. *Pogg. Ann.,* **1858**, *105*, 126-133.

[7]   Gans, R. Geolites and sililar compounds, their constitution and significance for thechnology and agriculture. Jahrb Kgl. Preuss. *Geol. Landesanstalt (Berlin),* **1905**, *25*, 179.

[8]   Girardi, F.; Pietra, R.; Sabbioni, E. Radiochemical separations by retention on ionic precipitate adsorption tests on 11 materials. *J. Radioanal. Chem.,* **1970**, *5*, 141.
        [http://dx.doi.org/10.1007/BF02513708]

[9]   Hellferich, F. *Ion-Exchange*; McGraw-Hill: New York, **1962**.

[10]   Gaines, G.L.; Thomas, H.C. Adsorption studies on clay minerals. 2. a formulation of the thermodynamics of exchange adsorption. *J. Chem. Phys.,* **1953**, *21*, 714-718.
        [http://dx.doi.org/10.1063/1.1698996]

[11]   Alberti, G.; Costantino, U.; Alluli, S.; Massucci, M.A. Crystalline insoluble acid salts of tetravalent metals. 14. forward and reverse sodium-potassium ion exchange isotherms on crystalline zirconium phosphate. *J. Inorg. Nucl. Chem.,* **1973**, *35*, 1339-1346.
        [http://dx.doi.org/10.1016/0022-1902(73)80208-8]

[12]   Alberti, G.; Costantino, U.; Pelliccioni, M. Crystalline insoluble acid salts of tetravalent metals. 13. ion exchange of crystalline zirconium phosphate with alkaline earth metal ions. *J. Inorg. Nucl. Chem.,* **1973**, *33*, 1327-1338.
        [http://dx.doi.org/10.1016/0022-1902(73)80207-6]

[13]   Clearfield, A.; Tuhtar, D.A. Mechanism of ion-exchange in zirconium-phosphates. 15. effect of crystallinity of exchanger on li^+-h^+ exchange of alpha-zirconium phosphate. *J. Phys. Chem.,* **1976**, *80*, 1296-1301.
        [http://dx.doi.org/10.1021/j100553a007]

[14]   Kullberg, L.; Clearfield, A. Mechanism of ion-exchange in zirconium-phosphates. 31. thermodynamics of alkali-metal ion-exchange on amorphous Zrp. *J. Phys. Chem.,* **1981**, *85*, 1578-1584.
        [http://dx.doi.org/10.1021/j150611a024]

[15]   Clearfield, A.; Kallnius, J.M. Mechanism of ion-exchange in zirconium-phosphates. 13. exchange of some divalent transition-metal ions on alpha-zirconium phosphate. *J. Inorg. Nucl. Chem.,* **1976**, *38*, 849-852.
        [http://dx.doi.org/10.1016/0022-1902(76)80369-7]

[16]   Clearfield, A.; Day, G.A.; Ruvaracr, A.; Milonjic, S. Ibid. On the mechanism of ion-exchange in zirconium-phosphates. 29. calorimetric determination of heats of k^+-h^+ exchange with alpha-zirconium phosphate. *J. Inorg. Nucl. Chem.,* **1981**, *43*, 165-169.
        [http://dx.doi.org/10.1016/0022-1902(81)80454-X]

[17]   Nancollas, G.H.; Tilak, B.V.K. Thermodynamics of cation exchange on semi-crystalline zirconium phosphate. *J. Inorg. Chem.,* **1969**, *31*, 3643.

[18]   Harkin, G.P.; Nancollas, G.H.; Paterson, R. The exchange of caesium and hydrogen ions on zirconium phosphate. *J. Inorg. Chem.,* **1964**, *26*, 305-310.

[19]   Larsen, E.M.; Vissers, D.R. The exchange Li^+, Na^+, K^+ with H^+ on zirconium phosphate. *J. Phys. Chem.,* **1960**, *64*, 1732-1736.
        [http://dx.doi.org/10.1021/j100840a031]

[20]   Nancollas, G.H.; Reid, D.S. Calorimetric studies of ion exchange on hydrous zirconia. *J. Inorg. Nucl. Chem.,* **1969**, *31*(1), 213-218.

[21]   Rawat, J.P.; Muktawat, K.P.S. Thermodynamics of ion-exchange on ferric antimonite. *J. Inorg. Nucl. Chem.,* **1981**, *43*, 2121-2128.
        [http://dx.doi.org/10.1016/0022-1902(81)80562-3]

[22]   Rawat, J.P.; Thind, P.S. Thermodynamics of cation-exchange in tantalum arsenate. *J. Indian Chem. Soc.,* **1980**, *57*, 819-822.

[23]   Varshney, K.G.; Singh, R.P.; Sharma, U. Thermodynamics of Ba^{2+}-H^+ and Sr^{2+}-H^+ exchange on zirconium(iv) phosphosilicate cation exchanger. *Proc. Indian Natl. Sci. Acad.,* **1985**, *51*, 726-734.

[24]   Varshney, K.G.; Singh, R.P.; Sharma, U. Thermodynamic of Ca(II)-H(I) and Mg(II)-H(I) exchanger on zirconium(iv) phosphosilicate cation exchanger. *Coll. Surf; (A).,* **1985**, *16*, 207-218.

[25]   Varshney, K.G.; Singh, R.P.; Rani, S. Thermodynamics of Ca^{2+}-H^+ and Mg^{2+}-H^+ exchanges on antimony(v) silicate cation exchanger. *Acta. Chem. Hung.,* **1984**, *115*, 403-413.

[26]   Herman, R.G.; Clearfield, A. Crystalline cerium(iv) phosphates. 2. ion-exchange characteristics with alkali-metal ions. *J. Inorg. Nucl. Chem.,* **1976**, *38*, 853-858.
       [http://dx.doi.org/10.1016/0022-1902(76)80370-3]

[27]   Rawat, J.P.; Singh, B. ion-exchange equilibria between alkali-metals and hydrogen-ions on iron(iii) antimonate, an inorganic-ion exchanger. *Bull. Chem. Soc. Jpn.,* **1984**, *57*, 862-865.
       [http://dx.doi.org/10.1246/bcsj.57.862]

[28]   Singh, R.P.; Varshney, K.G.; Rani, S. Adsorption thermodynamics of carbofuran on sandy clay loam and silt loam soils. *Ecotoxicol. Environ. Saf.,* **1985**, *10*(3), 309-313.
       [http://dx.doi.org/10.1016/0147-6513(85)90078-8] [PMID: 4092645]

[29]   Khan, A.A.; Niwas, R.; Bansal, O.P. A study on the adsorption behaviour of carbofuran on tin(IV) arsenophosphate H^+, Na^+ and Ca^{2+} forms. *J. Indian Chem. Soc.,* **1999**, *76*, 44-46.

[30]   Nachod, F.C.; Wood, W. The reaction velocity of ion exchange. *J. Inorg. Am. Chem. Soc.,* **1944**, *66*, 1350-1384.

[31]   Boyd, G.E.; Adamson, A.W.; Myers, L.S., Jr The exchange adsorption of ions from aqueous solutions by organic zeolites; kinetics. *J. Am. Chem. Soc.,* **1947**, *69*(11), 2836-2848.
       [http://dx.doi.org/10.1021/ja01203a066] [PMID: 20270838]

[32]   Sinha, P.K.; Panicker, P.K.; Amalraj, R.V.; Krishnasamy, V. Treatment of radioactive liquid waste containing cesium by indigenously available synthetic zeolites - a comparative-study. *Waste Manag.,* **1995**, *15*, 149-157.
       [http://dx.doi.org/10.1016/0956-053X(95)00014-Q]

[33]   Sinha, P.K.; Lal, K.B.; Panicker, P.K.; Krishnasamy, V. A comparative study on indigenously available synthetic zeolites for removal of strontium from solutions by ion-exchange. *Radiochim. Acta,* **1996**, *73*, 157-163.
       [http://dx.doi.org/10.1524/ract.1996.73.3.157]

[34]   Amphlett, C.B. Inorganic ion exchangers. *Elsevier,* **1964**, *541*, 3723.

[35]   Varshney, K.G.; Khan, A.A.; Maheshwari, A.; Anwar, S.; Sharma, U. Synthesis of a new thermally stable sn(iv) arsenosilicate cation exchanger and its application for the column chromatographic-separation of metal-ions. *Indian J. Chem. Technol.,* **1984**, *22*, 99-103.

[36]   Varshney, K.G.; Gupta, U. Tin(iv) antimonate as a lead-selective cation exchanger - synthesis, characterization, and analytical applications. *Bull. Chem. Soc. Jpn.,* **1990**, *63*, 1515-1520.
       [http://dx.doi.org/10.1246/bcsj.63.1515]

[37]   Nabi, S.A.; Rehman, N.; Farooqui, W.U.; Usmani, S. Synthesis, ion-exchange properties and analytical applications of a novel tin(iv)-sulphosalicylate exchanger. *Indian J. Chem.,* **1995**, *34*, 317-319.

[38]   Jain, A.K.; Singh, R.P.; Bala, C. Preparation and ion-exchange properties of stannic hexacyanoferrate(iii). *Chem. Anal.,* **1985**, *30*, 255-261.

[39]   Nabi, S.A.; Rao, R.A.K.; Siddiqui, A.R. Synthesis, ion-exchange properties and analytical applications of stannic selenoarsenate - comparison with other heteropolyacid salts. *J. Anal. Chem.,* **1982**, *311*,

503-506.

[40]   Nabi, S.A.; Siddiqui, Z.M.; Rao, R.A.K. Studies on Stannic Selenoarsenate (II). Separation of Uranium from numerous metal ions. *Sep. Sci. Technol.,* **1982**, *17*, 1681-1697.
[http://dx.doi.org/10.1080/01496398208055650]

[41]   Nabi, S.A.; Siddiqui, W.A. Sorption studies in dmso-nitric acid and dmso-acetic acid systems and separation of metal-ions on tin(iv)-iodophosphate. *J. Liq. Chromatogr.,* **1985**, *8*, 1159-1172.
[http://dx.doi.org/10.1080/01483918508067134]

[42]   Nabi, S.A.; Siddiqui, Z.M.; Farooqui, W.U. Synthesis, ion-exchange properties, and analytical applications of a novel tin(iv)-edta ion-exchanger. *Ibid,* **1982**, *55*, 2642-2641.

[43]   Varshney, K.G.; Khan, A.A.; Jam, J.B.; Varshney, S.S. Synthesis and ion-exchange properties of amine tin(iv) hexacyanoferrate(ii) and its use in the separation of cadmium(ii) from zinc(ii), manganese(ii), magnesium(ii) and aluminum(iii). *Indian J. Chem.,* **1982**, *21A*, 398-401.

[44]   Varshney, K.G.; Khan, A.A.; Jam, J.B. Varshney, S.S. Synthesis, Ion exchange behaviours and composition of silica based Sn(IV)hexacynoferrate(II). *Indian J. Chem. Technol.,* **1981**, *19*, 457-480.

[45]   Nabi, S.A.; Khan, A.M. Synthesis, ion exchange properties and analytical applications of stannic silicomolybdate: Effect of temperature on distribution coefficients of metal ions. *React. Funct. Polym.,* **2006**, *66*, 495-508.
[http://dx.doi.org/10.1016/j.reactfunctpolym.2005.10.002]

[46]   Rawat, J.P.; Ansari, A.A. Synthesis and ion exchange properties of sodium stannosilicate: A silver selective inorganic ion exchanger. *Bull. Chem. Soc. Jpn.,* **1990**, *63*, 1521-1525.
[http://dx.doi.org/10.1246/bcsj.63.1521]

[47]   Qureshi, M.; Kaushik, R.C. Synthesis and ion-exchange properties of crystalline stannic vanadophosphate. *Anal. Chem.,* **1977**, *49*, 165-168.
[http://dx.doi.org/10.1021/ac50009a049]

[48]   De, A.K.; Chowdhury, K. studies on thorium phosphate ion exchanger i. synthesis, properties and ion-exchange behaviour of thorium phosphate. *J. Chroniatogr.,* **1974**, *101*, 63-72.
[http://dx.doi.org/10.1016/S0021-9673(01)94731-4]

[49]   Qureshi, M.; Nabi, S.A. Synthesis and ion-exchange properties of thorium tungstate: separation of La^{3+} from Ba^{2+}, Sr^{2+}, Ca^{2+}, and Y^{3+} and of VO^{2+} from Fe^{3+} and Mn^{2+}. *J. Chem. Soc., (A).,* **1971**, 139-143.

[50]   De, A.K.; Chowdhury, K. Studies on synthetic inorganic ion exchangers-V: preparation, properties and ion-exchange behaviour of amorphous and crystalline thorium tungstate. *Talanta,* **1976**, *23*(2), 137-140.
[http://dx.doi.org/10.1016/0039-9140(76)80036-7] [PMID: 18961820]

[51]   Rawat, J.P.; Alam, M.; Aziz, H.M.A.; Singh, B. Zirconium–Bis(triethylamine): A New Class of Chelate Ion and Anion Exchanger. *Bull. Chem. Soc. Jpn.,* **1987**, *60*, 2619-2626.
[http://dx.doi.org/10.1246/bcsj.60.2619]

[52]   Khan, A.; Asiri, A.M.; Rub, M.A.; Azum, N.; Khan, A.A.P.; Khan, I. Review on composite cation exchanger as interdicipilinary materials in analytical chemistry. *Int. J. Electrochem. Sci.,* **2012**, *7*, 3854-3902.

[53]   Vernon, F.; Eccles, H. Chelating ion-exchangers containing n-substituted hydroxylamine functional groups: Part III. Hydroxamic acids. *Anal. Chim. Acta,* **1976**, *82*, 369-375.
[http://dx.doi.org/10.1016/S0003-2670(01)84614-6]

[54]   Jezorek, J.R.; Frieser, H. Metal-ion chelation chromatography on silica-immobilized 8-hydroxyquinoline. *Anal. Chem.,* **1979**, *51*, 366-373.
[http://dx.doi.org/10.1021/ac50039a011]

[55]   Hern, J.L. Report 1976, W79 00431, OWRT A030 WVA (2) Order No. PO 200179, pp. 70 (Eng.) Avail NTR From Gov. Rep. Announce Index (US) 79 (5), (1979). *CA (Edinb.),* **1979**, *91*(2), 73054z.

[56] Rawat, J.P.; Iqbal, M. Separation and Recovery of Some Metal Ions Using Pan Sorbed Zinc Silicate as Chelating Ion Exchanger. *J. Liq. Chromatogr.,* **1980**, *3*, 591-603.
[http://dx.doi.org/10.1080/01483918008059678]

[57] Sebesta, F.; Stefula, V. Composite ion exchanger with ammonium molybdophosphate and its properties. *J. Radioanal. Nucl. Chem.,* **1990**, *140*(1), 15-21.
[http://dx.doi.org/10.1007/BF02037360]

[58] Singh, D.K.; Darbari, A. Ligand exchange separation of metal ions on tetracycline hydrochloride sorbed on alumina. *Chromatographia,* **1986**, *22*(1), 88-90.
[http://dx.doi.org/10.1007/BF02257305]

[59] Saiqam, I. *Ph.D Thesis, D.C.E., Delhi (India),* **2000**, 84.

[60] Gupta, A.P.; Varshney, P.K. Studies on tetracycline hydrochloride sorbed zirconium tungstophosphate; La(III)-selective chelating ion exchanger. *React. Funct. Polym.,* **1996**, *31*, 111-116.
[http://dx.doi.org/10.1016/1381-5148(96)00045-4]

[61] Alberti, G.; Costantino, U.; Millini, R.; Vivani, R. Preparation, Characterization, and Structure of α-Zirconium Hydrogen Phosphate Hemihydrate. *J. Solid State Chem.,* **1994**, *113*, 289-295.
[http://dx.doi.org/10.1006/jssc.1994.1373]

[62] Alberti, G.; Casciola, M.; Dionigi, C.; Vivani, R. *Proceedings of International Conference on Ion-Exchange, ICIE '95*, Takamtsu, Japan**1995**.

[63] Costantino, U. *Inorganic Ion-Exchange Materials*; Clearfield, A., Ed.; CRC Press Inc.: Boca Raton, Florida, **1982**, p. 111.

[64] Clearfield, A.; Tindwa, R.M. On the mechanism of ion exchange in zirconium phosphates—XXI Intercalation of amines by α-zirconium phosphate. *J. Inorg. Nucl. Chem.,* **1979**, *41*, 871-878.
[http://dx.doi.org/10.1016/0022-1902(79)80283-3]

[65] Dines, N.B.; Giocomo, P.D.; Callahan, K.P.; Griffith, P.C.; Lane, R.H.; Cooksey, R.E. Chemically Modified Surfaces in Catalysis and Electrocatalysis *ACS Symposium Series 192, Ch. 12, Washington, D.C.,* **1982**.

[66] Varshney, K.G.; Naheed, S. Amine Sn(II) hexacyanoferrate(II) as an inorganic ion-exchanger. *J. Inorg. Nucl. Chem.,* **1977**, *39*, 2075-2078.
[http://dx.doi.org/10.1016/0022-1902(77)80550-2]

[67] Rawat, J.P.; Iqbal, M.; Alam, M. A New Modified Inorganic-Ion Exchanger For The Separation Of Anions. *Ann. Chim.,* **1985**, *75*, 87-99.

[68] Singh, D.K.; Bhatnagar, R.R. Darbari, A. Synthesis and Analytical Applications of a new chelating ion exchanger. *Indian J. Chem. Technol.,* **1986**, *24*, 25.

[69] Qureshi, S.Z.; Ahmed, I.; Khayer, M.R. Synthesis and physical studies on a new anion exchange material: Zirconium (IV) ethylenediamine and its application to the separation of $MoO_4{}^{2-}$ from other anionic species. *Ann. Chim. Sci. Mat.,* **1999**, *14*, 53-543.

[70] Hudson, M.J.; Castellon, E.R.; Sylvester, P. New Development in Ion-Exchange. *ICIE '91, Tokyo, Japan,* **1991**, p. 129.

[71] Wang, J.D.; Clearfield, A.; Pen, G.Z. Mat. Preparation of layered zirconium phosphonate/phosphate, zirconium phosphonate/phosphite and related compounds. *Chem. Phys.,* **1993**, *35*, 208-216.

[72] Tandon, S.; Pandit, B.; Chudesama, U. A new inorgano-organic ion exchanger; p-chlorophenol anchored to zirconium tungstate. *Transition. Met. Chem.,* **1996**, *21*, 7-10.
[http://dx.doi.org/10.1007/BF00166003]

[73] Malik, W.U.; Srivastava, S.K.; Kumar, S. Ion-exchange behaviour of pyridinium tungstoarsenate. *Talanta,* **1976**, *23*(4), 323-325.
[http://dx.doi.org/10.1016/0039-9140(76)80202-0] [PMID: 18961863]

[74]   Singh, D.K.; Darbari, A. Synthesis, ion-exchange properties and analytical applications of anilinium tin(IV) phosphate. *Bull. Chem. Soc. Jpn.,* **1988**, *61*, 1369-1373.
        [http://dx.doi.org/10.1246/bcsj.61.1369]

[75]   Singh, D.K.; Mehrotra, P. Synthesis, Ion-Exchange Properties, and Analytical Applications of Anilinium Zirconium(IV) Phosphate. *Bull. Chem. Soc. Jpn.,* **1990**, *63*, 3647-3652.
        [http://dx.doi.org/10.1246/bcsj.63.3647]

[76]   Nabi, S.A.; Islam, A.; Rahman, N. Synthesis, ion exchange properties and analytical applications of a semicrystalline ion exchange material: Zirconium (IV) sulphosalicylate. *Ann. Chim. Sci. Mat.,* **1997**, *22*, 463-473.

[77]   Islam, A. *Ph.D. Thesis, A.M.U., Aligarh (India),* **2000**, 98.

[78]   Malik, W.U.; Srivastava, S.K.; Kumar, S. Ion-exchange behaviour of pyridinium tungstoarsenate. *Talanta,* **1976**, *23*(4), 323-325.
        [http://dx.doi.org/10.1016/0039-9140(76)80202-0] [PMID: 18961863]

[79]   Chetverina, R.B. Effect of some conditions of synthesis of zirconium phosphate modified by sulfosalicylic acid on its ion exchange properties. *Zh. Prikl. Khim.,* **1977**, *50*, 188-190.

[80]   Khan, A.A.; Alam, M.M. Inamuddin. Preparation, characterization and analytical applications of a new and novel electrically conducting fibrous type polymeric–inorganic composite material: polypyrrole Th (IV) phosphate used as a cation-exchanger and Pb (II) ion-selective membrane electrode. *Mater. Res. Bull.,* **2005**, *40*, 289-305.
        [http://dx.doi.org/10.1016/j.materresbull.2004.10.014]

[81]   Khan, A.A.; Alam, M.M. Determination and separation of Pb^{2+} from aqueous solutions using a fibrous type organic-inorganic hybrid cation-exchange material: Polypyrrole thorium (IV) phosphate. *React. Funct. Polym.,* **2005**, *63*, 119-133.
        [http://dx.doi.org/10.1016/j.reactfunctpolym.2005.02.001]

[82]   Niwas, R.; Khan, A.A.; Vershney, K.G. Synthesis and ion exchange behaviour of polyaniline Sn(IV) arsenophosphate: a polymeric inorganic ion exchanger. *Coll. Sur. (A),* **1999**, *150*, 7-14.
        [http://dx.doi.org/10.1016/S0927-7757(98)00843-7]

[83]   Niwas, R.; Khan, A.A.; Vershney, K.G. Preparation and properties of styrene supported zirconium(IV) tungstophosphate: a mercury(II) selective inorganic–organic ion exchanger. *Indian J. Chem.,* **1998**, *37A*, 469-472.

[84]   Khan, A.A.; Niwas, R.; Alam, M.M. Ion-exchange kinetics on styrene supported zirconium(IV) tungstophosphate: An organic-inorganic type cation exchanger. *Indian J. Chem. Technol.,* **2002**, *9*, 256-260.

[85]   Pandit, B.; Chudasma, U. Synthesis, characterization and application of an inorgano organic material:p-chlorophenol anchored onto zirconium tungstate. *Bull. Mater. Sci.,* **2001**, *24*, 265-271.
        [http://dx.doi.org/10.1007/BF02704920]

[86]   Varshney, K.G.; Gupta, P.; Agrawal, A. *22nd National Conference in Chemistry '03, Indian Council of Chemists, I.I.T., Roorkee,* **2003**.

[87]   Khan, A.A.; Alam, I.M.M. Determination and separation of Pb^{2+} from aqueous solutions using a fibrous type organic-inorganic hybrid cation-exchange material: polypyrrole thorium(IV) phosphate. *React. Funct. Polym.,* **2005**, *63*, 119.
        [http://dx.doi.org/10.1016/j.reactfunctpolym.2005.02.001]

[88]   Khan, A.A.; Alam, I.M.M. Applications of Hg(II) sensitive polyaniline Sn(IV) phosphate composite cation-exchange material in determination of Hg^{2+} from aqueous solutions and in making ion-selective membrane electrode. *Sens. Act. Biol. Chem.,* **2006**, *120*, 10-18.

[89]   Strauss, S.D.; Puckorius, P.R. Cooling-water treatment for control of scaling, fouling, corrosion. *Power,* **1984**, *128*, S1-S24.

[90]   Mulik, J.D.; Sawicki, E. Ion chromatography. *Environ. Sci. Technol.,* **1979**, *13*, 804-809.
       [http://dx.doi.org/10.1021/es60155a014]

[91]   Miller, J.E.; Brown, N.E.; Krumkans, J.L.; Trudell, D.E.; Anthony, R.G.; Philip, C.V. *Science and
       Technology for disposal of Radioactive Wastes, Schulz, W.W*; Lombardo, N.J., Ed.; Plenum: New
       York, **1998**, p. 269.
       [http://dx.doi.org/10.1007/978-1-4899-1543-6_21]

[92]   Bond, A.H.; Dietz, M.L.; Sylvester, P.; Clearfield, A. *216th ACS National Meeting, Boston, M.A.,*
       1998.

[93]   Gordon, M.; Popvtzer, M.; Greenbaum, M.; MacArthur, M.; De Palma, J.R.; Maxwell, M.H. *Proc.
       Eur. Dialysis Transplant Assoc.,* **1968**, p. 86.

[94]   Clearfield, A.; Thakur, D.S. Zirconium and titanium phosphates as catalysts: a review. *Appl. Catal.,*
       1986, *26*, 1-26.
       [http://dx.doi.org/10.1016/S0166-9834(00)82538-5]

[95]   Meyer, K.H.; Strauss, W. Hclv. Permeability of membranes: VI. Passage of current through selective
       membranes. *Chim. Acta.,* **1940**, *23*, 795-800.
       [http://dx.doi.org/10.1002/hlca.19400230199]

[96]   Mindler, A.B.; Paulson, C. *National Meeting of the American Institute of Mining and Metallurgical
       Engineers,* Los Angeles, Calif, **1953**.

[97]   Shen, X.M.; Clearfield, A. Phase transitions and ion exchange behavior of electrolytically prepared
       manganese dioxide. *J. Solid State Chem.,* **1986**, *64*, 270-282.
       [http://dx.doi.org/10.1016/0022-4596(86)90071-X]

[98]   Bajaj, P.; Goyal, M.; Chavan, R.V. Thermal behavior of methacrylic acid–ethyl acrylate copolymers.
       J. Appl. Polym. Sci., **1994**, *51*, 423-433.
       [http://dx.doi.org/10.1002/app.1994.070510305]

[99]   Stein, E.W.; Clearfield, S.A.; Subramanian, M.A. Conductivity of group IV metal sulfophosphonates
       and a new class of interstratified metal amine-sulfophosphonates. *Solid State Ion.,* **1996**, *83*, 113-124.
       [http://dx.doi.org/10.1016/0167-2738(95)00215-4]

CHAPTER 6

The Importance of Iron oxides in Natural Environment and Significance of its Nanoparticles Application

Iqbal Ahmed[1,2,*], Kamisah Kormin[3], Rizwan Rajput[4], Muhammed H. Albeirutty[1,2], Zulfiqar Ahmad Rehan[5] and Jehan Zeb[6]

[1] *Center of Excellence in Desalination Technology, King Abdulaziz University, Jeddah 21589, Saudi Arabia*

[2] *Department of Mechanical Engineering, King Abdulaziz University, Jeddah 21589, Saudi Arabia*

[3] *Faculty of Management, University Technology Malaysia, 81310 Skudai, Johor, Malaysia*

[4] *Department of Chemistry, Government (MPL) Higher School Nawabshah, Sindh, Pakistan*

[5] *Department of Polymer Engineering, National Textile University Faisalabad, Faisalabad 37610, Pakistan*

[6] *Unit for Ain Zubaida Rehabilitation and Groundwater Research. King Abdulaziz University, Jeddah, Saudi Arabia, Jeddah 21589, Saudi Arabia*

Abstract: Iron oxide and its derivatives have been receiving importance and broad scale applications during the last two decades, due to their specific characteristics and use. There are many types of ferrous oxides have been characterized, whereas iron (III) oxide (Fe_2O_3), Fe(II)-deficient magnetite (Fe_2O_3, γ-Fe_2O_3) and ferrous-ferric oxide (Fe_3O_4) are, indeed, the relevant types of untainted ferrous oxides. Many hydroxides, such as ferric hydroxide ($Fe(OH)_3$) and oxyhydroxide ($Fe(O)OH$), are also of industrial importance, being precursors of pure or complex oxides. As for the compound iron oxides, ferrites are essential ferromagnetic materials. Among all the ferrites, the iron oxides and ferrite magnetic nanoparticles with appropriate surface chemistry are prepared either by wet chemical methods such as colloid chemical or sol-gel methods or by dry processes such as vapour deposition techniques. The conventional granulometric technique does not form nanoparticles (NP), and is often desired to assemble or pattern ferrous oxide nanoparticles to give magnetic or optogenetic functions. This review summarizes the comprehensive study of ferrous based nanomagnetic particles preparation together with a comparison of synthesis techniques and their characterization, particularly regarding magnetism properties, particle size and morphologies. Moreover, in this review, a recent study of modified microwave assisted system for nanomagnetic particles preparation has also been highlighted.

* **Corresponding author Iqbal Ahmed**: Department of Mechnical Engineering, Center of Excellence in Desalination Technology, King Abdulaziz University, Jeddah 21589, Saudi Arabia; Tel/Fax: +966-5- -3541984; E-mails: iqbalmouj@hotmail.com; iqbalmoujdin@gmail.com

Keywords: Applications, Iron oxide, Magnetic nanoparticles, Method.

INTRODUCTION

Historically, iron ore called Lodestone is the tremendously changeover metal oxides of scientific distinction. Iron ore minerals are the primary rock-composing components and turn out to be around five percent of our planet's cover, while ferrous oxides are of great significance for many of the properties and processes taking place in the ecosystems. The iron ores reserves held by 50 countries, and the only leading seven nations are producing for about three-quarters of total world production of iron ores [1, 2]. Table **1** summarizes the statistics of the global iron ore reserves as of 2015. All over the world, the fabrication of efficient iron ore, Australia and Brazil is the most prominent manufacturers and lead the world's iron ore exports. Australia processed about 824 million metric tons of iron ore in 2015, while Brazil's production came to an estimated 428 million metric tons [2]. Iron chemical element forms a pure substance iron [3]. Pure iron metal is a hard, dense and shiny metal with a greyish tinge. It is relative, ductile, easily forged, rolled and pulled in a letter to the wire, and magnetized and melts at 1539°S. Impurities can significantly alter the physical properties [3].

Table 1. Statistic of world iron ore reserves as of 2015 [1, 2].

Number	Iron Producer Countries	Millions Matric Tones	
		Iron Crude	Iron Contents
1	Australia	54000	24000
2	Russia	25000	14000
3	Brazil	23000	12000
4	China	23000	72000
5	United States	11500	3500
6	India	8100	5200
7	Ukraine	6500	2300
8	Canada	6300	2300
9	Sweden	3500	2200
10	Iran	2700	1500
11	Kazakhstan	2500	900
12	South Africa	1000	6500
13	Other Countries	18000	9500

Consequently, the most regularly used metal worldwide amongst all mineral metals is iron oxide [4 - 6]. The iron ore mines reserves are located in fifty countries. However, the most substantial part ~ 98% of iron ore cast-off in steel forging production. Principally, iron ores are shared word for oxides, hydroxides and oxy-hydroxides composed of Fe(II) and Fe(III) cations and O_2- and $OH-$ anions, respectively. The understanding and applications of iron oxide as an essential mineral originated thousands of years ago [3 - 6].

Importance of Iron Oxides Minerals in the Environment

Iron minerals holding significant amount of ferrous oxides remain in matrixed form to associate chemical compositions into silicates, carbonates, sulfides, nitrites, chlorides and phosphides but more often they are found in oxides form and are the most important sources of iron [3 - 5]. Table **2** summarizes the matrixes of oxide forms of iron ore minerals and designate the quartz, crystal species usually adapted as sources of iron. The greatest significantly castoff iron-based ores from which ferrous types of components can be straightforwardly leached out such as; magnetite (Fe_3O_4) contains 72% ferrous contents; hematite (Fe_2O_3) has ferrous materials about 72% and limonite ($2Fe_2O_3 \cdot 3H_2O$) which has about 60% ferrous contents. However, some of the least significant iron ores minerals such as; siderite ($FeCO_3$) that has 48.3% of ferrous materials and pyrite (FeS_2) has about 46.6% ferrous amount [7], where these amount of ferrous contents are in their refined states [8]. So far a lot of minerals metal oxides contain fifteen types of iron, however, amongst all the metals sources, iron metal can easily be extracted due to its natural form of oxide and hydroxides. There is also a group of FeIII-oxy-hydroxy-halides (generally based on salts) which are thoroughly linked to the pure oxides. A FeIII oxyhydroxy-sulfate [5], oxyhydroxy-nitrate and oxyhydroxy-chloride form belong to this group, respectively. The chloride form is in fact, the mineral akaganeite, although usually written as β-FeO(OH) or Fe^{3+}O(OH, Cl) other halogenides can also be incorporated into akaganeite.

Table 2. Composition of Iron ore minerals [3 - 6, 19, 24, 29, 32, 34, 42, 43]

Compositions	Compositions of iron oxide in minerals forms						
	Hematite	Magnetite	Maghemite	Limonite	Ilmenite	Ulvospinel	Goethite
FeO	-	32.28	-	-	-	-	-
Fe_2O_3	93.832	50.03	74.6	89.86	48.0	52.55	80.322
TiO_2	<0.02	7.265	13.0	-	47.6	10.750	-
Al_2O_3	0.894	5.980	1.8	5.206	0.2	4.750	3.395
SiO_2	0.584	0.068	0.3	3.822	< 0.02	1.00	0.450
MgO_2	<0.01	4.170	8.4	-	1.38	5.550	-
CaO	0.01	0.16	-	-	-	-	-

(Table 2) contd.....

Compositions	Compositions of iron oxide in minerals forms						
	Hematite	Magnetite	Maghemite	Limonite	Ilmenite	Ulvospinel	Goethite
Cr_2O_3	0.125	<0.05	-	-	0.08	-	-
V_2O_3	0.950	0.150	-	-	0.70	-	-
P_2O_5	0.042	-	-	0.118	-	0.290	0.225
MnO	0.01	0.220	0.9	-	-	0.5	-

The primary structural unit of all Fe" oxides is an octahedron, and every atom is adjacent around Fe either by six O or by both O and OH ions. Principally, all form of iron minerals are referred to as "iron oxides"; yet, merely a rare is utilized successfully as originates of iron. Iron oxides can be heated by using redox processes and cast-off as in metal iron in an economically feasible method [5]. Besides, the extracted iron oxides are playing an essential role in a variety of disciplines. These iron oxides may be required for investigations of their particular properties or used including real, environmental and industrial chemistry, corrosion science, mineralogy, geology, soil science, planetology, biology and medicine.

Classification of Iron Oxides Crystals

Iron ore crystals are frequently available as a form of magnetite and hematite. From the literature review and business magazine [1 - 6], it is observed that currently Australia and Brazil are the substantial producers and exporters of iron ore in the world and control the most considerable part of the world's iron ore reservoirs. Only Australia produce half of the world's iron ore exports, while Brazil exported around 23% of the world total iron exports. As of 2015 [1, 2], Brazil held a 12 billion metric tons of iron content and 23 billion metric tons of crude ore. Brazil produced 0.428 billion tons of iron ore in 2015 [2].

Nevertheless principally, iron oxides are accessible as several mineral forms such as,

  i. Hematite
 ii. Maghemite
iii. Magnetite
 iv. Goethite
  v. Limonite
 vi. Ilmenite
vii. Ulvospinel

Table **2** shows each iron ore mineral compositions. The most shared and distinctive properties of each crystal are their magnetical properties [9]. Usually, the nanoparticle is based on micron to submicron moieties that have a range of sizes from 1 nm to 100 nm as per the required application. Though there are models of NP some hundreds of nanometers in size prepared from inorganic or organic materials [9, 10], which have many novel properties compared with the bulk materials. In this way, magnetic NP has many unique magnetic properties such as superparamagnetic, high coercivity, low Curie temperature, high magnetic susceptibility, *etc.* Magnetic NP is of great interest to researchers from a broad range of disciplines, together with the magneticfluids, data storage, catalysis and bio-applications [12, 13]. Especially, magnetic ferro fluids are the applied study that has led to the integration of magnetic NP in a myriad of commercial applications [14].

Therefore, ferrous oxide nanoparticles hold available magnetic properties are permitting to their phase, shape and size [15]. Two aspects that control the magnetic behaviour of nanoparticles are size and surface anisotropy14. For example, Hyeon *et al.* [16] have studied the synthesis of highly crystalline and monodisperse γ-Fe_2O_3 nanocrystallites. They investigated a high-temperature (300°C) aging of iron−oleic acid metal complex, which was synthesized by the high-temperature disintegration of iron carbonyl ($Fe(CO)_5$) in the presence of oleic acid at 100 °C, was obtained to generate monodisperse iron nanoparticles. Hyeon *et al.* [17] have started growing in the obstructive temperature with an increase in particle size of iron oxide nanoparticles. Khurshid *et al.* [18] reported that the differences in the obstructive temperatures between spherical and cubic interchange-coupled FeO/Fe_3O_4 nanoparticles, with non-identical FeO: Fe_3O_4 ratios, have been synthesized by a high-temperature decomposition technique to study anisotropy reaction on their heating performance. X-ray diffraction (XRD) and transmission electron microscopy (TEM) unfold that the NP is composed of FeO and Fe_3O_4 parts, with an average size of 20 nm [17, 18]. Transformations of iron oxides nanoparticles also have capabilities in apparent anisotropy, and exchangeable preference holds to variances in the obstructive temperature amongst the cubic and spherical nanoparticles.

Therefore, the perspective of the primary study, Fe^{+++} oxide is an appropriate composite for the overall learning of polymorphism and the magnetic and structural phase transitions of nanoparticles. Thus, in Table **1** among all the minerals the hematite (α-Fe_2O_3), maghemite (β-Fe_2O_3, γ-Fe_2O_3 and ϵ-Fe_2O_3) and magnetite (Fe_3O_4), respectively are the most significant source of iron transition oxides.

i. Hematite (α-Fe₂O₃)

Hematite is of specific importance as an inexpensive, weak ferromagnetic, ecologically accessible and the most widespread Fe oxides in soils and sediments, respectively. As compared to magnetite and maghemite, the hematite (α-Fe$_2$O$_3$) is an ultimate steady iron oxide under normal conditions and frequently originate around the globe. Fig. (**1a**) shows an iron ore mineral of hematite, Fig. (**1b**), summarize the α-Fe$_2$O$_3$ transition, figure. 1c shows Mou *et al.* [20] crystal-phase and rhombohedral shape of Fe$_2$O$_3$ and Fig. (**1d**) revealed a cubic spinel single crystalline hematite (α-Fe$_2$O$_3$) phase, respectively. Hematite is the earlier iron oxide and is extensive in rock and soil and identified as ferric oxide, iron sesquioxide, red ochre, martite and kidney ore among others [19]. Hematite mineral ores have an identical and abundant source of the iron amount, which can provide up to about 70%-95 wt.% of the ore but it can also hold somewhat a considerable quantity of other transition metals, for example, aluminium (Al), chromium (Cr) and vanadium (V), respectively. Whereas, the hematite is an antiferromagnetic at temperatures lower than 677 °C *i.e.* the magnetic instant of α-Fe$_2$O$_3$ is relatively insignificant (~1emu/cc), although above the Morin point (-13 °C), it reveals a poor ferromagnetism [21, 22]. The hematite (alpha) has also ultimate numerous crystalline polymorphs structures such as rhombohedral, hexagonal, corundum and cubic spinel. Rollman *et al.* [21] investigated the first standards design of the structure and magnetic phases of hematite and calculated very general crystal structure of α-Fe$_2$O$_3$ as shown in Fig. (**1a**). Mou *et al.* [20] have comprehensively reviewed crystal segment and nature of crystal structures (Fig. **1b**) and oxygen anions packing arrays in a α-Fe$_2$O$_3$ nanoparticles for catalytic applications.

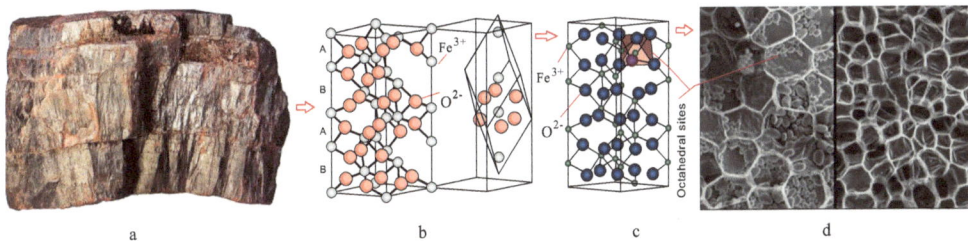

Fig. (1). Crystal structure of hematite (α-Fe$_2$O$_3$); a) shows hematite mineral iron transition [20], b) shows Rollmann *et al.* [21], hexagonal structure and magnetic phases of hematite, c) Mou *et al.* [20] crystal-phase and rhombohedral shape of Fe$_2$O$_3$, d) cubic spinel single-crystalline hematite (α-Fe$_2$O$_3$) phase.

Several researchers have studied the magnetic properties of hematite in complete form and the form of ultrafine nanoparticles and because of magnetic and biocompatibility characteristics, hematite magnetic nanoparticles find environmental, photoelectrochemical, sensors and medical applications [20 - 22]. Presently, the α-Fe_2O_3 photoelectrode has established significant accessibility such as cosmological energy transformation material because of its outstanding properties, for example, a lesser band gap (2.1 eV), great aversion to oxidization and cost-effective.

ii. Goethite (α-FeOOH)

In nature, Goethite, α-FeOOH, is one of the most abundant forms of iron oxides in global soils, sediments and ore deposits [4], as well as a typical weathering product in rocks of all types due to its high thermodynamic constancy. Organized with ferrihydrite (Fe8.2O8.5(OH)7.4·3H$_2$O) as an essential precursor of more stable and better-crystallized iron oxyhydroxides [23], these phases play crucial roles in numerous environments, from soils to weathered rocks, groundwater to the marine system [24]. Analysis of thermoremanent magnetization in goethite has confirmed that it contains poor ferromagnetic constituent. Thus, it should be considered that changes into hematite (-Fe_2O_3) between 453 K and 543 K through dehydrogenation have extensively been used in the preparation of magnetite and maghemite (Fe_2O_3). Agreed with the full magnetic carriage, the researchers made an adequate effort in the investigation of the metal cation substitutions in Fe(III)-oxyhydroxides and -oxides due to their importance in the industry and environment.

Naturally, goethite also contains various metal changes and thus, has an outstanding source of magnetic nanoparticles in industry, interpret the magnetic properties of goethite, precisely the influence of the crystallite size on these properties is essential for future technological development [25 - 29]. Fig. (**2**) shows most natural goethite mineral rock [29], octahedral crystal structures of goethite [28] and SEM image of hydrolysis goethite [30, 31], respectively. The first crystal structure of FeOOH was determined by Goldsztaub and Hoppe [26, 27] using X-ray diffraction photographic techniques [27]. The magnetic configuration of goethite with neutron powder diffraction on both natural and synthetic samples was examined by Forsyth *et al.* [32] and Szytula *et al.* [33] and found that it is anti-ferromagnetic below about 373 K (the Ne´el point). It was investigated by Gualtieri & Venturelli [34] and Nagai *et al.* [35], respectively, using synchrotron X-ray powder diffraction. However, despite both of its mineralogical and technological interest, no detailed structural information, such as anisotropic atomic displacement parameters, are available for goethite because of the lack of a single-crystal X-ray diffraction structure analysis.

The structure unit shared to all FeO and OH is the $FeO_3(OH)_3$ octahedron and Fig. (**2**) summarises crystal structure of goethite relating to the space arranged in specifically different ways. The Fe_3+O_6 octahedral share edges to the arrangement of dual chains successively analogous to c, which are supplementary linked to form a three-dimensional building by sharing vertices. There are two divergent O sites, O_1 and O_2, both coordinated to three Fe atoms, with O_2 additionally bonded to an H atom [36].

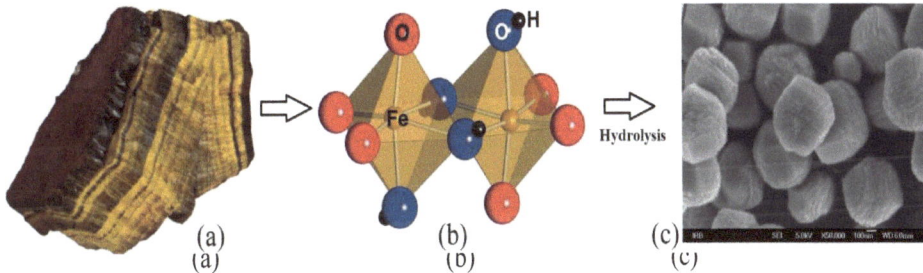

(a)
(b)
(c)
(a)
(b)
(c)

Fig. (2). (a) Most common goethite mineral rock [29]; (b) octahedral crystal structures of goethite [28]; (c) SEM image of hydrolysis goethite [30, 31].

iii. Limonite (Fe₂O₃.H₂O)

A limonite name such as "brown iron," is ferric oxide monohydrate ($Fe_2O_3.H_2O$) amorphous phase with no visible crystals, and does not exist as inactive forms because deposits are usually too small and too impure for use in modern metallurgy. It lives either as hematite or goethite in the form of yellowish brown with variable amounts of adsorbed water in its chemical structure. Boswell and Blanchard [37] first studied the cellular structure of limonite and revealed that the shape of the cellular pattern limonite box work is primarily governed by the fracture or cleavage pattern in here to the particular sulphide nodule before oxidation. Based on best of our knowledge not much work is studied on limonite.

iv. Maghemite (β-Fe₂O₃, γ-Fe₂O₃ and ε-Fe₂O₃)

Maghemite (γ-Fe_2O_3) is a second highest steady polymorph of iron oxide and has the same chemical composition as hematite, yet maghemite occurs in four polymorph forms (gamma, beta, epsilon) [38]. As compared to hematite, the maghemite has an identical crystalline spinel-like structure like magnetite (Fe_3O_4) and has an inverse spinel type space group with a = 8.351 Å and positions intertwined above the octahedral cation sites in the crystal lattice. Grau-Crespo [39] and workers were reported that the earliest evidence of a starting out from the consistency was investigated by Haul and Schoon in 1939 [40], and they revealed further replications in the powder diffraction form of maghemite synthesized by oxidizing magnetite. Moreover, another critical distinct of maghemite (γ-Fe_2O_3) with antiferromagnetic α-Fe_2O_3, reveals ferrimagnetic collation with a net magnetic moment (2.5 µB per formula unit) and greater magnetic examination temperature of -Fe$_2$O$_3$ is about 676.55 °C [39]. Moreover, the main distinguishing features of maghemite are the presence of vacancies in Fe position with symmetry reduction of 32O$_2^-$ ions; eight Fe^{3+} ions situated in tetrahedral coordination sites (A-sites) and sixteen Fe^{3+} ions in octahedral coordination sites (B-sites) per unit cell [41]. The other two polymorphs of meghemite have cubic bixbyite structure "beta" meghemite (β-Fe$_2$O$_3$), and orthorhombic structure "epsilon" meghemite (ε-Fe$_2$O$_3$) is ferromagnetic; while beta type Fe$_2$O$_3$ is a paramagnetic material [39, 40]. Figure 3 shows polymorphs structure of maghemite together with mineral ores. In maghemite, iron ions are distributed in the octahedral (Oh) and tetrahedral (Td) sites of the spinel structure, but maghemite differs from magnetite by the presence of cationic vacancies within the octahedral site [42]. The vacancies ordering scheme is closely related to the sample preparation method and result in symmetry lowering and possibly superstructures. The vacancies can be completely random or partly or entirely ordered. Several forms of nanoparticles have been produced and comprehensively investigated in current years [43]. To produce maghemite nanoparticles is not an easy job. However, several researchers reported that maghemite could be obtained by the dehydration of ferric hydroxide (γ-FeOOH) or by the oxidation of magnetite using solid or solution state route. While, several other different techniques are reported in previous studies for the synthesis of γ-Fe$_2$O$_3$ particles [44].

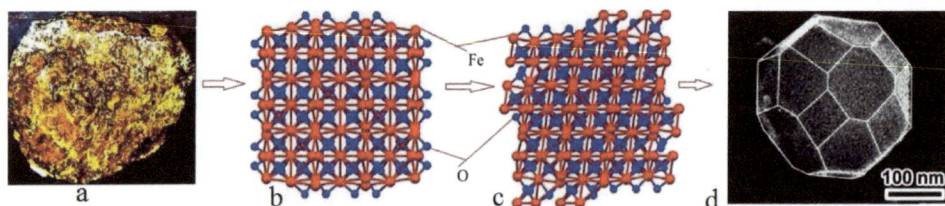

Fig. (3). Crystal structure of maghemite (Fe$_2$O$_3$); a) shows maghemite mineral iron transition metals, b) shows (Tuček *et al.*, 2015) cubic bixbyite structure and magnetic phases of β- Fe$_2$O$_3$, c) [42], crystal-phase and orthorhombic structure of ε-Fe$_2$O$_3$, d) Mou *et al.* [20] shows SEM photograph of γ-Fe$_2$O$_3$ nanoparticles.

v. Ilmenite

Ilmenite is denominated following the locality of the Ilmen Mountains, which are a part of the Southern Urals of Russia. This transition metal oxide (FeTiO$_3$) which has got fundamental importance in the investigation of rock magnetism and naturally existing mineral found in igneous rocks resulting from the higher layer down to depths and it is an essential industrial mineral in several deposits throughout the world [45]. Ilmenite is very similar in structure to hematite and is substantially the same as hematite with roughly half the iron replaced with titanium. However, ilmenite is a weakly magnetic iron-black or steel-gray mineral. Ilmenite is an economically essential mineral for the production of titania pigments. Depending on the pressure and temperature conditions, the polymorphs of $FeTiO_3$ crystallize in the ilmenite perovskite structures. Ilmenite is a stable phase at ambient conditions, while perovskite is stable at high pressure and room temperature. The solid solution of ilmenite and hematite is used to interpret the historical fluctuations in the Earth's magnetic field [46].

The magnetic properties of ilmenite ($FeTiO_3$) solid solutions are of specific significance in sustainable extraction *via* magnetic separation. The system is a solid solution between ilmenite ($FeTiO_3$) and hematite (Fe_2O_3), and it can be formed in the complete compositional range ($0 \leq x \leq 1$) due to the end-members having the same crystal structure and the cations having similar radii [47, 48]. The magnetic behavior of the solid solution is also of crucial importance to studies in which rock magnetization is used to interpret historical fluctuations. The earth's magnetic field containing for rocks displays self-reversed magnetization that is, the spontaneous magnetization occurring as the rock is cooled below the Curie temperature opposes the applied field [49]. The dependence of ilmenite's structure on temperature, pressure and composition is strongly coupled to its electronic, magnetic and optical properties. A detailed understanding of this coupling is of fundamental interest in the study of strongly interacting systems and also of great importance to some industrial processes.

In ilmenite, Fe and Ti form alternating bilayers with Fe-TiV-Ti-Fe (V shows vacant) ordering along the c-axis. Fig. (**4**) showed the ilmenite stone colour *i.e.,* grey or purple/black metallic and crystal structure and summarized that the Fe and Ti are octahedrally coordinated, where each octahedron shares one face with the adjacent octahedron of the other type cation and one face with an empty (vacant) octahedron along the (001)-direction. Within the layers, octahedra forms a honeycomb-like structure of edge-sharing octahedral. Concerning the magnetic arrangements, ferromagnetic (FM) and antiferromagnetic (AFM) coupling to the next adjacent layer (inter-bilayers) are considered. Different oxidation states are contemplated: AFM^{+2} (AFM-Fe^{2+}/Ti^{4+}), FM^{+2} (FM-Fe^{2+}/Ti^{4+}), AFM^{+3} ($AFMFe^{3+}/Ti^{3+}$) and FM^{+3} (FM-Fe^{3+}/Ti^{3+}). Within the bilayers, only FM coupling is taken into account between the atoms, which is experimentally confirmed by Kato *et al.* 1986 [49]. Additionally in their calculations [49, 50], an AFM arrangement is also investigated.

Ilmenite has considered a wide bandgap semiconductor with a measured bandgap of approximately 2.5 eV. 4,5 A productive unfold point for the interpretation of the electronic configuration of metamorphosis metal oxides is the ionic concept in which conventional charges are given to the ions. For $FeTiO_3$, this is tangled by the reality that there is two well-defined cation charge directing steady with O^{2-} anions, *i.e.*, $Fe^{3+}Ti^{4+}$ and $Fe^{3+}Ti^{3+}$. The charge shift excitation among these two conditions is well known and has been observed and modeled in some Fe and Ti-bearing crystalline. This resilience of the electronic structure of the Fe cation is robust to model using simulations established with factual capabilities. The delicate energy balance amongst the two charge shapes is strenuous to model, and the uninvolved task of a distinct charge to a position may lead to incorrect interactions.

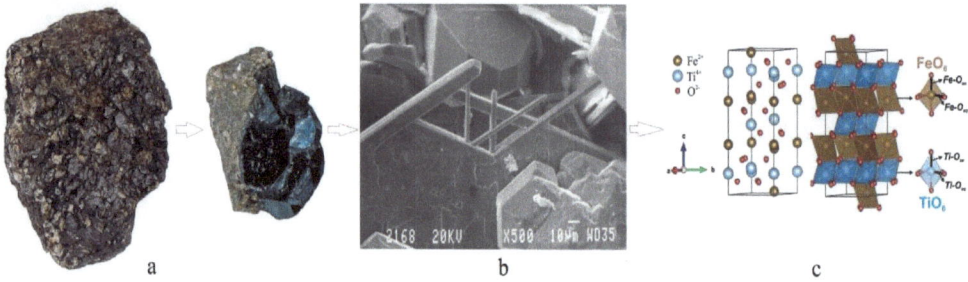

Fig. (4). The images of Ilmenite minerals and interpretation of octahedra ilmenite structure, a) mineral ore of ilmenite, b) SEM image of ore [50], c) crystal structure [51].

Moreover, the orbital occupancy of ilmenite is complicated, particularly the restrain d orbitals of the transition metal ions are not an obvious priori, and both a low spin and high spin states of the Fe ion are possible.

vi. Ulvospinel

Morgensen (1946) has discovered the natural ulvospinel (Fe_2TiO_4) and is a frequent accessory mineral in terrestrial metamorphic and vital igneous rocks being a ubiquitous magnetic mineral around the globe [52]. Since the first synthetic specimen of ulvospinel was prepared by Barthe and Posnjak [53], it has been broadly admitted that it is isomorphic with magnetite, that is, it causes the well-known spinel structure [54]. Principally from a composition of ulvospinel mineral ore covering "$xFe_2Ti_4-(1-x)Fe_3O_4$" (titanomagnetites) perspectives [55] regarding magnetic octahedral crystals where sufficient microstructures and the effect of extrinsic magnetic properties. Fig. (**5**) shows possible crystal structures of ulvospinel. Readman [56] has reported that the ulvospinel is contrary in position. Thus, spinel deliberated to have a cation scattering $Fe^{2++}(Fe^{2+}Ti^{4+})O_4$ position bracketed cations are situated on octahedral (B) sites [56]. If the magnitude and temperature dependence of the magnetic moments of the Fe^{2+} on octahedral and tetrahedral sites were equal and the spin arrangement precisely collinear, it would be expected that ulvospinel is antiferromagnetic.

The magnetic distinctive of ulvospinel minerals presents environmental field differences and provides evidence to geomagnetic develop. This metal ore contribution to the regions of remnant magnetization observed on the Martian surface argues for their significance in magnetic scrutinize efforts of the surface. The occurrence of spinodal decomposition in titanomagnetites may help explain large remnant magnetizations.

Nakamura and Fuwa, Hyperfine Interact: DOI 10.1007/s10751-013-0921-7

Fig. (5). ulvospinel minerals ore and octahedral crystals (a) high-temperature cubic phase and (b) low-temperature tetragonal phase.

Ulvospinel is critical to understand the magnetic properties and propitious metamorphosis of ulvospinel initial, afterward it is an edge member of the pseudo-binary system and maybe a phase present in spinodally decomposed titanomagnetite. Néel temperatures in ulvospinel differ linear in structure. Nevertheless, the exact locations of the cations are challenging to set accurately by X-ray diffraction because the similarity in scattering power of the iron and titanium atoms results in only small intensity differences in different cation arrangements [54]. Moreover, the intensities are too struck by a modification of the oxygen parameter. An accurate structural analysis should be possible by neutron diffraction because the effective scattering cross-sections for neutrons of iron and titanium ions are widely different [54]. However, Bosi *et al.* [57] reported that non-linear deviations in the crystal-chemical associations of the magnetite-ulvöspinel sequence could stand and clarified by an electron exchange reaction amongst Fe cations in entirely engaged sites. This response appears to be steady with a relaxation of the bonds to reduce the internal tension at reserved spinel space group symmetry.

Some studies of the properties of ulvospinel have been started mainly because of its relation to rock magnetism. Groschner and co-workers synthesized the anti-ferromagnetic of this mineral under argon to evaluate the role of an inert atmosphere on phase formation and magnetic properties of ulvospinel. They provided evidence that the part of environment on a promising phase transition and the magnetic properties of ulvospinel revealed the dependent transformation of ulvospinel with increasing temperature under the ambient atmosphere.

However, octahedral sites A and B are not equivalent, this would be surprising, so a type of weak ferrimagnetism or least L-type ferrimagnetism is expected. Indeed, ulvospinel is considered to possess a soft moment below about 120 Kelvin [58]. Groschner and co-workers [59] reported that the oxidation of ulvospinel to form metastable titanomaghemite at 300°C under atmospheric conditions. Lilova *et al.* [52] investigated thermodynamics of mixing and its dependence on cation distribution in magnetite-ulvöspinel (Fe_3O_4–Fe_2TiO_4) spinel solid solution using high-temperature oxide melt solution calorimetry and a range of structural and spectroscopic probes [52].

vii. Magnetite

Magnetite is one of the most significant iron minerals around the globe that is naturally magnetic. Technically, magnetite (Fe_3O_4) is a ferrimagnetic mineral and a member of the spinel group expressed as Fe_3O_4. Magnetite involving one part of FeO and one piece of Fe_2O_3 has a cubic inverse spinel structure with Fe3+ ions at A sites (tetrahedral coordination) and (Fe^{2+}, Fe^{3+}) ions at B sites (octahedral coordination and tetrahedral-octahedra layers). Fig. (**6**) summarizes crystal octahedral coordination and tetrahedral-octahedral structure of magnetite, and the ferrous type of the octahedral lattice sites due to higher ferrous crystal field stabilization energy [5, 6]. Magnetite crystals are seldom originating invisible clusters, and rather, they are more often dispersed into other surrounding rocks. Iron comes in two oxidation states, ferrous and ferric, and the components of magnetite crystals are primarily ferrous iron ions. Thus, magnetite and relatively oxidized magnetite have the darkest values due to the existence of FeII and FeIII causing intervalence charge transfer bands. The main differences between magnetite and maghemite, hematite, Goethite, and ulvospinel are the presence of FeII in magnetite and the presence of cation vacancies in maghemite [60]. Magnetite (Fe_3O_4) is a ferromagnetic mineral and a member of the spinel group. The ionic radius of FeII is more substantial than Fe^{III} so that the Fe^{II}-O bond is more prolonged and weaker than Fe^{III}-O bond [43]. Thus, it is one of a transition metal oxides in iron class and has been the subject of several types of research. Due to the unique magnetite properties, such as a faster breakdown of the Fe-O bond in acid solution, conversion into nanoparticles, magnetic nanoparticles, superconductivity and natural phase transition, respectively [43]. Amongst all above iron oxides type, the more significant iron oxides are hematite, maghemite and magnetite. It is noted that almost in all conventional methods use trivalent (Fe^{3+}) and divalent (Fe^{2+}) salt as the precursor for the synthesis process of iron nanoparticles. Before converting iron oxides into nanoparticles or supra nanomagnetic nanoparticles, to understand the behavior of iron physical and chemical properties are quite remarkable.

Fig. (6). (a) Pure rock of Magnetite iron; (b) opaque crystal of magnetite [61]; (c) SEM image of pure magnetite powder [62]; (d) crystal structure of magnetite [63].

Interaction of Iron in Natural Environment

Pure iron stable in air and under normal circumstances, its surface is covered with an oxide film, but its protective properties are flat, rust in a humid atmosphere and covered by a yellow-brown film which consists mainly of iron (III) hydroxide. Fig. (6) shows a pure rock of Magnetite iron, an opaque crystal of magnetite, the crystal structure of magnetite and SEM image of pure magnetite powder [61 - 63]. Film rust loses, porous and protects the metal from destruction, as ironwork in a dry atmosphere and humid place can become complete corrosion. The corrosion of iron is a fundamental principle and in the different environment (dry and wet). Moreover, pure iron has a strong interaction with strong oxidants (HCl_2, H_2S_4, HNO_3) formed compound ferric iron, and with others – bivalent Table **3** summarizes a general interaction behavior of iron oxides with various chemicals and in a different environment. In the following precursor, intermediate iron ores are behaved and interact as,

Table 3. Influence of Iro oxide reaction

Oxides	FeO iron (II) oxide	Fe_2O_3 iron (III) oxide	$Fe(OH)_2$ iron (II) hydroxide	$Fe(OH)_3$ iron (III) hydroxide
Physical properties	Black powder insoluble in water.	Light brown powder insoluble in water	Soluble in water	Soluble in water
Chemical properties: a) Cooperation with acidic conditions;	The basic oxide. It is quickly dissolved in acids to form salts of iron (II): $FeO + 2HCl = FeCl_2 + H_2O$; $FeO + H_2SO_4 = FeSO_4 + H_2O$. With nitric acid forms a salt of trivalent metal: $3FeO + 10NNO_3 = 3Fe(NO_3)_3 + NO + 5H_2O$.	Mild amphoteric properties dissolves in acids to form salts of iron (III), $Fe_2O_3 + 6HCl = 2FeCl_3 + 3H_2O$; $Fe_2O_3 + 3H_2SO_4 = Fe_2(SO_4)_3 + 3H_2O$. At high temperatures: $6Fe_2O_3 \rightarrow 4Fe_3O_4 + O_2$	Key properties. Easily dissolved in acids to form salts of iron (II) $Fe(OH)_2 + 2HCl = FeCl_2 + 2H_2O$.	Weak amphoteric properties, the weaker basis for $Fe(OH)_2$. Easily dissolved in acids to form salts of iron (III): $Fe(OH)_3 + 3HCl = FeCl_3 + 3H_2O$.
b) Interaction with alkalis.	not interact	When fused with alkali metal or alkaline carbonate salts formed metaferaty (II) alkali metals: $Fe_2O_3 + 2NaOH = 2NaFeO_2 + H_2O$; $Fe_2O_3 + Na_2CO_3 = 2NaFeO_2 + CO_2$.	repeatedly pluck $Fe(OH)_2$ throughout boiling with rigorous alkali soluble: $2NaOH + Fe(OH)_2 = Na_2Fe(OH)4$ (sulfate based tetrahidro soferat (II)	It dissolves in concentrated alkaline solutions to form salts: $Fe(OH)_3 + KOH = KFeO_2$ *(metaferat (III) potassium)*$+ 2H_2O$; $Fe(OH)_3 + NaOH \leftrightarrow NaFe(OH)_4$ *(tetrahidroksoferat (II) sodium)*. This salt is destroyed by altered concentrations of alkali to precursors.

(Table 3) contd.....

Oxides	FeO iron (II) oxide	Fe_2O_3 iron (III) oxide	$Fe(OH)_2$ iron (II) hydroxide	$Fe(OH)_3$ iron (III) hydroxide
Accomplishment	a) s iron (II) hydroxide: $Fe(OH)_2 = FeO + H_2O$ b) restoration of Fe2O3 hydrogen or carbon (II) oxide when heated: $Fe_2O_3 + CO = 2FeO + CO_2$; $Fe_2O_3 + H_2 = 2FeO + H_2O$	iron (III) hydroxide: $2Fe(OH)_3 = Fe_2O_3 + 3H_2O$.	The achievement of alkalis solution on iron (II) salts an ashy peeling residue that transforms its color to blue-green and then revolves brown due to the oxidation of Fe (II) to Fe (III): $2NaOH + FeCl_2 = Fe(OH)_2 \downarrow + 2NaCl$.	Performance alkali salts to iron (III) is formed as a reddish-brown sediment: $FeCl_3 + 3NaOH = Fe(OH)_3 \downarrow + 3NaCl$.
c) Disintegration when heated			$Fe(OH)_2 = FeO + H_2O$	$2Fe(OH)_3 = Fe_2O_3 + 3H_2O$.
d) Interaction with oxidants			a) oxidation in air: $4Fe(OH)_2 + O_2 + 2H_2O = 4Fe(OH)_3$ b) hydrogen peroxide oxidation: $2Fe(OH)_2 + H_2O_2 = 2Fe(OH)_3$	

a. **Interaction with Oxygen:**
     i. Heated to high temperatures, it burns in oxygen and air, becoming iron cinders:
$$3Fe + 2O_2 = Fe_3O_4$$
     ii. At oxidation in the presence of water vapor formed:
$$Fe_2O_3: 4Fe + 3O_2 = 2Fe_2O_3$$
     iii. Slow oxidation in dry air formed FeO:
$$2Fe + O_2 = 2FeO$$
     iv. In the hot state of the iron reacts with steam and make mix oxides:
$$3Fe + 4H_2O = Fe_3O_4 + 4H_2$$

b. **Interaction with Non-metals:**Iron reacts directly with most non-metals (halogens, sulfur, nitrogen, phosphorus, carbon, *etc.*). Many of these processes are exothermic, but for the reaction it should be hot iron.
   i. with chlorine and sulfur interacts with low heat:
      $2Fe + 3Cl_2 = 2FeCl_3$ (iron (III) chloride), $Fe + S = FeS$ (iron (II) sulfate);
   ii. at high temperature with other non-metals:
      $3Fe + C = Fe_3C$ (iron carbide), $2Fe + N_2 = 2FeN$ (iron (III) nitride), $2Fe + Si = Fe_2Si$ (iron silicide), $3Fe + 2P = Fe_3P_2$ (iron (II) phosphide)

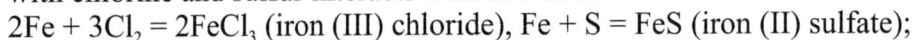

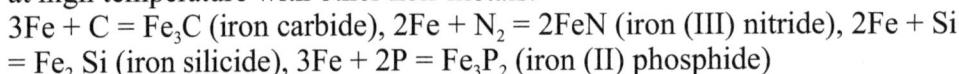

c. **Interaction with Acids:**
   i. In cooperation with diluted hydrochloric and sulfuric acids, it replaces the hydrogen with the formation of divalent salts of iron oxidation:
      $Fe + 2HCl = FeCl_2 + H_2$; $Fe + H_2SO_4(P) = FeSO_4 + H_2$
   ii. Iron at ordinary temperature does not react with concentrated sulfuric and nitric acids that formed on the surface film of insoluble compounds. Therefore, these acids can be stored in iron vessels. Diluted nitric acid dissolves iron in the cold:
      $4Fe + 10HNO_3 = 4Fe (NO_3)_2 + NH_4NO_3 + 3H_2O$;
   iii. The interaction with acids forms a mixture of salts of iron (II) and (III).
      $Fe_3O_4 + 8HCl = FeCl_2 + 2FeCl_3 + 4H_2O$;
      $Fe_3O_4 + 4H_2SO_4 = FeSO_4 + Fe_2(SO_4)_3 + 4H_2O$
   iv. In conjunction, nitric acid forms only one salt of $Fe(NO_3)_3$ as a result of oxidation of Fe (II) to Fe (III):
      $3Fe_3O_4 + 28HNO_3 = 9Fe(NO_3)_3 + NO + 14H_2O$
   v. When heated with concentrated sulfuric and nitric acids formed salt Fe^{3+}, depending on the concentration of acid can be formed in accordance with SO_2 (S), nitrogen oxides or free nitrogen:
      $2Fe + 6H_2SO_4 (k) = Fe_2 (SO_4)_3 + 3SO_2 + 6H_2O$;
      $Fe + 4HNO_3 (k) = Fe (NO_3)_3 + NO + 2H_2O$

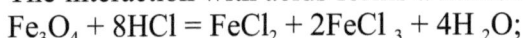

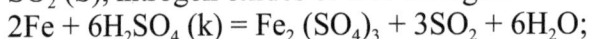

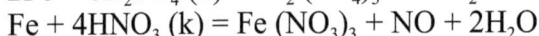

d. Interaction with Salts
   i. Iron salt solutions easily replace all metals that are among the stresses to the right of it, forming a compound of iron with a degree of oxidation +2:
      $Fe + Pb(NO_3)_2 = Fe (NO_3)_2 + Pb$;
      $Fe + CuSO_4 = FeSO_4 + Cu$

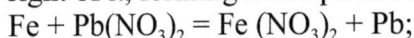

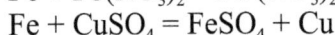

e. Interaction with Alkalis
   i. In the crushed state the iron reacts with hot concentrated solutions of alkalis to form complex salts:
      $Fe + 2NaOH + 2H_2O = Na_2Fe(OH)_4$ (tetrahidroksyferrat (II) sodium) $+ H_2$

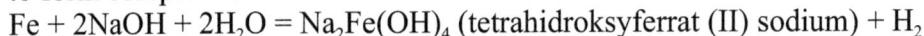

Classification of Iron-based Nano Materials Synthesis Methods

As summarized in Table **3**, the tendency of iron oxides can react or be modified with acids an bases. It can also be adjusted using either solid or solution techniques. Therefore, since last two decades, nanomagnetic materials are remarkably pursued in various engineering applications globally because of their superior thermal, physical and chemical properties [64]. This strategy of metal, with qualitative at dimensions of a nanometer, which creates novel structures, materials and mechanisms and Fig. (**7**) shows the general synthesis techniques of nanoparticles. Amongst all the methods, the chemical method is the most common particularly, in case of iron oxide nanoparticles. Moreover, iron oxides and nanomagnetic particles assure an exceptional improvement in several segments, for example, energy, materials synthesis, consumer commodity and medicine manufacturing [65 - 67]. The principal field of application to date is in electronics, photonics, pharmaceuticals, chemical synthesis and analysis, cosmetics and finishes for surfaces and textiles.

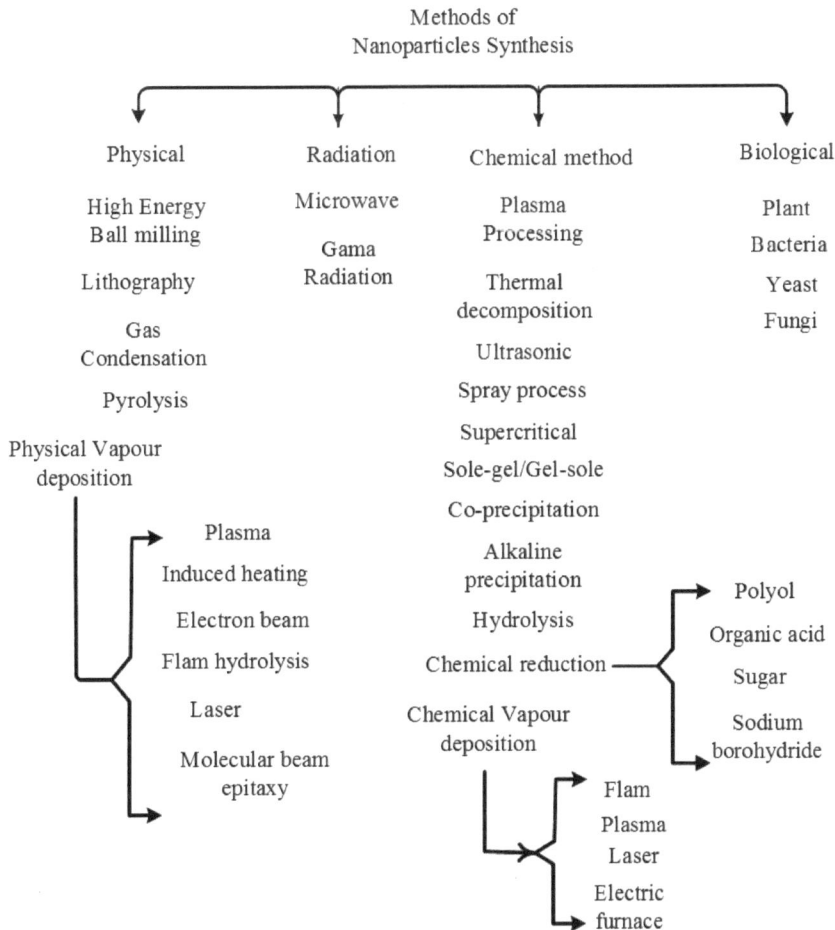

Fig. (7). Comprehensive synthesis techniques of nanoparticles.

The synthesis of numerous nanomagnetic particles with the size of 2 nm - 20 nm is very essential, because of their applications such as in multi-terabit in magnetic storage gadgets [68]. The particular magnetic effect of the nanoparticles appears primarily as long as the diminished sizes of obscure nanoparticles and present from interparticle interactions are imperceptible. The surfactant coating on magnetic nanoparticles prevents clustering due to steric repulsion. Dynamic adsorption and desorption of surfactant molecules towards particle surfaces during synthesis enable reactive species to be added to the thriving fragment. These nanoparticles perhaps scatter in many organic solvents and can be retrieved in powdered form by removing the solvent. However, during synthesis of metal oxide nanoparticles, there are still numerous issues concerning crystal structure, shape-control, uniformity of size control, particle size and alignment for device applications, respectively [69].

The influence of magnetic particles size behaviour of discrete nanomagnetic particles resulting in an innovative phenomenon, for instance, superparamagnetism, high infiltration field, high field inevitability and shifted loops after field cooling or superfluous anisotropy involvements [70]. Frenkel and Dorfman in 1930 [71], suggested that a particle of ferromagnetic material, below a critical particle size (<15 nm for the common elements), would consist of a single magnetic.

In the state crushed, the iron reacts with hot concentrated solutions of alkalis to domain. (Domains are the regions in which all the atomic moments point in the same direction so that within each domain the magnetization is saturated, attaining its maximum possible value) [72]. The magnetic behaviour of these particles above a certain temperature, *i.e.* the blocking temperature, is identical to that of atomic paramagnets (superparamagnetism). At blocking temperature, thermal energy of particles is greater than the energy of interaction of the particle moments with the applied magnetic field, resulting in the fluctuation of magnetic moments of the particles about the direction of magnetic field and the moments become disordered. Thus, high susceptibilities are involved [73].

As described above that iron ores as a form of ferrous oxide are established as distinctive patterns in nature. Amongst all iron minerals the hematite (α-Fe_2O_3), maghemite (γ-Fe_2O_3) and magnetite (Fe_3O_4), respectively are the mainly heterogeneous metal oxides ore minerals sources. Magnetite and maghemite, each of them based on single domains of size about 5–20 nm in diameter. Different type of ferrous oxide nonmagnetic particles can be identified regarding their atom differences and bulk counterparts in their chemical and physical characteristics [74]. Each type of ferrous oxide-based nanoparticle has assessed the critical surface area, quantum size and single magnetic state issues, respectively and assist to any dramatic change in magnetic properties resulting in quantum tunneling concerning magnetization and superparamagnetic phenomena. Stand on their absolute properties such as mechanical, thermal, physical, chemical and superparamagnetic nanoparticles, respectively offer a high potential for different applications [75, 76]. These utilizations interest nanomaterials of surface characteristics, shapes, specific sizes, and magnetic properties, correspondingly. Steep gradient magnetic separations (HGMS), ferrofluids, magnetic resonance imaging, fields of high-density data storage, electrode materials, catalysts and modified anti-corrosive coatings, wastewater treatment, bioseparations, and biomedicines, respectively are some applications discussed below [77 - 79].

Synthesis of Nanomagnetic Materials

Principally, qualitative steady state of iron oxides is based on the quantitative method for the nanomaterials. The preparation process plays a crucial role in determining the particle size and shape, size distribution, surface chemistry and therefore, the applications of the material [80]. The conditions necessary for the formation of magnetic particles are primarily the same as for non-magnetic particles, but some specific precautions are required because of high magnetic interactions among the particles. The necessary parameters are:

   i. Separation of the nucleation process from the growing process.
  ii. Protection of particles from aggregation.
 iii. A controlled supply of precursor materials.
 iv. Temperature and pH of the solution.

These parameters are intimately related, and sometimes it is difficult to separate them. Thus, the concentration of starting material, reaction temperature and pH in the solution must be optimized. Mahmoudi *et al.* [64] have reported a very comprehensive reviewed synthesis routes to attain significant crystallinity, polydispersity, proper shape, control of particle size and the magnetic properties, respectively. Several of them are examined below in Fig. (**8**). As concluded by Mahmoudi *et al.* [64], there is still more need to know about efficient synthesis routes of SPIONs. Moreover, after extensive reviewed by Mahmoudi and coworkers [64] in their perception, the recent research development of SPIONs still needs new techniques to overcome the conventional synthesis techniques of SPIONs [64].

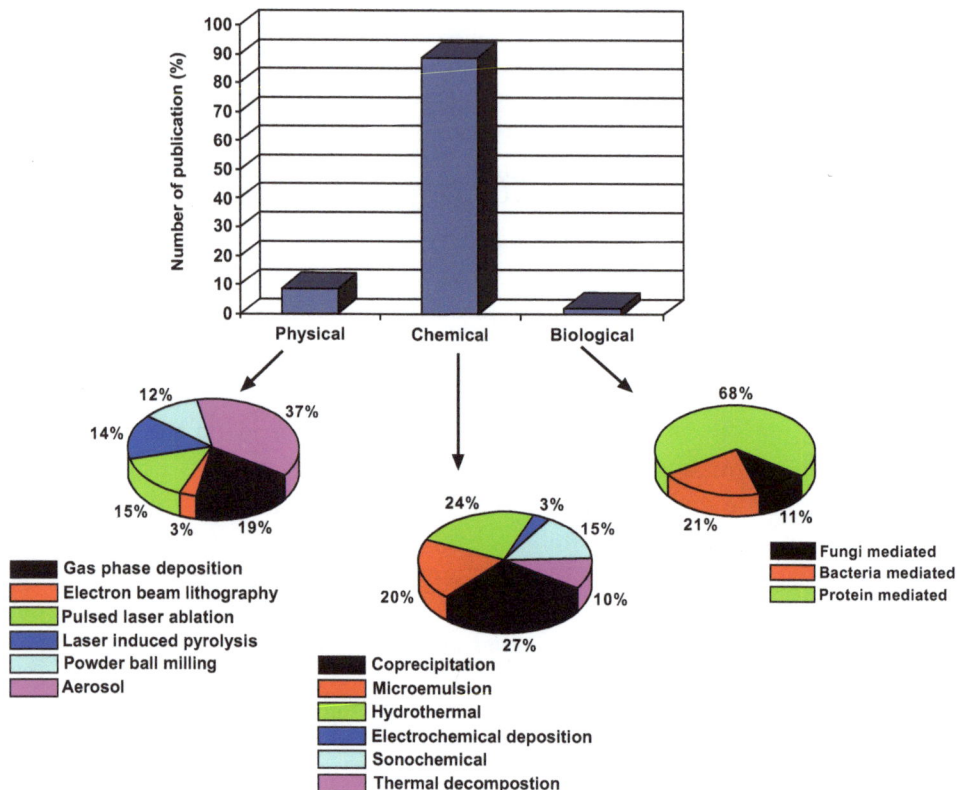

Fig. (8). A comparison of published work (up to date) on the synthesis of SPIONs by three different routes. Sources: Institute for Scientific Information. Adopted from ref [64].

Liquid Phase Methods

Liquid Phase enables the synthesis of nanomagnetic particles with complete features concerning pattern in an essential form and self-control in magnitude. Perhaps, this method is the most traditional techniques which propose a superior yield of nanomagnetic particles and surface treatments [81, 82]. Homogeneous precipitation reactions are used to come out with average size together with proper shape, *i.e.,* the methods that assume the separation of the nucleation and intensification of the nuclei [83, 84].

LaMer and Dinegar have proposed the classical model of nanoparticles development [85], whereabouts nuclei, so the approach is intended to develop size consistently by the dispersion of solutes from the solution to their surface as for as the last size is achieved. For monodispersity execution, nucleation should be evaded during the period of development. The most common method in the liquid phase is co-precipitation using aqueous solutions. However, some other solvent can also be used. The spherical magnetite particles size range about 30 nm to 100 nm can be achieved by the reaction of Fe(II) salt with a moderate oxidant (nitrate ions) and an alkali in aqueous solutions [83]. The particles phase and size are based on the pH of the solution, the counter ions present and the concentration of cations, respectively [86]. The mean size of the particles can be controlled (from 15 nm to 2 nm) by altering the pH and the ionic strength [87, 88].

Generally, because of the large surface-area to volume ratio of each nanoparticle are likely to aggregate and thus, reduce their surface energy [89]. By adding anionic surfactants as dispersing agents can be stabilized nanoparticles during the dissolution process [90 - 92]. Several studies have reported that for the achievement of better stabilization of nanoparticles during dispersion a protein coating onto the particles surface [93, 94] has a significant method such as starches [90, 93, 95] in the presence of polyelectrolytes or non-ionic detergents [94, 97]. The adsorption of such substances steadies the nanoparticles in the presence of the concentration of electrolyte that could or else be high enough for coagulation to occur [98]. Initially, Massart [98] has composed fabrication of superparamagnetic ferrous oxide particles using alkaline precipitation of $FeCl_2$ and $FeCl_3$.

Tavakoli and co-workers [100] have reported steady-state process for the development of magnetite (Fe_3O_4) nanoparticles and their diameter measured by XRD was 8 nm reported a roughly spherical. The support of this event was purposely examined to reveal the effect of the active and weak base such as NaOH CH_3NH_2, and ammonia [50, 51] as well as the pH value. Moreover, added cations such as Li^+, Na^+, K^+, $N(CH3)^{4+}$, CH_3NH^{3+}, NH^{4+} and the Fe^{2+}/Fe^{3+} ratio, respectively on the yield of the co-precipitation reaction and polydispersity and the diameter of the nanoparticles. When all of the set parameters are modulated, it is possible to obtain particles with a size ranging from 16.6 to 4.2 nm [100].

TWO-PHASE METHODS (MICROEMULSION)

The microemulsion specifically water-in-oil (w/o) based on nanosized water droplets dispersed in an oil phase and alleviated by surfactant molecules at the water/oil interface [101, 102]. By using the co-precipitation method to obtain and stabilized particles with have a broad size distribution; nevertheless, numerous other methods are currently being developed to produce nanoparticles with more uniform dimensions. The surfactant-stabilized nanoparticles that are typically in the range of 10 nm provide a confinement outcome that limits particle nucleation, agglomeration and growth [103]. The reverse micelle or emulsion technology's main advantage is the diversity of nanoparticles that can be obtained by varying the nature and amount of surfactant and co-surfactant, the oil phase or the reacting conditions.

Salazar-Alvarez [103] has revealed the process of ferrous oxide nanoparticles using inverse emulsions and this process is called nanoemulsion system. Technically, this reverse nanoemulsion system is based on AOT-BuOH/cHex/H_2O including the surfactant/water of molar ratio of 2.85 and a surfactant/co-surfactant molar ratio of 1 [103]. Various forms of surfactants, such as anionic, cationic, and non-ionic, respectively can be utilized. The only draw is their scale-up method, and injurious effects of residual surfactants on the properties of the particles are the main obstacles of microemulsion techniques. Salazar-Alvarez [103] has used the following synthetic method to prepare the nanoparticles;

- One nanoemulsion containing the iron source and another containing a solution of sodium hydroxide were mixed to form the magnetite nanoparticles.
- The nanoemulsion was lysed with acetone to remove the particles from the surfactant and washed several times with ethanol.
- The colloidal nanoparticles exhibit superparamagnetic behaviour with high magnetization values. The water and oil state usually consist of numerous dissolved constituents, and accordingly, the choice of the surfactant (and co-surfactant) is based on the system of physicochemical characteristics.

SOL-GEL TECHNIQUE

Sol-gel is a most common technique amongst the researchers for the preparation of nanoparticles. This process is based on the condensation and hydroxylation of molecular precursors in solution, instigating a "sol" of nanometric particles. The "sol" is then dried or "gelled" by solvent removal or by chemical reaction to obtain three-dimensional metal oxide network [104]. Technically, the sol-gel technique is sufficient for a wet route to the synthesis of nanostructured metal oxides [105 - 107]. Properties of a gel are very much dependent upon the structure created during the sol stage of the sol-gel process. The solvent used is water, but acid or base can also hydrolyze the precursors. Basic catalysis induces the formation of a colloidal gel, whereas acid catalysis yields a polymeric form of the gel [108]. At room temperature, these reactions are performed; further heat treatments are needed to acquire the final crystalline state [109, 110].

Several parameters are very influenced by the kinetic reaction such as; hydrolysis, condensation reactions, the rate reactions and of course the structure and properties of the gel are solvent, temperature, nature, the concentration of the salt precursors employed, agitation and pH, respectively [111 - 113]. The sequence of magnetic ordering during the sol-gel process, is highly based on the segments formed and the particle volume fraction, and is very sensitive to the size distribution and dispersion of the particles [114]. This method offers some advantages such as [114, 115].

i. According to experimental conditions, the possibility to obtain materials with a predetermined structure.
ii. The prospect to manage the particle size, monodispersity, pure amorphous phases, respectively.
iii. The homogeneity of the reaction products will organize the microstructure.
iv. The properties of the sol-gel matrix are the prospect to implant molecules, which sustain their immovability.

The aerogel composite has been prepared by the sol-gel method which is based on ferrous oxide-silica [116, 117] and forms to be the 2-3 array of consequence further reactive than traditional ferrous oxide. The escalation in reactivity was associated with the large surface area of ferrous oxide nanoparticles sustained on the silica aerogel [112, 118]. The tetraethyl orthosilicate and Fe (III) solutions as pecuniary precursors were fused in a typical polar aqueous solvent (alcohol), and the gels comprised a small number of days consequently were heated to fabricate the final materials [111, 119]. Sol-gel methods together contagion from by-products of reactions, in addition to the necessitate for post-treatment of the products.

SPRAY AND LASER PYROLYSIS

Spray and Laser pyrolysis have been established to be an efficient system for the straightforward and steady considerable production rate and of well-defined nanomagnetic particles under the exhaustive control of the experimental conditions [88].

In spray based pyrolysis process, a solution of iron salts and a squeeze reagent in a polar solvent is scattered into a sequence of reactors; where the polar solvent evaporates and the aerosol solute condenses [120]. The outcomes are the elementary size of the pure droplets and dried slag based on the particle. The maghemite particle size is about 5 to 60 nm with individual pattern have been procured using a variety of iron precursor salts in polar solution [121].

Laser pyrolysis of organometallic precursors [100, 122] based upon the resonant interaction between laser photons and sensitizer or reactant, fairly one gaseous types. A sensitizer is an energy exile carrier that is miffed by the absorption of CO_2 laser radiation and expels the absorbed energy to the reactants by collision [123]. The method correlates heating a crafty mixture of gasses with a steady wave CO_2 laser to start and prolong a chemical reaction as far as a critical concentration of nuclei reaches in the reaction zone, and homogeneous nucleation of particles occurs [124]. The nucleated particles formed during the reaction are entrained by the gas stream and are collected at the exit [88].

GAS/AEROSOL METHODS

Gas/aerosol phase technique has a highly qualitative product yield but the average rate of product yield is typically little adequate. The factors like gas-phase contamination, the concentration of oxygen and the heating time need to be controlled accurately to receive neat products. The equipments used in this system are also excessive.

POLYOLS METHOD

Polysol method for nanoparticles synthesis particularly for metal oxide nanoparticles is a well promising method achieve size-controlled nanoparticles with distinct shapes, and that could be utilized in biomedical applications [125]. By leading the kinetics of the precipitation, definite shape and size can be obtained from non-agglomerated metal particles. Better control of the metal particles can be achieved as average size by seeding the reactive medium with unidentified particles (heterogeneous nucleation). In this way, uniform particles result together with nucleation and growth steps can be separated entirely, and ferrous oxide nanoparticles around 0.10 micrometer (μm/length) can be obtained by disproportionation of ferrous hydroxide in organic media [126].

Typically polyethylene glycol (PEG) proposes attractive properties: such as they proceed as solvents capable of dissolving inorganic compounds such as;

- Due to their comparatively elevated boiling points,
- Due to their high dielectric steady form,
- An extensive operating temperature range (from 25 °C to the boiling point) for preparing inorganic compounds [127].

PEG has numerous hydroxyl groups such as based solvents (polyols) thus, PEG-based polyols also provide dropping agents as well as stabilizers to manage particle enlargement and avoid interparticle aggregation. Joseyphus *et al.* [128] have measured numerous aspects leading the production yield of iron particles similar to the type of polyols, hydroxyl ion concentration, ferrous ion concentration, iron salts and reaction temperature, respectively. They established out the yield and size of iron particles varied depending upon the reduction potential of the polyols.

HYDROTHERMAL REACTION METHODS

Typically, the hydrothermal reactions are carried out in aqueous media in reactors or autoclaves. During hydrothermal reactions, the temperature might be more than 200 °C, and the pressure can be higher than 1.379 MPa. Hydrothermal techniques depend upon the tendency of dehydrating metal salts on exalted circumstances and water to hydrolyze. However, the very low solubility of the resulting metal oxides in water at these conditions, particle size and morphology dominion is one of the leads of hydrothermal techniques and as well as to generate supersaturation [129]. Additional post-processing steps are needed, to accomplish the engineering of particle surfaces. It is also effortless to scale up.

Hao and Teja [129] have conducted a very comprehensive analysis of variables which influence on nanoparticle size and its morphology such as the hydrothermal technique. During their study, they investigated the precursor amount (wt.% or vol.), residence time (minutes) and temperature (°C), and it was concluded that;

i. The Higher amount of precursor would enhance the particle size and size distribution.
ii. The impact of residence time had a more significant on average particle size than feed amount.
iii. At short residence times, the monodispersed particles can be produced.

Xu and Teja [130] have also employed the continuous hydrothermal techniques to generate polyvinyl alcohol (PVA) coated iron oxide nanoparticles. PVA was chosen as the coating material because it has the required solution properties in water and it holds many isolated hydroxyl functional groups, which can absorb and complex with metal ions. The results of their experiment showed that particles with uniform shape and narrow particle size distribution were obtained in the presence of PVA. The average particle size decreased with the increasing PVA concentration when the residence time was of in the order of 2s and became nearly independent of PVA concentration, when the residence time was 10s or higher [130].

SONOLYSIS

Ferrous oxide nanoparticles can be synthesized. Several researchers [132 - 134] used this approach for decomposition of organometallic precursors, organic coupling agents, or structural hosts, polymers, respectively to limit the nanoparticle growth. Ultrasonic irradiation caused cavitation in an aqueous medium, where the collapse of microbubbles growth, and formation, occurred [133, 134]. Throughout sonolysis system, the cavitation can provoke more than 1.8 MPa pressure temperature about 5000 °C which can enable many unusual chemical reactions to occur [135, 136]. Thermally, induced processes contribute nanoparticles in crystalline form in many cases. Besides that, ultrasonically assisted reactions produced an amorphous yield that is generated in collapsing cavitation bubbles as large cooling rates (1010 K/s) to prevent their crystallization during quenching, thus calling for heat-treatment after synthesis [137, 138].

MICROWAVE IRRADIATION

Microwave irradiation assisted techniques have been known as volumetric heat since the early 1940s and have been used successfully in the chemical (edible and non-edible) industry. Microwave irradiation assisted technology in chemistry has received substantial interest in recent years, as its use has been started in material synthesis and preparative chemistry since 1986 [139]. The most significant advantage of microwave irradiation is that it can heat a substance uniformly leading to more homogeneous nucleation and a shorter crystallization time compared through a glass or plastic reaction container, with those for conventional heating, and this is beneficial to the formation of uniform colloidal materials. Norihito Kijima [140] reported synthesizing ultrafine α-Fe_2O_3 nanoparticles with a highly narrow distribution of microwave heating. As compared with other previously published iron oxides, ultrafine α-Fe_2O_3 nanoparticles showed a significantly high electrochemical performance because of their uniformity and size [11].

Application of Magnetic Nanoparticles

Magnetite is a very promising iron oxide because of its proved biocompatibility. Magnetic nanoparticles of iron oxides exhibit excellent drug delivery system, applications and drug targeting to tumors as well as anticancer agents [4, 141]. The use of magnetic nanoparticles for drug delivery must address issues such as drug-loading capacity, desired release profile, aqueous dispersion stability, biocompatibility with cells and tissue, and retention of magnetic properties after modification with polymers or chemical reaction [143 - 145]. Although para-magnetic capture is an excellent diagnostic instrument, their significance and relevance have been restricted by the magnetic nanoparticles particle technology and the design of the separator [146]. Fig. (**9**) shows a typical configuration utilized in nano-biomaterials applied to medical or biological problems [147]. For High Gradient Magnetic Separations (HGMS), these particles are tailored to fit particular metal or molecules, which can be removed from the water or waste slurries. In a process, a liquid phase containing magnetic particles passed through a matrix of wires magnetized by an applied magnetic field. The particles are held onto the wires but can be released if the magnetic field is cut off [147].

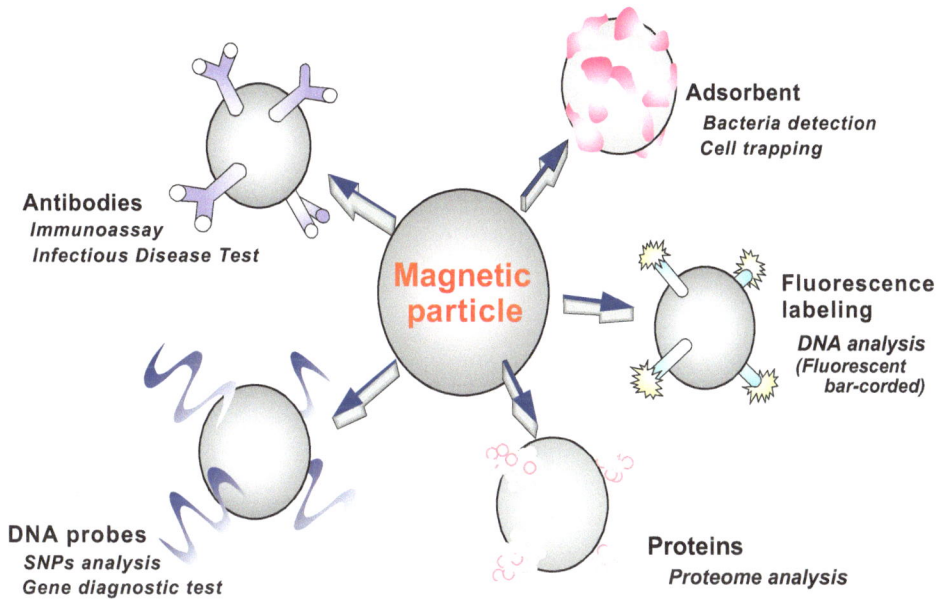

Fig. (9). Typical configurations utilized in nano-biomaterials applied to medical or biological problems. Adapted from ref [147].

Fig. (**10**) shows a conventional high-gradient magnetic separation capability. The application of magnetic nanoparticles for industrial water particularly Fe_3O_4, γFe_2O_3 and αFe_2O_3 based have been reported as competent, economical and treatment process almost hundred years old. Moreover, lately nanomagnetic particles environmental friendly substitute of other treatment materials, from the perspectives of both resource conservation and ecological remediation [64, 148]. Maghemite is useful in recording and data storage applications [149]. In data storage applications, the particles must have a stable, switchable magnetic state that is not affected by temperature fluctuations [4]. The particles should exhibit both high coercivity and high remanence, and they should be uniformly small, and resistant to corrosion, friction and temperature changes. It is doped with cobalt to improve its coercivity and storage capacity.

Ferrofluids employ nanoparticles of Magnetite. These are the colloids usually consisting of 10 nm nanoparticles, coated with a surfactant and suspended in a liquid such as a transformer oil. Ferrofluids have no net magnetic moment except when it is under the influence of an applied field. An external magnet is, therefore, able to trap the fluid in a particular location to act as a seal. Ferrofluids prove useful in optical switches and tunable diffraction gratings because of their impressive magnetic field subject optical anisotropy property [149]. Hematite and magnetite have been used as catalysts for some industrially essential reactions, which include the desulfurization of natural gas synthesis, the high-temperature water gas shift reaction and in the Haber process. They are also involved in the oxidation of alcohols and the large-scale manufacture of butadiene [4, 150, 151]. All three forms of magnetic iron oxide are commonly used in synthetic pigments in paints, ceramics and porcelain [4]. They possess some desirable attributes for these applications because they display a range of colours with pure hues and high tinting strength. They are also stable and highly resistant to acids and alkalis.

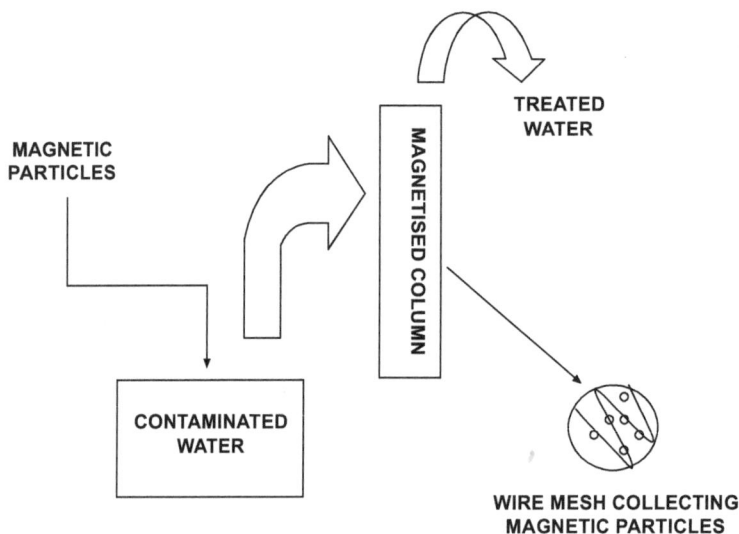

Fig. (10). A typical high-gradient magnetic separation facility. Adapted from ref [148].

Commercial iron oxide nanoparticles of Maghemite (Endorem® and Resovit®) have been used as contrast agents in NMR imaging. Endorem® and Resovit® are for location and diagnosis of brain and cardiac infarcts, liver lesions or tumors, respectively where the magnetic nanoparticles tend to accumulate at higher levels due to the differences in tissue composition and endocytotic uptake processes [89]. Especially, promising results have been detected in the improvement of sensitivity of detection and delineation of pathological structures, such as primary and metastatic brain tumors, inflammation and ischemia [152].

Spinel ferrites of the type $M_2+Fe_2 3O_4$ (M=Mn, Mg, Zn, Ni, Co, Cd, *etc.*) also attract the research interest because of their versatile, practical applications. The special requests in the electronics are as the magnetron, electric motors, electric guitar pickups, loudspeakers and electronic sensors, respectively. Also transformers, pulse transformers, rotating transformers, actuators, TWT amplifiers, communication, vehicle signage, shelf and bin marking, craft, hobby, toys, industrial automation equipment, catalysts and adsorptive materials [153 - 156].

CONCLUSION AND PERSPECTIVES

The synthesis of ferrous oxide nanoparticles or nanomagnetic particles, covering an extensive range of compositions and tunable sizes, has made substantial progress, especially over the past decade. The primary concern of recent research ferrous oxide nanoparticle synthesis is exploring new high-tech techniques to improve the conventional ones to obtain reproducible/ stable superparamagnetic nanoparticles with the required pattern and uniform size. Moreover, biocompatibility in the biological environment and saturation magnetization including colloidal cohesion concerning, respectively. Several kinds of monodisperse spherical nanocrystals with controllable particle sizes and compositions have been fabricated by a broad range of synthetic chemical procedures: sol-gel, co-precipitation, microemulsion aerosol, respectively Moreover, polyols processes, decomposition process *via* hydrothermal metal-surfactant complexes and sonolysis assisted method. The recent technique, *i.e.*, microwave assisted irradiation has also been studied. As this approach is environmental friendly, the microwave assisted preparation technique is appropriating importance in other fields of research and a domestic microwave can be used for research purpose as well. A significant problem for all methods is the design of magnetic nanoparticles with sufficient surface coatings that offer optimum performance *in vitro* and *in vivo* biological applications. Thus, future studies should also aim to address different challenges faced in iron oxide magnetic nanoparticles synthesis and applications. Additional problems include scale-up, toxicity, and safety of large-scale particle production cost and processes.

CONSENT FOR PUBLICATION

Not applicable.

CONFLICT OF INTEREST

The author (editor) declares no conflict of interest, financial or otherwise.

ACKNOWLEDGEMENTS

I am thankful to the Center of Excellence in Desalination Technology, King Abdulaziz University Jeddah for the working environment and moral support. I would also like to appreciate Kamisah Kormin from the Faculty of Business

Management, University Technology Malaysia for her kind assistance.

REFERENCES

[1]   U.S. Geological Survey (USGS). http://minerals.usgs.gov/minerals/pubs/commodity/iron_ore/index.html

[2]   Statista. *Iron ore mine production worldwide by country from 2010 to 2015 (in million metric tons),* **2015**.https://www.statista.com/statistics/267381/world-reserves-of-iron-ore-by- country/

[3]   Royal Society of Chemistry (RSC). http://www.rsc.org/periodic-table/element/26/iron

[4]   Cornell, R.M.; Schwertmann, U. *Introduction to the Iron Oxides*; Wiley-CH Verlag GmbH & Co. KGaA, **2004**.

[5]   Cornell, R.M.; Schwertmann, U. *The iron oxides: structure, properties, reactions, occurrences, and uses*; , **2003**.
        [http://dx.doi.org/10.1002/3527602097]

[6]   Cornell, R. *M and Schwertmann U. The iron oxides*; VCH Press: Weinheim, Germany, **1996**.

[7]   Zboril, R.; Mashlan, M.; Petridis, D. Iron(III) Oxides from thermal processes synthesis, structural and magnetic properties, mössbauer spectroscopy characterization, and applications. *Chem. Mater.,* **2002**, *14*(3), 969-982.
        [http://dx.doi.org/10.1021/cm0111074]

[8]   Joan, J.K.; Muumbo, A.M.; Makokha, A.B.; Kimutai, S.K. Characterization of selected mineral ores in the eastern zone of Kenya: case study of mwingi north constituency in kitui county. *Int. J. Mining Eng. Mineral. Proc.,* **2015**, *4*(1), 8-17.
        [http://dx.doi.org/10.5923/j.mining.20150401.02]

[9]   Kittel, C. Physical Theory of Ferromagnetic domains. *Rev. Mod. Phys.,* **1949**, *21*, 541.
        [http://dx.doi.org/10.1103/RevModPhys.21.541]

[10]   Mehmood, S.H. Magnetic anisotropy in fine magnetic particles. *J. Magn. Magn. Mater.,* **1993**, *118*, 359-364.
        [http://dx.doi.org/10.1016/0304-8853(93)90439-9]

[11]   Haneeda, K. Recent advances in the magnetism of fine particles. *Can. J. Phys.,* **1987**, *65*, 1233.
        [http://dx.doi.org/10.1139/p87-198]

[12]   Wu, W.; Wu, Z.; Yu, T.; Jiang, C.; Kim, W.S. Recent progress on magnetic iron oxide nanoparticles: synthesis, surface functional strategies and biomedical applications. In: *Sci. Technol. Adv. Mater*; , **2015**; 16, p. 43.

[13]   Drbohlavova, J.; Hrdy, R.; Adam, V.; Kizek, R.; Schneeweiss, O.; Hubalek, J. Preparation and properties of various magnetic nanoparticles. *Sensors (Basel),* **2009**, *9*(4), 2352-2362.
        [http://dx.doi.org/10.3390/s90402352] [PMID: 22574017]

[14]   Ramimoghadam, D.; Bagheri, S.; Abd Hamid, S.B. Progress in electrochemical synthesis of magnetic iron oxide nanoparticles. *J. Magn. Magn. Mater.,* **2014**, *368*, 207-229.
        [http://dx.doi.org/10.1016/j.jmmm.2014.05.015]

[15]   Jeevanandam, G.S. Synthesis of self-assembled prismatic iron oxide nanoparticles by a novel thermal decomposition route. *Royal Society of Chemistry,* **2013**, *3*, 189-200.
        [http://dx.doi.org/10.1039/C2RA22004K]

[16]   Hyeon, T.; Lee, S.S.; Park, J.; Chung, Y.; Na, H.B. Synthesis of highly crystalline and monodisperse maghemite nanocrystallites without a size-selection process. *J. Am. Chem. Soc.,* **2001**, *123*(51), 12798-12801.
        [http://dx.doi.org/10.1021/ja016812s] [PMID: 11749537]

[17]   Hyeon, T. Chemical synthesis of magnetic nanoparticles. *Chem. Commun. (Camb.),* **2003**, (8), 927-

934.
[http://dx.doi.org/10.1039/b207789b] [PMID: 12744306]

[18]   Khurshid, H.; Alonso, H.; Nemati, J.; Phan, Z.; Mukherjee, M.H. Anisotropy effects in magnetic hyperthermia: A comparison between spherical and cubic exchange-coupled FeO/Fe_3O_4 nanoparticles. *J. Appl. Physics,* **2015**, *117* 17A337.

[19]   Mahmed, N. *Ph.D. Thesis. Development of Multifunctional Magnetic Core Nanoparticles. Aalto University,* **2013**.http://urn.fi/URN:ISBN:978-952-60-5106-2

[20]   Mou, X.; Wei, X.; Li, Y.; Shen, W. Tuning crystal-phase and shape of Fe_2O_3 nanoparticles for catalytic applications. *CrystEngComm,* **2012**, *14*(16), 5107-5120.
[http://dx.doi.org/10.1039/c2ce25109d]

[21]   Rollmann, G.; Rohrbach, A.; Entel, P.; Hafner, J. First-principles calculation of the structure and magnetic phases of hematite. *Phys. Rev. B,* **2004**, *69*(16), 165107.
[http://dx.doi.org/10.1103/PhysRevB.69.165107]

[22]   Cao, H.; Wang, G.; Zhang, L.; Liang, Y.; Zhang, S.; Zhang, X. Shape and magnetic properties of single-crystalline hematite (α-Fe_2O_3) nanocrystals. *ChemPhysChem,* **2006**, *7*(9), 1897-1901.
[http://dx.doi.org/10.1002/cphc.200600130] [PMID: 16881086]

[23]   Michel, F.M.; Barrón, V.; Torrent, J.; Morales, M.P.; Serna, C.J.; Boily, J.F.; Liu, Q.; Ambrosini, A.; Cismasu, A.C.; Brown, G.E., Jr Ordered ferrimagnetic form of ferrihydrite reveals links among structure, composition, and magnetism. *Proc. Natl. Acad. Sci. USA,* **2010**, *107*(7), 2787-2792.
[http://dx.doi.org/10.1073/pnas.0910170107] [PMID: 20133643]

[24]   Jambor, J.L.; Dutrizac, J.E. Occurrence and Constitution of Natural and Synthetic Ferrihydrite, a Widespread Iron Oxyhydroxide. *Chem. Rev.,* **1998**, *98*(7), 2549-2586.
[http://dx.doi.org/10.1021/cr970105t] [PMID: 11848971]

[25]   Steen Mørup, D.E.M.; Frandsen, C.; Christian, R.H.B. Mikkel, Hansen. F. Experimental and theoretical studies of nanoparticles of antiferromagnetic materials. *J. Phys. Condens. Matter,* **2007**, *19*(21), 213202.
[http://dx.doi.org/10.1088/0953-8984/19/21/213202]

[26]   Goldsztaub, M.S. *Bull. Soc. Fr. Mineral.,* **1935**, *58*, 6-67.

[27]   Hoppe, W. Uber die Kristallstruktur von a-AlOOH (Diaspor) und a-FeOOH (Nadeleisenerz). *Z. Kristallogr.,* *941*(103), 73-89.

[28]   Song, X. *Ph.D. Thesis, Surface and Bulk Reactivity of Iron Oxyhydroxides: A Molecular Perspective. (Ph.D.), Umeå University,* **2013**.

[29]   Mindate.org. *GoethiteUSAMichigan, Marquette Co., Marquette iron range, Champion Typical acicular crystals forming botyroidal masses in iron ore.,* http://www.mindat.org/photo-340368.html

[30]   Ristic, M.; Music, S.; Godec, M. Properties of -FeOOH, -FeOOH and -Fe_2O_3 particles precipitated by hydrolysis of Fe^{3+} ions in perchlorate containing aqueous solutions. *J. Alloys Compd.,* **2006**, *417*, 292-299.
[http://dx.doi.org/10.1016/j.jallcom.2005.09.043]

[31]   Ristić, M.; Musić, S. Precipitation by forced hydrolysis of Fe^{3+} ions in perchlorate-containing aqueous solutions. *Presentation: Poster at E-MRS Fall Meeting 2007, Symposium A,* **2007**.

[32]   Forsyth, J.B.; Hedley, I.G.; Johnson, C.E. The magnetic structure and hyperfine field of goethite (a'-FeOOH). *J. Phys. C. Proc. Phys. Soc.,* **1968**, *1*, 179-188.

[33]   Szytua, A.; Burewicz, A.; Dimitrijevic, Z.; Krasnicki, S.; Rzany, H.; Todorovio, J.; Wanic, A.; Wolski, W. Neutron Diffraction Studies of a-FeOOH. *Phys. Status Solidi,* **1968**, *26*, 429.
[http://dx.doi.org/10.1002/pssb.19680260205]

[34]   Gualtieri, A.; Venturelli, P. Access Denied *in situ* study of the goethite-hematite phase transformation by real time synchrotron powder diffraction. *Am. Mineral.,* **1999**, *84*, 895-904.

[http://dx.doi.org/10.2138/am-1999-5-624]

[35]   Nagai, T.; Kagi, H.; Yamanaka, T. Variation of hydrogen-bonded O…O distances in goethite at high pressure. *Am. Mineral.,* **2003**, *88*, 1423-1427.
[http://dx.doi.org/10.2138/am-2003-1005]

[36]   Yang, H.; Lu, R.; Downs, R.T.; Costin, G. Goethite, [alpha]-FeO(OH), from single- crystal data. *Acta Crystallogr. Sect. E Struct. Rep. Online,* **2006**, *62*(12), i250-i252.
[http://dx.doi.org/10.1107/S1600536806047258]

[37]   Boswell, P.F.; Blanchard, R. Cellular structure in limonite. *Econ. Geol.,* **1929**, *XXIV*, 8.

[38]   Kluchova, K.; Zboril, R.; Tucek, J.; Pecova, M.; Zajoncova, L.; Safarik, I.; Mashlan, M.; Markova, I.; Jancik, D.; Sebela, M.; Bartonkova, H.; Bellesi, V.; Novak, P.; Petridis, D. Superparamagnetic maghemite nanoparticles from solid-state synthesis - their functionalization towards peroral MRI contrast agent and magnetic carrier for trypsin immobilization. *Biomaterials,* **2009**, *30*(15), 2855-2863.
[http://dx.doi.org/10.1016/j.biomaterials.2009.02.023] [PMID: 19264355]

[39]   Grau-Crespo, R.; Al-Baitai, A.Y.; Saadoune, I.; De Leeuw, N.H. Vacancy ordering and electronic structure of γ-Fe$_2$O$_3$ (maghemite): a theoretical investigation. *J. Phys. Condens. Matter,* **2010**, *22*(25), 255401.
[http://dx.doi.org/10.1088/0953-8984/22/25/255401] [PMID: 21393797]

[40]   Haul, R.; Schoon, T. *Z. Phys. Chem.,* **1939**, *44B*, 216.
[http://dx.doi.org/10.1515/zpch-1939-0117]

[41]   De-Boer, C.B.D.; Mark, J. Unusual thermomagnetic behaviour of haematites: neoformation of a highly magnetic spinel phase on heating in air. *Geophys. J.,* **2001**, *144*(2), 481-494.
[http://dx.doi.org/10.1046/j.0956-540X.2000.01333.x]

[42]   Tuček, J.; Machala, L.; Ono, S.; Namai, A.; Yoshikiyo, M.; Imoto, K.; Tokoro, H.; Ohkoshi, S.; Zbořil, R. Zeta-Fe$_2$O$_3$--A new stable polymorph in iron(III) oxide family. *Sci. Rep.,* **2015**, *5*, 15091.
[http://dx.doi.org/10.1038/srep15091] [PMID: 26469883]

[43]   Chirita, M.; Grozescu, I. Fe$_2$O$_3$ – Nanoparticles, Physical Properties and Their Photochemical And Photoelectrochemical Applications. *Chem. Bull. "POLITEHNICA. Univ. (Timisoara),* **2009**, *54*(68), 1.

[44]   Serna, C.J.; Morales, M.P. 2. Maghemite (γ-Fe$_2$O$_3$): A versatile magnetic colloidal material. In: *Surface and Colloid Science*; Matijevic, C.; Borkovec, M., Eds.; Kluwer Academic/Plenum Publishers, **2018**; 17, pp. 27-81.

[45]   Wilson, N.C.; Muscat, J.; Mkhonto, D.; Ngoepe, P.E.; Harrison, N.M. Structure and properties of ilmenite from first principles. *Phys. Rev. B,* **2005**, *71*(7), 075202.
[http://dx.doi.org/10.1103/PhysRevB.71.075202]

[46]   Nabi, S.H.S. Ph.D. Thesis, Microscopic origin of magnetism in the hematite-ilmenite system. *Ludwig-Maximilians University,* **2010**.

[47]   Putnis, A. *Introduction to Mineral Sciences*; Cambridge University Press, **1992**.

[48]   Charilaou, M. Ph.D. Thesis. Magnetic thermodynamics in the FeTiO$_3$–Fe$_2$O$_3$ system: Experiment and modeling. **2012**. ETH Zurich, 2012 (DISS. ETH Nr. 20244)

[49]   Kato, H.; Yamaguchi, Y.; Yamada, M.; Funahashi, S.; Nakagawa, Y.; Takei, H. Neutron scattering study of magnetic excitations in oblique easy-axis antiferromagnet FeTiO$_3$. *J. Phys. C Solid State Phys.,* **1986**, *19*, 6993-7011.
[http://dx.doi.org/10.1088/0022-3719/19/35/013]

[50]   Wilson, N.C.; Muscat, J.; Mkhonto, D.; Ngoepe, P.E.; Harrison, N.M. Structure and properties of ilmenite from first principles. *Phys. Rev. B,* **2005**, *71*, 075202.
[http://dx.doi.org/10.1103/PhysRevB.71.075202]

[51]   Ribeiro, R.A.P.; de Lazaro, S.R. Structural, electronic and elastic properties of FeBO$_3$ (B 1/4 Ti, Sn, Si, Zr) ilmenite: a density functional theory study. *RSC Advances,* **2014**, *4*, 59839.

[http://dx.doi.org/10.1039/C4RA11320A]

[52]   Lilova, K.I.; Pearce, C.I.; Gorski, C. 3 Rosso, K.M.; Navrotsky, K.M. Thermodynamics of the magnetite-ulvöspinel (Fe_3O_4-Fe_2TiO_4) solid solution. *Am. Mineral.,* **2012**, *97*, 1330-1338.
[http://dx.doi.org/10.2138/am.2012.4076]

[53]   Barth, T.F.W.; Posnjak, E. Spinel structures: with and without variate atom quipoints. *Z. Kristallogr.,* **1932**, *82*, 325-341.

[54]   Forster, R.H.; Hall, E.O. A neutron and X-ray diffraction study of ulvospinel, Fe_2TiO_4. *Acta Crystallogr.,* **1965**, *18*(5), 857-862.
[http://dx.doi.org/10.1107/S0365110X65002104]

[55]   Groschner, C.; Lan, S.; Wise, A.; Leary, A.; Lucas, M.S.; Park, C.; McHenry, M.E. The role of atmosphere on phase transformations and magnetic properties of ulvospinel. *IEEE Trans. Magn.,* **2013**, *49*(7), 4273-4276.
[http://dx.doi.org/10.1109/TMAG.2013.2250928]

[56]   Xu, R. R.W. Mag Magnetic Properties of Ulvospinel. *Phys. Earth Planet. Inter.,* **1978**, (16), 196-199.

[57]   Bosi, F.; Hålenius, U.; Skogby, H. Crystal chemistry of the magnetite- ulvöspinel series. *Am. Mineral.,* **2009**, *94*, 181-189.
[http://dx.doi.org/10.2138/am.2009.3002]

[58]   Nakamura, S.; Fuwa, A. Local and dynamic Jahn-Teller distortion in ulvöspinel Fe_2TiO_4. *Hyperfine Interact.,*
[http://dx.doi.org/10.1007/s10751-013-0921-7]

[59]   Groschner, C.; Lan, S.; Wise, A.; Leary, A.; Lucas, M.S.; Park, C.; Laughlin, D.E. Diaz- Michelena, M.; McHenry, M.E. The role of atmosphere on phase transformations and magnetic properties of ulvospinel. *IEEE Trans. Magn.,* **2013**, *49*, 7.
[http://dx.doi.org/10.1109/TMAG.2013.2250928]

[60]   Noh, J.; Osman, O.; Aziz, S.G.; Winget, P.; Brédas, J. A density functional theory investigation of the electronic structure and spin moments of magnetite. *Sci. Technol. Adv. Mater.,* **2014**, *15* 044202 (8pp)
[http://dx.doi.org/10.1088/1468-6996/15/4/044202]

[61]   http://www.sandatlas.org/magnetite/

[62]   Hua, Q.; Huang, W. Chemical etching induced a shape change of magnetite microcrystals. *J. Mater. Chem.,* **2008**, *18*(36), 4286-4290.
[http://dx.doi.org/10.1039/b807212d]

[63]   Martin, F.; Arno, S.; Matthias, S. Ab initio study of the half-metal to metal transition in strained magnetite. *New J. Phys.,* **2007**, *9*(1), 5.
[http://dx.doi.org/10.1088/1367-2630/9/1/005]

[64]   Mahmoudi, M.; Sant, S.; Wang, B.; Laurent, S.; Sen, T. Superparamagnetic iron oxide nanoparticles (SPIONs): development, surface modification and applications in chemotherapy. *Adv. Drug Deliv. Rev.,* **2011**, *63*(1-2), 24-46.
[http://dx.doi.org/10.1016/j.addr.2010.05.006] [PMID: 20685224]

[65]   Tari, A.; Chantrell, R.W.; Charles, S.W.; Popplewell, J. The magnetic properties and stability of a ferrofluid containing Fe_3O_4 particles. *Physica B+C,* **1979**, *97*(1), 57-64.
[http://dx.doi.org/10.1016/0378-4363(79)90007-X]

[66]   Poizot, P.; Laruelle, S.; Grugeon, S.; Dupont, L.; Tarascon, J.M. Nano-sized transition-metal oxides as negative-electrode materials for lithium-ion batteries. *Nature,* **2000**, *407*(6803), 496-499.
[http://dx.doi.org/10.1038/35035045] [PMID: 11028997]

[67]   Mahmoudi, M.; Sant, S.; Wang, B.; Laurent, S.; Sen, T. Superparamagnetic iron oxide nanoparticles (SPIONs): development, surface modification and applications in chemotherapy. *Adv. Drug Deliv. Rev.,* **2011**, *63*(1-2), 24-46.

[http://dx.doi.org/10.1016/j.addr.2010.05.006] [PMID: 20685224]

[68] O'Handley, R.C. *Modern Magnetic Materials: Principles and Applications*; Wiley, **2000**.

[69] Rudolf, H.; Silvio, D.; Robert, M.; Matthias, Z. Magnetic particle hyperthermia: nanoparticle magnetism and materials development for cancer therapy. *J. Phys. Condens. Matter,* **2006**, *18*(38), S2919.
[http://dx.doi.org/10.1088/0953-8984/18/38/S26]

[70] Xavier, B.; Amílcar, L. Finite-size effects in fine particles: magnetic and transport properties. *J. Phys. D Appl. Phys.,* **2002**, *35*(6), R15.
[http://dx.doi.org/10.1088/0022-3727/35/6/201]

[71] Doefman, J.F.J. Spontaneous and induced magnetisation in ferromagnetic bodies. *Nature,* **1930**, *126*, 274-275.
[http://dx.doi.org/10.1038/126274a0]

[72] Owens, F.J.C.P.P., Jr *The Physics and Chemistry of Nanosolids*; Wiley, **2008**.

[73] Bean, C.P.; Livingston, J.D. Superparamagnetism. *J. Appl. Phys.,* **1959**, *30*(4), S120-S129.

[74] Babes, L.; Denizot, B.; Tanguy, G.; Jallet, P.; Jallet, P. Le Jeune JJ. Synthesis of iron oxide nanoparticles used as mri contrast agents: a parametric study. *J. Colloid Interface Sci.,* **1999**, *212*(2), 474-482.
[http://dx.doi.org/10.1006/jcis.1998.6053] [PMID: 10092379]

[75] Arbab, A.S.; Bashaw, L.A.; Miller, B.R.; Jordan, E.K.; Lewis, B.K.; Kalish, H.; Frank, J.A. Characterization of biophysical and metabolic properties of cells labeled with superparamagnetic iron oxide nanoparticles and transfection agent for cellular MR imaging. *Radiology,* **2003**, *229*(3), 838-846.
[http://dx.doi.org/10.1148/radiol.2293021215] [PMID: 14657318]

[76] Goya, G.F.; Berquó, T.S.; Fonseca, F.C.; Morales, M.P. Static and dynamic magnetic properties of spherical magnetite nanoparticles. *J. Appl. Phys.,* **2003**, *94*(5), 3520-3528.
[http://dx.doi.org/10.1063/1.1599959]

[77] Denizli, A.; Say, R. Preparation of magnetic dye affinity adsorbent and its use in the removal of aluminium ions. *J. Biomater. Sci. Polym. Ed.,* **2001**, *12*(10), 1059-1073.
[http://dx.doi.org/10.1163/15685620152691850] [PMID: 11853378]

[78] Suber, L.; Foglia, S.; Maria Ingo, G.; Boukos, N. Synthesis, and structural and morphological characterization of iron oxide–ion-exchange resin and –cellulose nanocomposites. *Appl. Organomet. Chem.,* **2001**, *15*(5), 414-420.
[http://dx.doi.org/10.1002/aoc.163]

[79] Chen, D.H.; Huang, S.H. Fast separation of bromelain by polyacrylic acid-bound iron oxide magnetic nanoparticles. *Process Biochem.,* **2004**, *39*(12), 2207-2211.
[http://dx.doi.org/10.1016/j.procbio.2003.11.014]

[80] Jeong, U.; Teng, X.; Wang, Y.; Yang, H.; Xia, Y. Superparamagnetic colloids: controlled synthesis and niche applications. *Adv. Mater.,* **2007**, *19*(1), 33-60.
[http://dx.doi.org/10.1002/adma.200600674]

[81] Li, Y.S.; Church, J.S.; Woodhead, A.L.; Moussa, F. Preparation and characterization of silica coated iron oxide magnetic nano-particles. *Spectrochim. Acta A Mol. Biomol. Spectrosc.,* **2010**, *76*(5), 484-489.
[http://dx.doi.org/10.1016/j.saa.2010.04.004] [PMID: 20452273]

[82] Majewski, P.; Thierry, B. Functionalized Magnetite Nanoparticles-Synthesis, Properties, and Bio-Applications. *Crit. Rev. Solid State Mater. Sci.,* **2007**, *32*(3-4), 203-215.
[http://dx.doi.org/10.1080/10408430701776680]

[83] Sugimoto, T.; Matijevic, E. Formation of uniform spherical magnetite particles by crystallization from ferrous hydroxide gels. *J. Coll. and Interf. Sci.,* **1980**, *74*(1), 227-243.

[http://dx.doi.org/10.1016/0021-9797(80)90187-3]

[84] Tang, B.; Yuan, L.; Shi, T.; Yu, L.; Zhu, Y. Preparation of nano-sized magnetic particles from spent pickling liquors by ultrasonic-assisted chemical co-precipitation. *J. Hazard. Mater.,* **2009**, *163*(2-3), 1173-1178.
      [http://dx.doi.org/10.1016/j.jhazmat.2008.07.095] [PMID: 18762377]

[85] LaMer, V.K.; Dinegar, R.H. Theory, Production and Mechanism of Formation of Monodispersed Hydrosols. *J. Am. Chem. Soc.,* **1950**, *72*(11), 4847-4854.
      [http://dx.doi.org/10.1021/ja01167a001]

[86] Chastellain, M.; Petri, A.; Hofmann, H. Particle size investigations of a multistep synthesis of PVA coated superparamagnetic nanoparticles. *J. Colloid Interface Sci.,* **2004**, *278*(2), 353-360.
      [http://dx.doi.org/10.1016/j.jcis.2004.06.025] [PMID: 15450454]

[87] Tartaj, P.; Morales, M.P.; González-Carreño, T.; Veintemillas-Verdaguer, S.; Serna, C.J. Advances in magnetic nanoparticles for biotechnology applications. *J. Magn. Magn. Mater.,* **2005**, *290–291*(Part 1), 28-34.
      [http://dx.doi.org/10.1016/j.jmmm.2004.11.155]

[88] Pedro, T.; María del Puerto, M.; Sabino, V.V.; Teresita, G.C.; Carlos, J.S. The preparation of magnetic nanoparticles for applications in biomedicine. *J. Phys. D Appl. Phys.,* **2003**, *36*(13), R182.
      [http://dx.doi.org/10.1088/0022-3727/36/13/202]

[89] Kim, D.K.; Zhang, Y.; Voit, W.; Rao, K.V.; Muhammed, M. Synthesis and characterization of surfactant-coated superparamagnetic monodispersed iron oxide nanoparticles. *J. Magn. Magn. Mater.,* **2001**, *225*(1–2), 30-36.
      [http://dx.doi.org/10.1016/S0304-8853(00)01224-5]

[90] Kim, D.K.; Mikhaylova, M.; Wang, F.H.; Kehr, J.; Bjelke, B.; Zhang, Y.; Muhammed, M. Starch-Coated Superparamagnetic Nanoparticles as MR Contrast Agents. *Chem. Mater.,* **2003**, *15*(23), 4343-4351.
      [http://dx.doi.org/10.1021/cm031104m]

[91] Kim, D.K.; Mikhaylova, M.; Zhang, Y.; Muhammed, M. Protective Coating of Superparamagnetic Iron Oxide Nanoparticles. *Chem. Mater.,* **2003**, *15*(8), 1617-1627.
      [http://dx.doi.org/10.1021/cm021349j]

[92] Lin, C.L.; Lee, C.F.; Chiu, W.Y. Preparation and properties of poly(acrylic acid) oligomer stabilized superparamagnetic ferrofluid. *J. Colloid Interface Sci.,* **2005**, *291*(2), 411-420.
      [http://dx.doi.org/10.1016/j.jcis.2005.05.023] [PMID: 16009367]

[93] Mohapatra, S.; Pramanik, N.; Mukherjee, S.; Ghosh, S.K.; Pramanik, P. A simple synthesis of amine-derivatised superparamagnetic iron oxide nanoparticles for bioapplications. *J. Mater. Sci.,* **2007**, *42*(17), 7566-7574.
      [http://dx.doi.org/10.1007/s10853-007-1597-7]

[94] Lee, H.; Lee, E.; Kim, D.K.; Jang, N.K.; Jeong, Y.Y.; Jon, S. Antibiofouling polymer-coated superparamagnetic iron oxide nanoparticles as potential magnetic resonance contrast agents for *in vivo* cancer imaging. *J. Am. Chem. Soc.,* **2006**, *128*(22), 7383-7389.
      [http://dx.doi.org/10.1021/ja061529k] [PMID: 16734494]

[95] Bergemann, C.; Müller-Schulte, D.; Oster, J.; Brassard, L.; Lubbe, A.S. Magnetic ion-exchange nano- and microparticles for medical, biochemical and molecular biological applications. *J. Magn. Magn. Mater.,* **1999**, *194*, 45.
      [http://dx.doi.org/10.1016/S0304-8853(98)00554-X]

[96] Rochelle, M.; Cornell, U. The Iron Oxides: Structure, Properties, Reactions, Occurrences and Uses. In: *Completely Revised and Extended Edition (Second ed.)*; Wiley.m, **2006**.

[97] Laurent, S.; Forge, D.; Port, M.; Roch, A.; Robic, C.; Vander Elst, L.; Muller, R.N. Magnetic iron oxide nanoparticles: synthesis, stabilization, vectorization, physicochemical characterizations, and

biological applications. *Chem. Rev.,* **2008**, *108*(6), 2064-2110.
[http://dx.doi.org/10.1021/cr068445e] [PMID: 18543879]

[98]   Massart, R. Preparation of aqueous magnetic liquids in alkaline and acidic media. *IEEE Trans. Magn.,* **1981**, *17*(2), 1247-1248.
[http://dx.doi.org/10.1109/TMAG.1981.1061188]

[99]   Gribanov, N.M.; Bibik, E.E.; Buzunov, O.V.; Naumov, V.N. Physico-chemical regularities of obtaining highly dispersed magnetite by the method of chemical condensation. *J. Magn. Magn. Mater.,* **1990**, *85*(1), 7-10.
[http://dx.doi.org/10.1016/0304-8853(90)90005-B]

[100] Tavakoli, A.; Sohrabi, M.; Kargari, A. A review of methods for synthesis of nanostructured metals with emphasis on iron compounds. *Chem. Pap.,* **2007**, *61*(3), 151-170.
[http://dx.doi.org/10.2478/s11696-007-0014-7]

[101] Pang, Y.X.; Bao, X. Aluminium oxide nanoparticles prepared by water-in-oil microemulsions. *J. Mater. Chem.,* **2002**, *12*(12), 3699-3704.
[http://dx.doi.org/10.1039/b206501k]

[102] Capek, I. Preparation of metal nanoparticles in water-in-oil (w/o) microemulsions. *Adv. Colloid Interface Sci.,* **2004**, *110*(1-2), 49-74.
[http://dx.doi.org/10.1016/j.cis.2004.02.003] [PMID: 15142823]

[103] Salazar-Alvarez, G. PhD Thesis. Characterisation and Applications of Iron Oxide Nanoparticles. *KTH Materials Science and Engineering, Stockholm, Sweden,* **2004**. http://kth.diva-portal.org/smash/get/diva2:14857/FULLTEXT01

[104] Dai, Z.; Meiser, F.; Möhwald, H. Nanoengineering of iron oxide and iron oxide/silica hollow spheres by sequential layering combined with a sol-gel process. *J. Colloid. Interface. Sci.,* **2005**, *288*(1), 298-300.

[105] Durães, L.; Costa, B.F.O.; Vasques, J.; Campos, J. Portugal, APhase investigation of as-prepared iron oxide/hydroxide produced by sol-gel synthesis. *Mater. Lett.,* **2005**, *59*(7), 859-863.
[http://dx.doi.org/10.1016/j.matlet.2004.10.066]

[106] Ismail, A.A. Synthesis and characterization of $Y_2O_3/Fe_2O_3/TiO_2$ nanoparticles by sol-gel method. *Appl. Catal. B,* **2005**, *58*(1-2), 115-121.
[http://dx.doi.org/10.1016/j.apcatb.2004.11.022]

[107] Lam, U.T.; Mammucari, R.; Suzuki, K.; Foster, N.R. Processing of Iron Oxide Nanoparticles by Supercritical Fluids. *Ind. Eng. Chem. Res.,* **2008**, *47*(3), 599-614.
[http://dx.doi.org/10.1021/ie070494+]

[108] Liu, X Q.; Tao, S.W.; Shen, Y.S. Preparation and characterization of nanocrystalline α-Fe_2O_3 by a sol-gel process. *Sensors and Actuators B: Chemical.,* **1997**, *40*(2), 161-165.

[109] Kijima, N.; Yoshinaga, M.; Awaka, J.; Akimoto, J. Microwave synthesis, characterization, and electrochemical properties of α-Fe_2O_3 nanoparticles. *Solid State Ion.,* **2011**, *192*(1), 293-297.
[http://dx.doi.org/10.1016/j.ssi.2010.07.012]

[110] Cannas, C.; Gatteschi, D.; Musinu, A.; Piccaluga, G.; Sangregorio, C. Structural and magnetic properties of fe2o3 nanoparticles dispersed over a silica matrix. *J. Phys. Chem. B,* **1998**, *102*(40), 7721-7726.
[http://dx.doi.org/10.1021/jp981355w]

[111] Ennas, G.; Musinu, A.; Piccaluga, G.; Zedda, D.; Gatteschi, D.; Sangregorio, C.; Spano, G. Characterization of iron oxide nanoparticles in an fe₂o₃–sio₂ composite prepared by a sol–gel method. *Chem. Mater.,* **1998**, *10*(2), 495-502.
[http://dx.doi.org/10.1021/cm970400u]

[112] Tavakoli, A.A.; Sohrabi, M.; Kargari, A. A review of methods for synthesis of nanostructured metals with emphasis on iron compounds. *Chem. Pap.,* **2007**, *61*(3), 151-170.

[http://dx.doi.org/10.2478/s11696-007-0014-7]

[113] da Costa, G.M.; De Grave, E.; de Bakker, P.M.A.; Vandenberghe, R.E. Synthesis and characterization of some iron oxides by sol-gel method. *J. Solid State Chem.*, **1994**, *113*(2), 405-412.
[http://dx.doi.org/10.1006/jssc.1994.1388]

[114] Raileanu, M.; Crisan, M.; Petrache, C.; Crisan, D.; Jitianu, A.; Zaharescu, M.; Predoi, D.; Kuncser, V.; Filoti, G. Sol-Gel FexOy – SiO$_2$ nanocomposites. *Rom. J. Phys.*, **2005**, *50*(5-6), 595.

[115] Tadić, M.; Marković, D.; Spasojević, V.; Kusigerski, V.; Remškar, M.; Pirnat, J.; Jagličić, Z. Synthesis and magnetic properties of concentrated α-Fe$_2$O$_3$ nanoparticles in a silica matrix. *J. Alloys Compd.*, **2007**, *441*(1–2), 291-296.
[http://dx.doi.org/10.1016/j.jallcom.2006.09.099]

[116] Xu, Z.Z.; Wang, C.C.; Yang, W.L.; Fu, S.K. Synthesis of superparamagnetic Fe$_3$O$_4$/SiO$_2$ composite particles *via* sol-gel process based on inverse miniemulsion. *J. Mater. Sci.*, **2005**, *40*(17), 4667-4669.
[http://dx.doi.org/10.1007/s10853-005-3924-1]

[117] Deng, Y.H.; Wang, C.C.; Hu, J.H.; Yang, W.L.; Fu, S.K. Investigation of formation of silica-coated magnetite nanoparticles *via* sol-gel approach. *Colloids Surf. A Physicochem. Eng. Asp.*, **2005**, *262*(1-3), 87-93.
[http://dx.doi.org/10.1016/j.colsurfa.2005.04.009]

[118] Wang, C.T.; Ro, S.H. Nanocluster iron oxide-silica aerogel catalysts for methanol partial oxidation. *Appl. Catal. A Gen.*, **2005**, *285*(1-2), 196-204.
[http://dx.doi.org/10.1016/j.apcata.2005.02.029]

[119] Bruni, S.; Cariati, F.; Casu, M.; Lai, A.; Musinu, A.; Piccaluga, G. Solinas, SIR and NMR study of nanoparticle-support interactions in a Fe$_2$O$_3$-SiO$_2$ nanocomposite prepared by a Sol-gel method. *Nanostruct. Mater.*, **1999**, *11*(5), 573-586.
[http://dx.doi.org/10.1016/S0965-9773(99)00335-9]

[120] Pecharrománn, C.; González-Carreño, T.; Iglesias, J.E. The infrared dielectric properties of maghemite, γ-Fe$_2$O$_3$, from reflectance measurement on pressed powders. *Phys. Chem. Miner.*, **1995**, *22*(1), 21-29.
[http://dx.doi.org/10.1007/BF00202677]

[121] González-Carreño, T.; Morales, M.P.; Gracia, M.; Serna, C.J. Preparation of uniform γ-Fe$_2$O$_3$ particles with nanometer size by spray pyrolysis. *Mater. Lett.*, **1993**, *18*(3), 151-155.
[http://dx.doi.org/10.1016/0167-577X(93)90116-F]

[122] Morjan, I.; Alexandrescu, R.; Soare, I.; Dumitrache, F.; Sandu, I.; Voicu, I.; Martelli, S. Nanoscale powders of different iron oxide phases prepared by continuous laser irradiation of iron pentacarbonyl-containing gas precursors. *Mater. Sci. Eng: C.*, **2003**, *23*(1–2), 211-216.
[http://dx.doi.org/10.1016/S0928-4931(02)00269-2]

[123] Dumitrache, F.; Morjan, I.; Alexandrescu, R.; Ciupina, V.; Prodan, G.; Voicu, I.; Soare, I. Iron-iron oxide core-shell nanoparticles synthesized by laser pyrolysis followed by superficial oxidation. *Appl. Surf. Sci.*, **2005**, *247*(1-4), 25-31.
[http://dx.doi.org/10.1016/j.apsusc.2005.01.037]

[124] Pedro Tartaj, P.; De Jonghe, L.C. Preparation of nanospherical amorphous zircon powders by a microemulsion-mediated process. *J. Mater. Chem.*, **2000**, *10*(12), 2786-2790.
[http://dx.doi.org/10.1039/b002720k]

[125] Schweiger, C.; Pietzonka, C.; Heverhagen, J.; Kissel, T. Novel magnetic iron oxide nanoparticles coated with poly(ethylene imine)-g-poly(ethylene glycol) for potential biomedical application: synthesis, stability, cytotoxicity and MR imaging. *Int. J. Pharm.*, **2011**, *408*(1-2), 130-137.
[http://dx.doi.org/10.1016/j.ijpharm.2010.12.046] [PMID: 21315813]

[126] Fievet, F.; Lagier, J.P.; Blin, B.; Beaudoin, B.; Figlarz, M. Homogeneous and heterogeneous nucleations in the polyol process for the preparation of micron and submicron size metal particles. *Solid State Ion.*, **1989**, *32*, 198-205.

[http://dx.doi.org/10.1016/0167-2738(89)90222-1]

[127] Jézéquel, D.; Guenot, J.; Jouini, N.; Fiévet, F. Submicrometer zinc oxide particles: Elaboration in polyol medium and morphological characteristics. *J. Mater. Res.,* **1995**, *10*(01), 77-83.
[http://dx.doi.org/10.1557/JMR.1995.0077]

[128] Joseyphus, R.J.; Kodama, D.; Matsumoto, T.; Sato, Y.; Jeyadevan, B.; Tohji, K. Role of polyol in the synthesis of Fe particles. Original Research article. *Magn. Magn. Mater.,* **2007**, *310*(2), 2393.
[http://dx.doi.org/10.1016/j.jmmm.2006.10.1132]

[129] Hao, Y.; Teja, A.S. Continuous hydrothermal crystallization of α–Fe$_2$O$_3$ and Co$_3$O$_4$ nanoparticles. *J. Mater. Res.,* **2003**, *18*(02), 415-422.
[http://dx.doi.org/10.1557/JMR.2003.0053]

[130] Xu, C.; Teja, A.S. Continuous hydrothermal synthesis of iron oxide and PVA-protected iron oxide nanoparticles. *J. Supercrit. Fluids,* **2008**, *44*(1), 85-91.
[http://dx.doi.org/10.1016/j.supflu.2007.09.033]

[131] Osuna, J.; de Caro, D.; Amiens, C.; Chaudret, B.; Snoeck, E.; Respaud, M.; Fert, A. Synthesis, characterization, and magnetic properties of cobalt nanoparticles from an organometallic precursor. *J. Phys. Chem.,* **1996**, *100*(35), 14571-14574.
[http://dx.doi.org/10.1021/jp961086e]

[132] Park, S.J.; Kim, S.; Lee, S.; Khim, Z.G.; Char, K.; Hyeon, T. Synthesis and magnetic studies of uniform iron nanorods and nanospheres. *J. Am. Chem. Soc.,* **2000**, *122*(35), 8581-8582.
[http://dx.doi.org/10.1021/ja001628c]

[133] Roshan, H.A.; Vaezi, M.R.; Shokuhfar, A.; Rajabali, Z. Synthesis of iron oxide nanoparticles *via* the sonochemical method and their characterization. *Particuology,* **2011**, *9*(1), 95-99.
[http://dx.doi.org/10.1016/j.partic.2010.05.013]

[134] van Wijngaarden, L. Mechanics of collapsing cavitation bubbles. *Ultrason. Sonochem.,* **2016**, *29*, 524-527.
[http://dx.doi.org/10.1016/j.ultsonch.2015.04.006] [PMID: 25890856]

[135] Suslick, K.S. *Sonochemistry Kirk-Othmer Encyclopedia of Chemical Technology*; New York John Wiley and Sons, Inc, **1998**, 26, pp. 516-541.

[136] Wang, X.; Chen, G.; Guo, W. Sonochemical degradation kinetics of methyl violet in aqueous solutions. *Molecules,* **2003**, *8*(1), 40.
[http://dx.doi.org/10.3390/80100040]

[137] Mason, T.J.; Lorimer, J.P. *General Principles Applied Sonochemistry,* **2003**, , 25-74.

[138] Suslick, K.S.; Didenko, Y.; Fang, M.M.; Hyeon, T.; Kolbeck, K.J.; McNamara, W.B.; Wong, M. Acoustic cavitation and its chemical consequences. Philosophical Transactions of the Royal Society of London. Series A. *Math. Phys. Eng. Sci.,* **1751**, *1999*(357), 335-353.
[http://dx.doi.org/10.1098/rsta.1999.0330]

[139] Gedye, R.; Smith, F.; Westaway, K.; Ali, H.; Baldisera, L.; Laberge, L.; Rousell, J. The use of microwave ovens for rapid organic synthesis. *Tetrahedron Lett.,* **1986**, *27*(3), 279-282.
[http://dx.doi.org/10.1016/S0040-4039(00)83996-9]

[140] Kijima, N.; Yoshinaga, M.; Awaka, J. Akimoto, j. Microwave synthesis, characterization, and electrochemical properties of α-Fe$_2$O$_3$ nanoparticles. *Solid State Ion.,* **2011**, *192*, 293-297.
[http://dx.doi.org/10.1016/j.ssi.2010.07.012]

[141] Cullity, B.D.; Graham, C.D. *Introduction to magnetic materials*; IEEE Press, and John Wiley & Sons, Inc, **2009**.

[142] Gupta, A.K.; Berry, C.; Gupta, M.; Curtis, A. Receptor-mediated targeting of magnetic nanoparticles using insulin as a surface ligand to prevent endocytosis. *IEEE Trans. Nanobioscience,* **2003**, *2*(4), 255-261.

[http://dx.doi.org/10.1109/TNB.2003.820279] [PMID: 15376916]

[143] Gupta, A.K.; Wells, S. Surface-modified superparamagnetic nanoparticles for drug delivery: preparation, characterization, and cytotoxicity studies. *IEEE Trans. Nanobioscience,* **2004**, *3*(1), 66-73.
[http://dx.doi.org/10.1109/TNB.2003.820277] [PMID: 15382647]

[144] Jain, T.K.; Morales, M.A.; Sahoo, S.K.; Leslie-Pelecky, D.L.; Labhasetwar, V. Iron oxide nanoparticles for sustained delivery of anticancer agents. *Mol. Pharm.,* **2005**, *2*(3), 194-205.
[http://dx.doi.org/10.1021/mp0500014] [PMID: 15934780]

[145] Berry, C.C.; Charles, S.; Wells, S.; Dalby, M.J.; Curtis, A.S.G. The influence of transferrin stabilised magnetic nanoparticles on human dermal fibroblasts in culture. *Int. J. Pharm.,* **2004**, *269*(1), 211-225.
[http://dx.doi.org/10.1016/j.ijpharm.2003.09.042] [PMID: 14698593]

[146] Han, K.H.; Frazier, A.B. Paramagnetic capture mode magnetophoretic microseparator for high efficiency blood cell separations. *Lab Chip,* **2006**, *6*(2), 265-273.
[http://dx.doi.org/10.1039/B514539B] [PMID: 16450037]

[147] http://www.biomagneticsolutions.com/magnetic-separations/

[147] Ngomsik, A.F.; Bee, A.; Draye, M.; Cote, G.; Cabuil, V. Magnetic nano- and microparticles for metal removal and environmental applications: a review. *C. R. Chim.,* **2005**, *8*(6–7), 963-970.
[http://dx.doi.org/10.1016/j.crci.2005.01.001]

[148] Kim, Y.G.; Song1, J.B.; Yang, D.G.; Lee, J.S.; Park, Y.J.; Kang, D.H and Lee, H.G.. Effects of filter shapes on the capture efficiency of a superconducting high-gradient magnetic separation system. *Supercond. Sci. Technol.,* **2013**, (26) 085002 (7pp).

[149] Frank, J.; Owens, C.P.P. *The Physics and Chemistry of Nanosolids*; Wiley, **2008**.

[150] Azhar Uddin, M.; Tsuda, H.; Wu, S.; Sasaoka, E. Catalytic decomposition of biomass tars with iron oxide catalysts. *Fuel,* **2008**, *87*(4–5), 451-459.
[http://dx.doi.org/10.1016/j.fuel.2007.06.021]

[151] Li, C.; Shen, Y.; Jia, M.; Sheng, S.; Adebajo, M.O.; Zhu, H. Catalytic combustion of formaldehyde on gold/iron-oxide catalysts. *Catal. Commun.,* **2008**, *9*(3), 355-361.
[http://dx.doi.org/10.1016/j.catcom.2007.06.020]

[152] Roberts, T.P.L.; Chuang, N.; Roberts, H.C. Neuroimaging: do we really need new contrast agents for MRI? *Eur. J. Radiol.,* **2000**, *34*(3), 166-178.
[http://dx.doi.org/10.1016/S0720-048X(00)00197-2] [PMID: 10927159]

[153] Deraz, N.M. Size and crystallinity-dependent magnetic properties of copper ferrite nano- particles. *J. Alloys Compd.,* **2010**, *501*(2), 317-325.
[http://dx.doi.org/10.1016/j.jallcom.2010.04.096]

[154] Sugimoto, M. The Past, Present, and Future of Ferrites. *J. Am. Ceram. Soc.,* **1999**, *82*(2), 269-280.
[http://dx.doi.org/10.1111/j.1551-2916.1999.tb20058.x]

[155] Sugimoto, T.; Matijević, E. Formation of uniform spherical magnetite particles by crystallization from ferrous hydroxide gels. *J. Colloid Interface Sci.,* **1980**, *74*(1), 227-243.
[http://dx.doi.org/10.1016/0021-9797(80)90187-3]

[156] Sun, Z.; Liu, L.; Jia, D.Z.; Pan, W. Simple synthesis of $CuFe_2O_4$ nanoparticles as gas- sensing materials. *Sens. Actuators B Chem.,* **2007**, *125*(1), 144-148.
[http://dx.doi.org/10.1016/j.snb.2007.01.050]

CHAPTER 7

Potential of Nanoparticles for the Development of Polymeric Membranes

Zulfiqar Ahmad Rehan[1,*], Iqbal Ahmed[2,3], L Gzara[2], Tanveer Hussain[1] and **Enrico Drioli[4]**

[1] *Department of Polymer Engineering, National Textile University Faisalabad, Faisalabad 37610, Pakistan*

[2] *Center of Excellence in Desalination Technology, King Abdulaziz University, Jeddah 21589, Saudi Arabia*

[3] *Department of Mechanical Engineering, King Abdulaziz University, Jeddah 21589, Saudi Arabia*

[4] *Institute on Membrane Technology (ITM-CNR), c/o University of Calabria, via P. Bucci 17/C, 87030 Rende, (CS), Italy*

Abstract: Since the last century, polymeric membrane technology has been offering economical and efficient solutions for emissions control and energy security. However, polymeric membranes in separation and purification environment, are facing challenges such as correct material selection, polymer modification for the resolution of permeability, selectivity, low resistance to fouling and low mechanical strength. Among all these problems, the three types of fouling *i.e.*, inorganic, organic and biological are the critical applications of membrane issue. Technically, organic and inorganic fouling can be controlled or minimized but the practical approaches to control biological fouling, particularly the membranes, used in the seawater desalination application are still facing issues, and the limitations to control bio-fouling. Recently, nanoparticles have shown real potential for the improvement of the membrane life. This chapter gives a comprehensive review of the critical aspects of the preparation and use of polymeric membranes in combination with standard additives and their influence on membrane permeability and selectively. The recent development of polymeric membranes loaded with nanoparticles against bio-fouling has also been reviewed in this chapter.

Keywords: Development, Fouling, Nanoparticles.

---

* **Corresponding author Zulfiqar Ahmad Rehan**: Department of Polymer Engineering, National Textile Research Center, National Textile University Faisalabad, Faisalabad 37610, Pakistan; Tel: 092-041-9230081; Email: zarehan@ntu.edu.pk

Sher Bahadar Khan (Ed.)

INTRODUCTION

Freshwater scarcity and fossil fuel energy consumption cause environmental problems and sustainability issues yet, in water and energy-rich countries. Globally, the rapid expansion of economic activity is demanding the growth of energy expenditure and deficiency of soft water is increasing with time, and by coming decades, this extension will become worst specially the availability of freshwater [1]. Fig. (**1**) gives a comprehensive representation of the freshwater scenario. Over the earlier period, membrane technology has become a necessary separation technology, particularly for seawater desalination and is an encouraging novel way to manage and overcome the freshwater thread in the future. Principally, pressure driven membrane is works at ambient conditions and no hazardous chemicals are required during freshwater production with simple separation processes, well-understood operating parameters and environmental friendliness. Therefore, thanks to the recent development of energy efficient membrane process such as pressure restarted osmosis process (PRO) and forward osmosis (FO) process in seawater desalination industry [2].

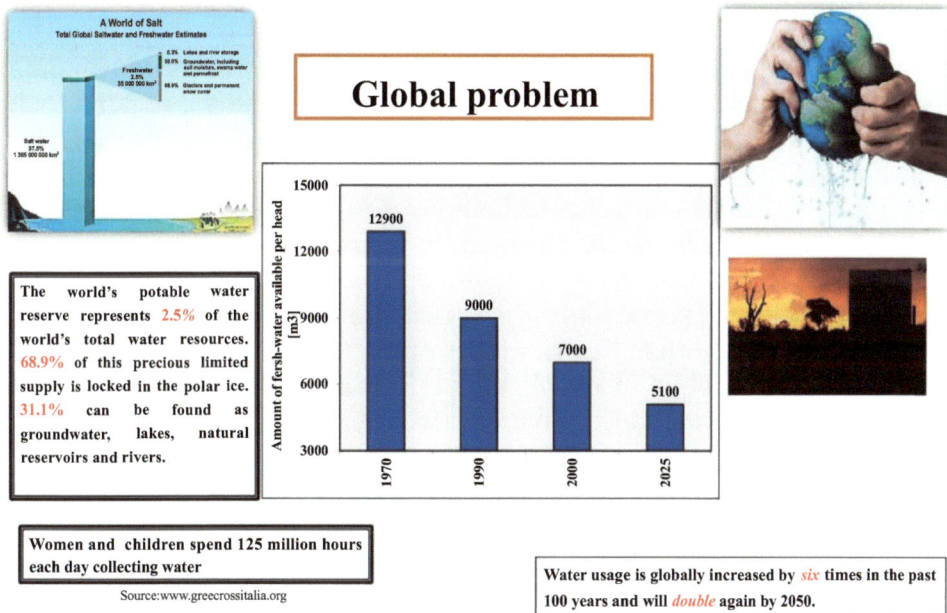

Fig. (1). A comprehensive scenario of portable water in world.

The first commercial membrane was invented in 1960s, right after the discovery of semipermeable membranes. Several types of organic and inorganic polymeric materials have been explored together with efficient fabrication techniques of the polymeric membrane. With considerable progress in polymeric membrane materials, various kinds of membrane applications in separation and purification industry are available commercially such as RO, nanofiltration (NF), ultrafiltration (UF), microfiltration (MF), dialysis, electrodialysis, pervaporation (PV), membrane distillation (MD), *etc.* The membranologists are still exploring best membrane materials and energy efficient membrane process for particular application. Recently, some of the state-of-the-art new methods such as membrane PRO, FO, reverse electrodialysis (RED), membrane contractor, catalytic membrane reactors and fuel cell membranes, respectively have been extensively explored and have reported a high prospect for application in filtration, separation and purification process industry [3 - 5].

In separation and purification process, the membrane process is more reliable regarding product yield without the hazardous by-product, and is a dominant technology especially in saline water, wastewater, food and pharmaceutical as well as the electronics industry, respectively. As compared to other separation and purification process, membrane technology is still facing a steady state performance issues concerning time regarding permeate and rejection rate due to the fouling phenomenon [5].

Their membrane systems in water treatment process are facing most common kinds of fouling such as crystalline fouling, particulate matters, organic fouling, colloidal fouling and biological fouling due to the protein in feed, respectively [6]. The membranes in water application, the particulate matters and colloidal fouling persisted the prime reason for the declining of permeate and rejection even after eras of significant membrane development in industrial wastewater management [7]. Due to the wetting affinity of a pressure driven membrane having a drawback to be fouled closer, particularly biological fouling or either due to the accumulation of an additional barrier layer or due to a failure of the barrier, the consequence is a reduction of membrane performance. Other process parameter and operational conditions also have an ascendancy on the degree of fouling. Terrible type membrane fouling may need massive amounts of chemical cleaning or replacement of membrane. As a result, processing expenditures of plant maintenance are therefore, more significant than before [8].

Membrane fouling and biofouling originate due to the affection of solutes and microbial growth onto the surface of the membrane, respectively. Fig. (**2**) shows typical fouling on the membrane surface. The implementation of fouling and biofouling inhibitors is also a supplementary obstruction of the membrane pores thwarting the permeate from transporting through the membrane, hence elevates the trans-membrane pressure and lowering the productivity. Technically, the membrane biofouling phenomenon is a dynamic process due to the colonization of microbial activity at the membrane surface [9]. Once this microbial colonization indelibly adhered, these organisms in the direction of generating extracellular polymeric secretions (EPS) involved polysaccharides, lipoproteins, proteins, glycoproteins, and other bio-macromolecules, respectively [10]. The development of fully grown biofilm is due to the amassing of EPS and reproduction of bacteria [11 - 14].

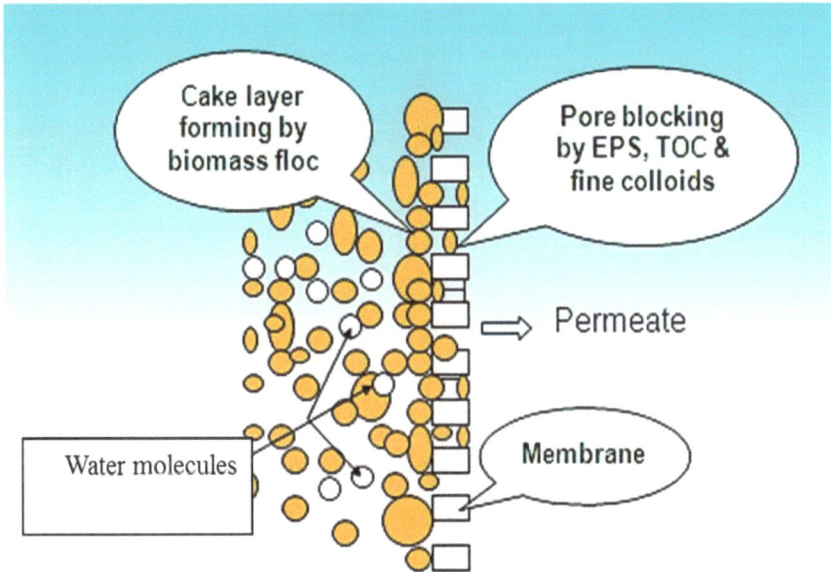

Fig. (2). Disadvantages of fouling and bio-fouling.

Perhaps, several ways are accessible to settle down the fouling problem, for instance, pre-treatment, the modifications of membrane surface, integration of ultrasonic, membrane cleaning using chemical and physical processes. Membrane cleaning by physical methods like backwashing (when permeate is applied to flush the membrane backward) and lessening (when no filtration takes place) has been integrated into standard functional approaches [15]. However, the most critical field in the research is membrane surface modification and has become one of the besieged for membranologists. It has been believed that developing more hydrophilic membranes could reduced membrane fouling [16]. The most common methods for the modification of water treatment membrane surface are blending [17], grafting [18], surface chemical reaction [19, 20] and nanoparticle incorporation [27]. The existence of diaphanously of inorganic nanoparticles dispersion in the polymeric matrix has been established to be very advantageous to settle down the membrane performance for a broad spectrum of processes, collection from gas separation and pervaporation [22, 23], to ultrafiltration, microfiltration and nanofiltration [7, 24].

NANOCOMPOSITE MEMBRANES

In the current prospect, the meaning of nano-composite membranes is the combination of nanoparticles (NPs) and polymeric membranes. Nanocomposite membranes are classified as,

  i. Thin-film nanocomposite membranes (TFNC)
 ii. Mixed matrix nanocomposite membranes (MMNM)

Thin-film nanocomposite membranes (TFNC) fabricate by using dip-coating or deposition techniques, and nanoparticles are endorsed to self-construct towards membrane surface. Kim *et al.* [28] have applied interfacial polymerization (IP) of 1,3-benzenediamine (C_6H_{4-1},3-$(NH_2)_2$) and 1,3,5-Benzenetricarbonyl trichloride ($C_6H_3(COCl)_3$) for the fabrication of polyamide thin film composite (TFC) membrane onto the PSF support. Likewise, Li *et al.* [29] have used sol-gel method for the fabrication of self-assembly of NPs in the PES polymer matrix, in which tetrabutyl titanate ($Ti(OCH_2CH_2CH_2CH_3)_4$) was applied as NPs precursor. While, Wang *et al.* [30] have used the similar technique with PSF polymer, it was observed in both the studies that decrease in contact angle due to hydrophilicity induced by NPs.

Mixed matrix nanocomposite membranes (MMNM) hence, is also known as nanoparticles-blend membranes prepared by dispersing nanoparticles onto the polymer dope solution before membrane casting. Inorganic micro-materials such as silica, zirconia, and alumina, respectively have been used formerly, and all these micro-materials are used mostly as fillers that enhanced the efficiency of membrane process by improving the permeate as well as improving the salt rejection and the thermal, mechanical and chemical stabilities [31 - 36].

However, most of the inorganic nanoparticles (NPs), *e.g.*, silver, cobalt, Zinc oxide, copper oxide, TiO_2, *etc.* used in membrane modification are either metal or metal oxide form. The significant of using metal oxide NPs is well-developed cost effectiveness synthesis techniques and perfection in hydrophilicity of polymeric membrane. To evaluate the polymeric membrane fouling performance recently, several studies have been issued in the works unfolding studies on membranes modifications by using self-assembled NPs such as dip coating, sol-gel, blended, impregnated or embedded, pressurized deposition techniques, respectively [61 - 72]. Yu *et al.* reported that the hydrophilicity of SiO_2 NPs is due to the hydrogen bonding between the hydroxyl group of water and SiO_2, which repels hydrophobic foulant protein and BSA molecule, resulting in excellent antifouling ability and higher permeate recovery of the SiO_2/PVC nanocomposite UF membrane prepared by NIPS method [55]. Rabiee *et al.* reported the anti-fouling performance of TiO_2/EPVC composite membrane by performing flux recovery experiment, in which pure water flux as well as the flux of model foulants solution (500 ppm bovine serum albumin solution), was calculated in another cycle at different time intervals [62].

The method of fabrication, polymer material and performance of these membranes is summarized in Table **2**. It was also investigated that ZnO nanoparticles blended with polymer matrix increase the hydrophilicity, improve antifouling performance and flux recovery [72]. Some of the published work related to the incorporation of ZnO NP in different polymers is described in Table **3**. The metal oxides NPs such as TiO_2 and ZnO, which are photo-catalytic, hydrophilic and anti-bacterial, and silver NPs which possess an effective anti-bacterial and hydrophilic properties were used for antifouling membranes [79].

Commercialization of NPs based polymeric membranes cannot considered without continuously emerging novel NPs with unique morphologies and properties. Such as Ag-based NPs (Table **1**). TiO_2 NPs for membrane fabrication composite membranes could be applied for the anti-bacterial persistence, while TiO_2-based NPs-membranes bargain the applications in dye degradation and removal as well as bacterial disinfection. Contrasting photo-catalytic TiO_2-, SiO_2- or Al_2O_3- based NPs, carbon nanotechnology has fascinated more consideration, since the past periods because of its surprisingly different properties compared with their pre-cursors and macro-sized or micro-sized counterparts.

Table 1. TiO_2 NPs for membrane fabrication.

Polymer Material	Technique/Method	Performance as Rejection	Ref.
PVDF	Blending	BSA, dye	[37 - 39]
Polysulfone, PVDF	Blending	Activated sludge	[40]
Polyethersulfone	sol-gel	BSA	[29, 41]
PES coated with Polyamide	Self-fabricated	$MgSO_4$	[42]
TFC Polyamide	Self-fabricated	Sodium chloride	[28]
Sulfonated Polyethersulfone	Self-fabricated	Activated sludge	[43, 44]
Polyethersulfone	Self-fabricated	PEG-5000	[45]
Polyethersulfone	Self-fabricated and blending	Milk (non-skimmed)	[46]
PVDF/Poly (styrene-alt-maleic anhydride)	Self-fabricated	BSA	[47]
Polyethersulfone	Blending	NaCl	[48, 49]

(Table 1) contd.....

Polymer Material	Technique/Method	Performance as Rejection	Ref.
PVC Emulsion	Blending	BSA	[50]
PVDF/ (SPES)	Self-fabricated	BSA	[51]
Polyethersulfone	Blending	BSA, Whey	[52 - 54]
Polysulfone	sol-gel	BSA	[55]
(SPES)/PVDF	Blending	BSA	[56]
PES coated with PVA	Self-fabricated	NaCl	[57]
Polysulfone /Chitosan	Blending	NaCl	[58]
Polysulfone	Blending	BSA, humic acid	[59, 60]

Table 2. SiO$_2$ NPs for membrane fabrication.

Polymer Material	Technique	Performance as Rejection	References
Polyethersulfone	Blending	BSA	[73]
PVDF/Glycidyl methacrylate	Blending	NaCl	[74]
PVDFHF	sol-gel	BSA	[75]
Polysulfone laminated with polyamide	Dip-coating of commercial as well as as-synthesized NPs	Isopropanol, dioxane, NaCl	[76]
PVC	Blending	Waste water of sewage treatment plant, BSA	[55]
PSf/PVA	Blending	Na$_2$SO$_4$	[77]
PVC	sol-gel	BSA	[78]

Table 3. ZnO and Miscellaneous NPs for membrane fabrication.

Polymer Material	Technique	Performance as Rejection	References
ZnO			
Polyethersulfone	Blending	BSA, humic acid	[81, 82]
PVDF	Blending	Sodium alginate, humic acid, BSA	[72, 83, 84]
PSf	Blending	Oleic acid	[85]
PVDF	Dip-coating of pre-treated PVDF as well as blending	Copper ion, BSA	[86]
Polyethersulfone	sol-gel	BSA	[87]

Carbon nanotubes as a form of one-dimensional nanomaterial carbon nanotube (CNT) and an allotrope of carbon, multiwall carbon nanotubes (MWCNTs) are acknowledged for their outstanding surface mechanical, thermal, adsorption and chemical properties, respectively [80]. The primary significance of using CNTs, especially MWCNTs, is their natural scale-up, higher purity and low production cost (estimated to be $100/kg) reported in [88] and $1/kg published in [89]. Recently, the graphene oxides which are two-dimensional nanomaterials are another form of carbon nanomaterial invented after CNTs. However, the graphene oxides hold remarkably high surface area to weight ratio giving rise to outstanding thermal, mechanical and of course, chemical stabilities. Technically, graphene oxide (GO) is the oxidized form of graphene and is preferred over graphene due to its hydrophilic nature, owing to the presence of polar hydroxyl and carboxyl groups. Various studies have reported different synthesis techniques of GO; however, Hummers' method [90] and its improved version [91] are the most widely used methods. Some of the applications of carbon-based nanocomposite membranes in recent years are listed in Table **4**.

Table 4. **Carbon-based nano-materials for membrane modification.**

Polymer Material	Technique/ Material	Performance of Membranes	References
Polysulfone	Blending/Modified MWCNTs	BSA,PVP-55K, PEO-100K, PEG-20K,	[92, 98]
Brominated polyphenylene oxide	Blending/ MWCNTs	Egg albumin	[93]
Polypropylene and PES	Surface coating/ MWCNTs	Dyes	[94]
Polyethersulfone	MWCNTs/blending	BSA	[95]
Polyacrylonitrile	Blending/ MWCNTs	–	[96]
PES coated with chitosan	Blending with chitosan /MWCNTs	NaCl, Na_2SO_4, $MgSO_4$	[97]
Polyethersulfone	Blending/ MWCNTs	Sodium alginate	[98]
PVDF	blending/GO	BSA, yeast, NOM	[99 - 102]
PVDF	blending/GO-Oxidized MWCNTs	BSA	[103]
PVDF	Blending/GO-MWCNT	BSA	[104]
Polyethersulfone	blending/GO	Dye, milk powder solution	[105]
PVDF	Functionalized GO/blending	BSA	[106]
Polysulfone	rGO&rGO-PANI/blending	NaCl	[107]

Zhao *et al.* [101] have reported the anti-fouling behaviour of polysulfone UF isocyanate-treated GO (iGO) membranes and using BSA and ovalbumin as model foulants for filtration experiment. It was established that the significant value of negative zeta potential and augmented hydrophilicity were responsible for the anti-fouling effect. Innovative nanomaterials are being developed that are produced from the composites of different types of nanoparticles, *viz.* GO–TiO_2, TiO_2–SiO_2 and Ag–SiO_2, respectively. These are referred, in the present context, as nanoparticle (NP) composites and Table **4** has summarised various types of NPs for the synthesis of antifouling membranes.

NANOPARTICLE COMPOSITE MEMBRANES

To combat drinking water scarcity and energy crisis for sustainable development, there is an increasing demand for innovative membrane materials. Several chemical and material science researchers have been worked out on exclusively unique types of nanoparticle composites membranes with enhanced properties compared to NPs, such as coating of NPs over microspheres, decorating NPs with some other nanomaterials, incorporating NPs with some other nanomaterial and functionalization of NPs, respectively. Since some decades, Silver NPs are known for the antibacterial or bactericidal effect [111]. TiO_2 has been exploited enough for its photocatalysis and photo-degradation applications. Table **5** summarised nanoparticles synthesis techniques and its use for polymeric membranes modification and their form.

Table 5. Nanoparticle composites for membrane modification.

Membrane Material	Method/ Composite Material	Feed for Performance	References
Polyamide on porous PSf membrane	sol-gel & self-fabricated/ TiO_2–SiO_2	1.0 g/liter aqueous solution PEG 600	[112]
Polysulfone	Blending/ Cerium doped nano-silica particles	emulsion (Oil-in-water)	[113]
Polysulfone	Blending/Sulfated Y-doped zirconia	emulsion (Oil-in-water)	[114, 115]
Polyethersulfone	Blending/Silver decorated sodium zirconium phosphate	1000 mg/L BSA solution in phosphate buffer	[116]
Polysulfone	Blending/ Acid modified MWCNT-Ag	2.0 g/liter Na_2SO_4 and NaCl solution	[117]
Cellulose Acetate	self-fabricated/ TiO_2-GO	20 ppm humic acid in water	[118]
Polyethersulfone	Blending/ Ag–SiO_2 microsphere	500 mg/L PEG20000 and 1 g/L BSA	[119, 120]
Polysulfone	self-fabricated/ TiO_2-GO	50 ppm solution of methylene blue	[121]
Polysulfone	blending/ GO– SiO_2	1000 mg/L BSA solution	[122]
Alumina support with PVA coating	self-fabricated & sol-gel/ Ag–TiO_2	500 ppm PEG 20000	[123]
PVDF	blending/ rGO– TiO_2	500 mg/L BSA solution	[124, 125]

MEMBRANE CLASSIFICATION

Membranes are categorized by the types and nature of the materials, the membrane geometry, morphology, preparation method, separation regime and processes [126, 127]. A representative membrane classification is shown in Fig. (**3**). Technically, membranes are grouped into organic (polymeric) and inorganic membranes based on the membrane materials, the organic polymeric membranes, including amorphous, glassy, crystalline and rubbery are suitable for the membrane manufacturing. Polymeric composite membrane synthesis methods involve phase inversion, coating, stretching and interfacial reaction, respectively. Among these preparing methods, the phase inversion is the primary approach to prepare most of the polymeric membranes.

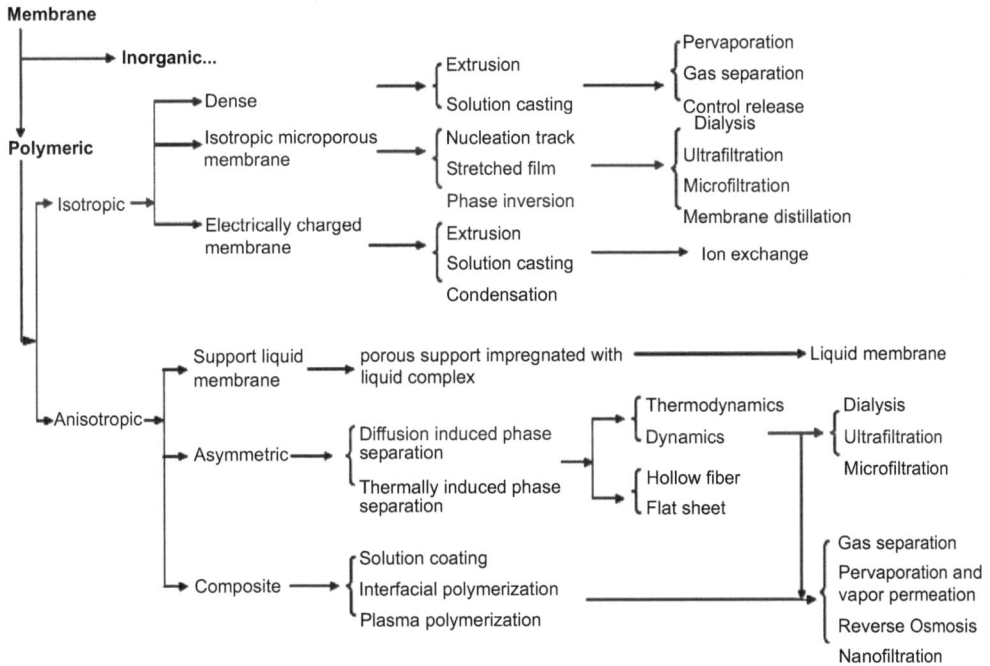

Fig. (3). Schematic diagram of membrane classification [128].

MEMBRANE MATERIALS

The polymeric type of membrane materials are based on such a polymer (NPS composite membrane), glass, ceramic, carbon and metal to make composite membranes or liquid/liquid, liquid/solid, liquid/gas, gas/gas and crystallizer application respectively. Currently, the industrial grade membranes are primarily prepared from polymeric materials. For synthetic polymeric membrane constituents, the polymers should validate substantial tensile strength, thermal stability over a wide range of temperature and possess chemical stability over a range of pH. Besides, they can also be processed into a flat sheet or hollow fiber membranes, according to the required application.

The first commercial membranes were based on cellulose acetate and polysulfone. Afterward, other polymeric materials were exploited for membranes such as polyamide (PA), polyimide (PI), polyetherimide, polyacrylonitrile (PAN), polysulfone (PSf), polyethersulfones (ES), polyvinylidene fluorides (PVDF) and polycarbonate (PC), respectively [129]. Here, we will discuss Polyethersulfone in detail.

Polysulfone and Polyethersulfone

The amorphous type polymeric materials, which are usually economical with significant selectivity and capable of being processed are one kind of ruling content in the membrane separation technology particularly, in pressure driven and gas separation type of membranes. Currently, available polymeric materials offer the ultimate promise of both academic research and industrial application. The amorphous sulfone polymer family includes polyphenyl sulfone, polysulfone and polysulfone blends, polyethersulfone and its blends. These polymers are characterized as ultra-efficient resins because they offer exceptional performance in specific areas include polyamide-imide, aromatic polyketone and PVDF for membranes production [130]. The chemical structures of polysulfone and polyethersulfone are shown in Fig. (4).

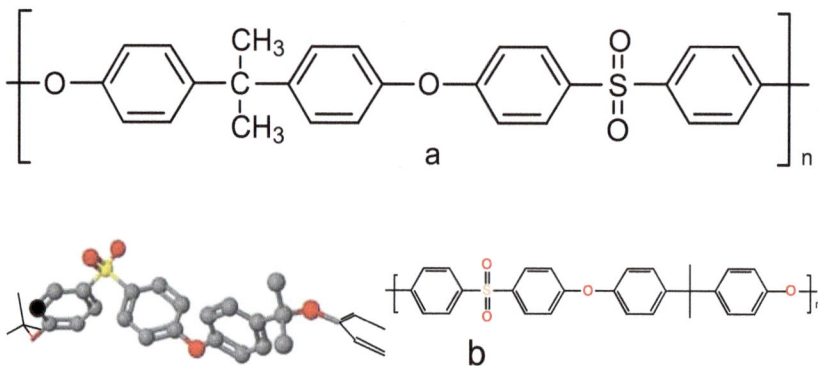

Fig. (4). Chemical structure of Polysulfone (a) and structure of polyethersulfone (b).

Polyethersulfone (PES) and PSF have been used by many researchers in the preparation of hollow fiber or flat sheet membranes as membrane-forming material. Polyethersulfone is a closely related derivative of polysulfone which is entirely devoid of aliphatic hydrocarbon groups and has a higher glass transition temperature of 230°C. Thus, many membranologists prefer to use PES due to its lower hydrophobicity, when compared to other polymers commonly used for membrane preparation [131]. The excellent hydrolytic reliability that distinguished PSF from other thermoset or thermoplastics is a result of the sustainability of aqueous hydrolysis of both the phenylene sulfone and ether groups.

PES consists of phenylene ring structures connected with sulfone ether linkages (-O-) OR groups (SO_2) in the backbone structure to make a polymer. The sulfone groups shift to form the polymer stiff with a high glass transition temperature (230°C) and, together with the ring structures, tend to make the polymer chemically resistant [130, 131]. The ether linkages, derived from the polymer's precursor, make it more flexible and processable.

THE SOLVENT CHOICE OF POLYMERIC MEMBRANES

The escalation of polymeric membrane solvent is significant for NCM and TFC membrane manufacturing as it affects the following separation performance. In asymmetric membranes prepared by phase separation methods, the macrophase separation can be perceived for blended membranes cast from polymer/solvent solution [131, 132]. The selections of solvent are principally imperative in the nucleophilic aromatic substitution polymerization reaction. Before the choice of aprotic or non-aprotic solvent for selected polymer solution, the necessities of the chosen solvent for the polymer have three significant factors such as:

a. It is essential that the solvent does not undergo any reaction with the reactants prominent to side products.
b. The solvation dispersion command of the solvent must be such that all of the reactants and products have sufficient solubility to react with each other.
c. The solvent should be able to aid in the dissociation of the nucleophilic anion from the metal cation associated with it. Nucleophilic substitution reactions are frequently carried out in polar aprotic solvents as opposed to aprotic solvents or either hydroxyl group or amine group based solvent. The intention is that the nucleophilicity of the nucleophile is exceedingly reliant on solute/solvent interactions.

Protic solvents will exceedingly solvate the nucleophile, thereby minimizing its tendency to react with the activated carbon; examples are acetone, alcohol and water. The relative nucleophilicity of anion changes occurred because of the absence of protic solvents, and it is suggested that, presumably due to the removal of solvation consequences [133]. The most commonly cast-off solvents are N-methyl pyrrolidone (NMP), dimethylacetamide (DMAc), dimethylformamide (DMF), dimethylsulfoxide (DMSO), tetrahydrofuran THF, *etc*. The precipitations of non-solvent bath are often water or a mixture of water and solvent.

Fig. (**5**) summarised some of most common polar aprotic solvents used in polymer dissolution for the fabrication of membrane. These solvents not only enhance the nucleophilicity of the nucleophile by strongly solvating the cations associated with the nucleophiles but they can also reduce the solvation of the nucleophiles [133].

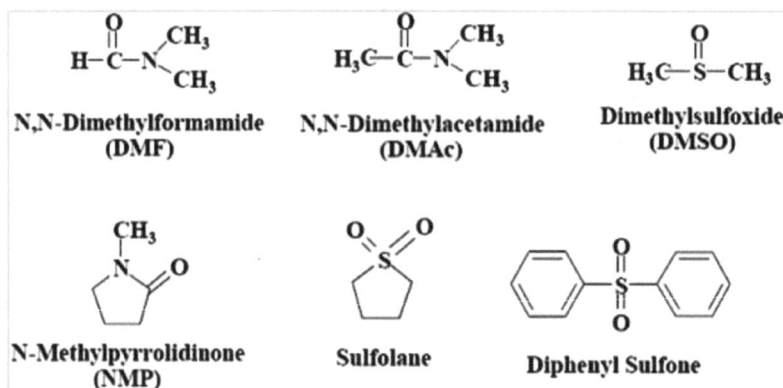

Fig. (5). Examples of aprotic polar solvent [134].

However, several ones or more solvent pairs consequence in altered membrane morphology and as well as the performance of the polymeric membranes. Swier *et al*. have used several different solvents for the fabrication of blends of sulfonated polyetheretherketone (SPEKK)/PES membranes and investigated the morphology by changing the casting solvent and temperature [134]. The membrane cast from NMP shows normal morphology in comparison with the membrane prepared from DMAc and the DMAc based membranes appeared to possess a more co-continuous morphology cast at the same casting temperature. It is thus, reasonable to hypothesize that the interaction of polymer and solvent during casting, as well as by residual solvent in the as-cast membrane will be strongly influenced on to the membrane properties, which are correlated to membrane morphology. For symmetric (dense) membrane also, fabrication conditions can markedly affect permeation properties [135]. Almost all the classes of polymeric membrane, performance strongly depends on both the intrinsic material properties and the membrane synthesis circumstances.

Strathmann reported filtration properties and porosities of membranes prepared from a polyimide-DMAc-water and DMAC-benzene mix solvent system [136]. Several studies revealed that by changing the proper solvent, the polymeric membrane morphology could be controlled. The morphological transformations seem to be directed by solvent volatility pooled with polymer-solvent interactions, which appear to be correlated to the conductivity difference between the SPES–NMP and SPES–DMAc membranes. Yun *et al.* described that the interaction force amongst solvent and polymer resulted in the differentiation of thermodynamic state and gelation kinetics of the casting solution, and further changed the membrane morphology and performance [137]. According to the solubility parameter principle, the polymer is more accessible to dissolve in a solvent, if solubility parameters of the polymer are closer to that of the solvent.

PHASE INVERSION MEMBRANES

Strathmann reported filtration properties and porosities of membranes prepared from a polyimide-DMAc-water and DMAC-benzene mix solvent system [136]. Several studies said that by using an appropriate solvent for casting solution, the membrane morphology could be controlled [131 - 137]. The morphological changes appear to be governed by solvent volatility combined with polymer-solvent interactions, which seem to be correlated to the conductivity difference between the SPES-NMP and SPES-DMA membranes. Yun *et al.* reported that the interaction force between solvent and polymer resulted in the differentiation of thermodynamic state and gelation kinetics of the casting solution, and further changed the membrane morphology and performance [137]. According to the solubility parameter principle, if solubility parameters of the polymer were closer to that of the solvent, the polymer was more comfortable to dissolve in this solvent.

 i. Thermally induced phase separation (TIPS): The precipitation is induced by decreasing the temperature of the polymer solution.
 ii. Diffusion induced phase separation (DIPS): Diffusional mass exchange leads to a change in the local composition of the polymer film, and the precipitation is influenced by approaching a polymer solution to a vapor or liquid. Fig. (**6**) summarised three kinds of techniques which have been developed to reach DIPS.
 iii. Immersion precipitation: The cause of rapid precipitation of the casting solution from the top surface downward will have occurred, of the casting solution is immersed into a non-solvent bath, and the non-solvent from the coagulation bath diffuses into the casting solution and the solvent diffuses from the casting solution into the non-solvent bath.
 iv. Vapor adsorption: The adsorption of the non-solvent will cause the

precipitation of the casting solution, and the casting solution should be contained to a vapor containing non-solvent and some inner or ordinary gases such as helium, nitrogen, or air.

v. Solvent evaporation: The polymer dope solution should be synthesized with a volatile solvent such as acetone, dichloromethane, tetrahydrofuran, *etc.* and a less or no volatile non-solvent and a polymer. Preferential evaporation of the volatile solvent will produce meta- or unstable compositions and precipitation will be induced [139]. In the process of membrane manufacturing, the solvent evaporation, vapor adsorption and immersion precipitation may occur instantaneously. The membrane is formed using one of the several methods.

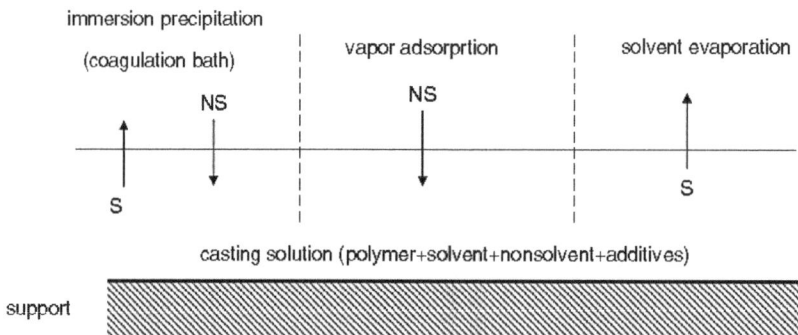

Fig. (6). Schematic representation of three DIPS processes (S: solvent; NS: non-solvent) [128].

The final morphology of the fibers and membranes obtained will vary greatly, depending on the selection and properties of the materials and the process conditions. The phase inversion process is induced by either TIPS or DIPS to change the thermodynamic properties of the dope solution. In DIPS, immersion precipitation or NIPS (non-solvent induced phase inversion) is most significant [128].

Immersion precipitation

Immersion precipitation prepares most commercially available membranes. In this process, a polymer dope solution is to fabricate as thin film support or extruded through a die casting and is successively immersed in a non-solvent bath. Fig. (7) summarised a general process of membranes made by precipitation in a non-solvent. The polymer solution, which may consist of non-solvent to improve pore construction, is forthwith coagulated while interacting with a bulk non-solvent phase containing one or more non-solvents.

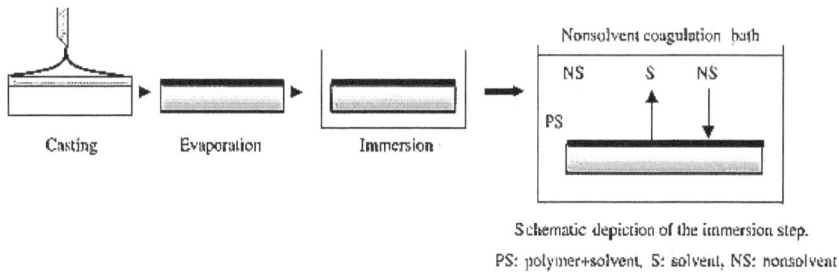

Schematic depiction of the immersion step.
PS: polymer+solvent, S: solvent, NS: nonsolvent

Fig. (7). Schematic illustration of the immersion precipitations phase inversion process [131].

Kesting distinguished between a wet or combined evaporation process, a dry or complete evaporation process and a thermal process [130]. This difference is established in the technique used to accomplish the phase inversion process; that is the foundation of the precipitation and solidification of the polymer. Individually, differences of the phase inversion process deliberated about the mechanism of membranes formation, membranes structural characteristics (anisotropy *versus* asymmetry or skinning, and structural irregularities) and membranes production consideration. Strathmann explored the details mechanism of membranes formation using phase inversion method and comprehensively highlighted the factors which are influences during the process.

The primary process variables considered are the selectivity of a suitable polymer and its concentration, the selection of solvent or solvent concentration, the choice of precipitant or precipitant system and precipitation temperature. It is also demonstrated that through proper control of the process variables, the same membrane structure can be prepared from various polymers and that the same polymer can be used to develop multiple membrane structures (CA/DMAc, PA/DMSO, PSf/DMF and mix solvent system; PA/DMAc + benzene). The rate precipitation (precipitant; formic acid, methanol, water, acetone, acetic acid and glycerol) is shown to have a direct influence on the structure of the porous support layer. Strathmann also briefly discussed the effects of evaporation before precipitation and impact of post-formation treatment such as annealing [140].

CONCEPT OF ASYMMETRIC MEMBRANES BY DRY/WET PHASE INVERSION PROCESS

Dry processing

Dry processing is the most straightforward technique for membrane formation. In this process, the solvent is evaporated from the polymer dope solution until either vitrification or phase separation occurs. If the polymer dope solution stays in the unchanging region of the phase diagram until the point of vitrification, a dense, homogenous structure will be formed. It can be achieved by using a pure binary dope consisting of polymer and a volatile solvent. If a ternary solution comprising of thermoplastic polymer, aprotic solvent (must tend to disperse polymer altogether), and non-solvent is used, evaporation of the solvent will typically lead to phase separation and an asymmetric structure. It is defined as a real dry phase separation process [140].

Wet Processing

In wet processing, phase separation is induced by placing a nascent membrane in a non-solvent quench medium. The diffusional exchange of solvent and non-solvent will influence an unstable composition, causing the homogeneous phase to separate and form a two-phase structure. The construction of asymmetric membranes formed as integrally skinned by the wet phase inversion process has been attributed to a disproportionate polymer scattering at the onset of liquid-liquid (L-L) phase separation in the immersed film [141, 142]. Thus, the following significant aspect of the manufacturing of the surface thickness is a more substantial indigenous polymer amount in the peripheral region associated with that in the sublayer of a tailored film [143]. Two unique concepts have been proposed for the asymmetric polymer distribution, and it's the causes of at the onset of L-L phase separation:

1. Higher concentration of thermoplastic polymer is the consequence of solvent abstraction from the beyond the field of the fabricated film by evaporation before the precipitation phase.
2. A significant thermoplastic polymer amount (weight %) in the membrane top place is the result of the exceptionally prompt solvent discharge relative to the non-solvent inflow occurring during the initial stages of the precipitation phase.

Baker *et al.* demonstrated experimentally that without an evaporation step, *i.e.*, by direct immersion a high flux membranes and integrally skinned asymmetric membranes can be designed [144]. Other studies revealed that the skin layer thickness and its porosity could be controlled by along with an evaporation step in the wet phase inversion process [145 - 148]. Ideal asymmetric membranes for gas separations, as considered here, must meet the following requirements:

1. The skin layer needs to be deformity free to assure that the gas carrier is restrained entirely by a solution/diffusion mechanism to achieve the maximum selectivity.
2. The skin layer should be as thin as possible to maximize the membrane productivity.
3. The substructure should not afford any resistance to gas transport but

accommodate adequate mechanical strength to back the exquisite skin layer during high-pressure operation.

The consolidation of ultrathin and defect-free skin layers is not achieved for membranes made by the wet phase inversion process. It has even been deliberate that membranes fabricate by traditional phase inversion technique always contain defects due to incomplete coalescence of the skin layer [149].

Dry/Wet Phase Inversion Process

The dry/wet process has been shown to produce thin and defect-free skin layers in some glassy polymers in flat sheet geometry [150]. A later investigation by Pesek [151] demonstrated that the dry/wet process could be successfully implemented in hollow fiber geometry, albeit under purely academic conditions. Through by dry/wet phase, inversion technique the integrally skinned asymmetric membranes can also be fabricated. In dry/wet the phase separation is induced in the furthermost part of the membrane casting at the same time as an evaporation step, whereas L-L phase segregation in the bulk film happens subsequently during a precipitation phase. Pinnau and Koros have investigated a very comprehensive study of phase inversion processes such as wet, dry and dry/wet and fabricated polysulfone membranes using several types of solvent and nonsolvent [152]. Also, they were also used multicomponent materials and presented a physically important mechanism for the formation of ultrathin and defect-free skin layers of membranes for gas separation application using phase inversion processes [150]. Also, they have summarized the empirically developed rules for the formation of optimized asymmetric membranes made by the dry/wet phase inversion process. Koros *et al.* [150, 152] have established clearly that the physical processes which appeared throughout the evaporation part were of decisive importance for the skin layer formation of asymmetric membranes made by dry/wet phase inversion [152]. Fig. (**8**) summarised the comprehensive studies of dry/wet phase inversion technique (evaporation) for an exemplary ternary casting system used for the preparation of membranes made by is illustrated schematically in [150].

Fig. (**8**) shows several theoretical points during a polymeric membrane formation, the point 'A' gives the composition of the initially stable polymer solution. Several membranologist has established this theory that in case of higher vapor pressure solvent as compared to the non-solvent component, it is reasonable as a first approximation to presuppose that only solvent is removed from the outermost region of the cast film during the initial stages of the evaporation process [150 - 153].

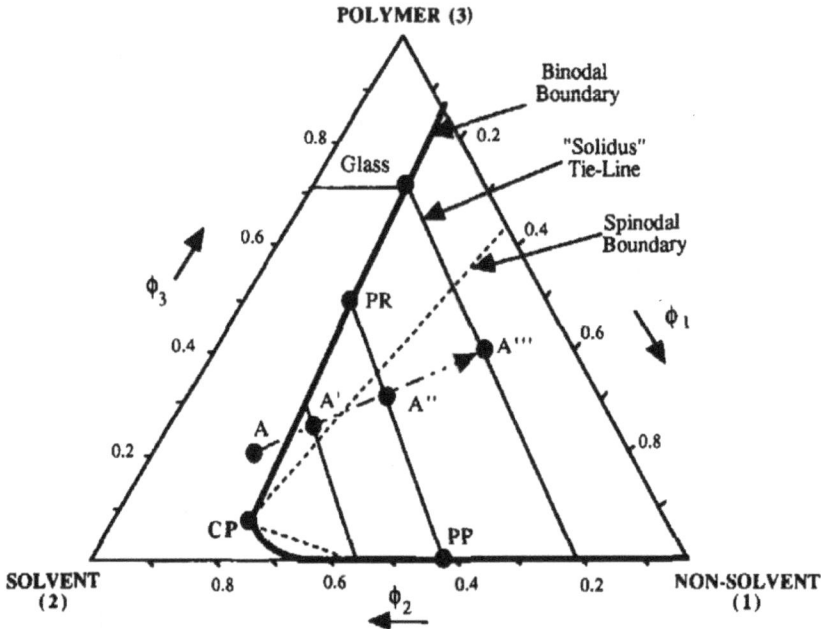

Fig. (8). Schematic representation of possible physical events occurring during vaporation of the solvent component from a ternary casting solution [150].
Note: PP = polymer poor; PR = polymer rich; CP = critical point.

In this case, the evaporation curve is given by an unvarying ratio of the polymer volume fraction, $\phi3$, to the non-solvent volume fraction, $\phi1$. In such type of ideal condition during the forced-convective evaporation process, as a consequence of the only just stable polymer dope solution, the phase separation will be inclined to take place straight away in the cast film. The consequential physical occurrences which investigate the polymeric membrane construction process during the dry phase inversion stage will strongly depend on the evaporation kinetics (*i.e.*, the velocity of the gas stream blown across the membrane surface and the evaporation time). Conversely, high gas flow velocities and longer evaporation times may show the way to an average composition "A" in the tangential membrane region. Since composition "A" lives deep in the precarious position, L-L phase separation will take place by spinodal disintegration. Fig. (**8**) shows a good agreement that in case of L-L phase separation by spinodal disintegration will direct to a much upper polymer concentration in the polymer-rich phase as compared to that obtained by the corresponding nucleation and growth mechanism. Spinodal deteriorates L-L phase segregate morphology incorporate of a highly conventional, bicontinuous network of the polymer-rich and the poor polymer phase [150 - 155].

Furthermore, the theoretical description Fig. (**8**), has established that the consequent of wet precipitation step transforms the highly plasticized, but homogeneous surface layer immediately into a mostly solvent-free glass surrounding only the equilibrium quantity of the precipitation medium. As the precipitation approach saturates across the homogeneous surface layer, solvent and non-solvent contained in the underlying transition layer can move by counter diffusion into the miscible precipitation bath. It is essential to note that the vitrification of the nodular transition layer should happen as expeditiously as possible to avoid any loss of the interconnectivity and porosity [156]. The use of a thermodynamically stable non-solvent such as methanol or water guarantees that the bi-continuous transition layer solidifies about instantaneously, as recently demonstrated experimentally." Furthermore, the use of a powerful quench medium leads to instantaneous L-L phase separation in the bulk of the membrane, for the regions which were not phase inverted during the evaporation process. As often perceived for membranes made by wet phase inversion, instantaneous L-L phase separation results in an Open-cell, sponge-like substructure [157].

CONCLUSION

The composite types of membranes tailoring techniques and application has established distinct properties regarding chemical, physical, biological, respectively and recent studies show engaging for the extensive variety of innovative development in numerous pitch of fields. This review has focused nanoparticles based membranes for pressure-driven membranes such as microfiltration (MF), ultrafiltration (UF), nanofiltration (NF), reverse osmosis (RO) and process development, respectively. This study has also highlighted the importance of nanoparticles or multicomponent materials on thin film composite membrane (TFC), electrodialysis, gas separation and pervaporation membranes respectively. Moreover, this review somewhat lay a hand on the influence of nanoparticles on the improvement of membrane performance and minimizing their fouling tendency in the pressure-driven membrane. Generally, by add-in and integration an appropriate amount of nanoparticles inside polymer solution or on to a thin selective polymeric membrane layer, both may perhaps fabricate composite membranes with superior performance characteristics as weigh against to the traditional membrane process. Several researchers have reported, that nanoparticles based composite membranes not only outstanding antifouling defiance and better antibacterial outcome but also probably overcome the arrangement consequence among solute selectivity and water flux or permeability.

CONSENT FOR PUBLICATION

Not applicable.

CONFLICT OF INTEREST

The author (editor) declares no conflict of interest, financial or otherwise.

ACKNOWLEDGEMENT

The authors thank full of Centre of Excellence in Desalination Technology (CEDT), King Abdulaziz University Jeddah to provide a platform and moral support to accomplish this work.

REFERENCES

[1]   Shannon, M.A.; Bohn, P.W.; Elimelech, M.; Georgiadis, J.G.; Mariñas, B.J.; Mayes, A.M. Science and technology for water purification in the coming decades. *Nature,* **2008,** *452*(7185), 301-310.
        [http://dx.doi.org/10.1038/nature06599] [PMID: 18354474]

[2]   Logan, B.E.; Elimelech, M. Membrane-based processes for sustainable power generation using water. *Nature,* **2012,** *488*(7411), 313-319.
        [http://dx.doi.org/10.1038/nature11477] [PMID: 22895336]

[3]   Drioli, E.; Giorno, L. Comprehensive membrane science and engineering.*Basic Aspects in Polymeric Membrane Preparation, Strathmann, H., Giorno, L., Drioli, E*; Elsevier Science: Amsterdam, **2010,** Vol. 1, pp. 91-110.

[4]   Cui, Z.L.; Drioli, E.; Lee, Y.M. Recent progress in fluoro-polymers for membranes. *Prog. Polym. Sci.,* **2014,** *39*(1), 164-198.
        [http://dx.doi.org/10.1016/j.progpolymsci.2013.07.008]

[5]   Długołęcki, P.; Dabrowska, J.; Nijmeijer, K.; Wesslinga, M. Ion conductive spacers for increased power generation in reverse electrodialysis. *J. Membr. Sci.,* **2010,** *347*(1), 101-107.
        [http://dx.doi.org/10.1016/j.memsci.2009.10.011]

[6]   Flemming, H.C. Reverse osmosis membrane biofouling. *Exp. Therm. Fluid Sci.,* **1997,** *14*(4), 382.
        [http://dx.doi.org/10.1016/S0894-1777(96)00140-9]

[7]   Kilduff, J.E.; Mattaraj, S.; Pieracci, J.P.; Belfort, G. Photochemical modification of poly(ether sulfone) and sulfonated poly(sulfone) nanofiltration membranes for control of fouling by natural organic matter. *Desalination,* **2000,** *132*(1-3), 133-142.
        [http://dx.doi.org/10.1016/S0011-9164(00)00142-9]

[8]   Baker, J.S.; Dudley, L.Y. Biofouling in membrane systems - a review. *Desalination,* **1998,** *118*(1), 81-89.
        [http://dx.doi.org/10.1016/S0011-9164(98)00091-5]

[9]   Kochkodan, V.M.; Sharma, V.K. Graft polymerization and plasma treatment of polymer membranes for fouling reduction: a review. *J Environ Sci Health A Tox Hazard Subst Environ Eng,* **2012,** *47*(12), 1713-1727.
        [http://dx.doi.org/10.1080/10934529.2012.689183] [PMID: 22755517]

[10]   Mansouri, J.; Harrisson, S.; Chen, V. Strategies for controlling biofouling in membrane filtration systems: challenges and opportunities. *J. Mater. Chem.,* **2010,** *20*(22), 4567-4586.
        [http://dx.doi.org/10.1039/b926440j]

[11]   Cheng, G.; Zhang, Z.; Chen, S.; Bryers, J.D.; Jiang, S. Inhibition of bacterial adhesion and biofilm formation on zwitterionic surfaces. *Biomaterials,* **2007,** *28*(29), 4192-4199.
        [http://dx.doi.org/10.1016/j.biomaterials.2007.05.041] [PMID: 17604099]

[12]   Costerton, J.W.; Cheng, K.J.; Geesey, G.G.; Ladd, T.I.; Nickel, J.C.; Dasgupta, M.; Marrie, T.J. Bacterial biofilms in nature and disease. *Annu. Rev. Microbiol.,* **1987,** *41*, 435-464.

[http://dx.doi.org/10.1146/annurev.mi.41.100187.002251] [PMID: 3318676]

[13]   Banerjee, I.; Pangule, R.C.; Kane, R.S. Antifouling coatings: recent developments in the design of surfaces that prevent fouling by proteins, bacteria, and marine organisms. *Adv. Mater.,* **2011**, *23*(6), 690-718.
[http://dx.doi.org/10.1002/adma.201001215] [PMID: 20886559]

[14]   Misdan, N.; Ismail, A.F.; Hilal, N. Recent advances in the development of (bio) fouling resistant thin film composite membranes for desalination. *Desalination,* **2016**, *380*(1), 105-111.
[http://dx.doi.org/10.1016/j.desal.2015.06.001]

[15]   Le-Clech, P.; Chen, V.; Fane, T.A.G. Fouling in membrane bioreactors used in wastewater treatment. *J. Membr. Sci.,* **2006**, *284*(1), 17.
[http://dx.doi.org/10.1016/j.memsci.2006.08.019]

[16]   Viero, A.F.; Sant'Anna, G.L., Jr; Nobrega, R. The use of polyetherimide hollow fibres in a submerged membrane bioreactor operating with air backwashing. *J. Membr. Sci.,* **2007**, *302*(1), 127.
[http://dx.doi.org/10.1016/j.memsci.2007.06.036]

[17]   Wang, Y.Q.; Su, Y.L.; Sun, Q.; Ma, X.L.; Jiang, Z.Y. Generation of anti-biofouling ultrafiltration membrane surface by blending novel branched amphiphilic polymers with polyethersulfone. *J. Membr. Sci.,* **2006**, *286*(1-2), 228.
[http://dx.doi.org/10.1016/j.memsci.2006.09.040]

[18]   Asatekin, A.; Menniti, A.; Kang, S.; Elimelech, M.; Morgenroth, E.; Mayes, A.M. Antifouling nanofiltration membranes for membrane bioreactors from self-assembling graft copolymers. *J. Membr. Sci.,* **2006**, *285*(1), 81.
[http://dx.doi.org/10.1016/j.memsci.2006.07.042]

[19]   Yang, S.; Liu, Z. Preparation and characterization of polyacrylonitrile ultrafiltration membranes. *J. Membr. Sci.,* **2003**, *222*(1), 87.
[http://dx.doi.org/10.1016/S0376-7388(03)00220-5]

[20]   Wang, M.; Wu, L.; Gao, C. The influence of phase inversion process modified by chemical reaction on membrane properties and morphology. *J. Membr. Sci.,* **2006**, *270*(1), 154.
[http://dx.doi.org/10.1016/j.memsci.2005.06.051]

[21]   Yu, S.; Zuo, X.; Bao, R.; Xu, X.; Wang, J.; Xu, J. Effect of SiO_2 nanoparticle addition on the characteristics of a new organic–inorganic hybrid membrane. *Polymer (Guildf.),* **2009**, *50*(2), 553-559.
[http://dx.doi.org/10.1016/j.polymer.2008.11.012]

[22]   Pandey, P.; Chauhan, R.S. Membranes for gas separation. *Prog. Polym. Sci.,* **2001**, *26*(6), 853-893.
[http://dx.doi.org/10.1016/S0079-6700(01)00009-0]

[23]   Ji, W.; Sikdar, S.K. A pollution reduction methodology for chemical process simulators. *Ind. Eng. Chem. Res.,* **1996**, *35*, 4128.
[http://dx.doi.org/10.1021/ie9601108]

[24]   Genne, I.; Kuypers, S.; Leysen, R. Effect of the addition of ZrO_2 to polysulfone based UF membranes. *J. Membr. Sci.,* **1996**, *113*, 343.
[http://dx.doi.org/10.1016/0376-7388(95)00132-8]

[25]   Schaep, J.; Vandecasteele, C.; Leysen, R. W. Salt retention of Zirfon® membranes. *Separ. Purif. Tech.,* **1998**, *14*(1), 127.
[http://dx.doi.org/10.1016/S1383-5866(98)00067-7]

[26]   Bottino, A.; Capannelli, G.; D'Asti, V.; Piaggio, P. Preparation and properties of novel organic-inorganic porous membranes. *Separ. Purif. Tech.,* **2001**, *269*(1), 22-23.

[27]   Aerts, P.; Greenberg, A.R.; Leysen, R.; Krantz, W.B.; Reinsch, V.E.; Jacobs, P.A. The influence of filler concentration on the compaction and filtration properties of Zirfon®-composite ultrafiltration membranes. *Separ. Purif. Tech.,* **2001**, *663*(1), 22-23.

[28]   Kim, S.H.; Kwak, S.Y.; Sohn, B.H.; Park, T.H. Design of TiO_2 nanoparticle self-assembled aromatic polyamide thin-film-composite (TFC) membrane as an approach to solve biofouling problem. *J. Membr. Sci.,* **2003**, *211*(1), 157-165.
[http://dx.doi.org/10.1016/S0376-7388(02)00418-0]

[29]   Li, X.; Fang, X.; Pang, R.; Li, J.; Sun, X.; Shen, J.; Han, W.; Wang, L. Self-assembly of TiO_2 nanoparticles around the pores of PES ultrafiltration membrane for mitigating organic fouling. *J. Membr. Sci.,* **2014**, *467*(1), 226-235.

[30]   Yang, Y.; Wang, P. Preparation and characterizations of a new PS/TiO_2 hybrid membranes by sol-gel process. *Polymer (Guildf.),* **2006**, *47*, 2683-2688.
[http://dx.doi.org/10.1016/j.polymer.2006.01.019]

[31]   Wara, N.M; Francis, L.F.; Velamakanni, B.V. Addition of alumina to cellulose acetate membranes. *J. Membr. Sci.,* **1995**, *104*(1), 43-49.
[http://dx.doi.org/10.1016/0376-7388(95)00010-A]

[32]   Genne, I.; Kuypers, S.; Leysen, R. Effect of the addition of ZrO_2 to polysulfone based UF membranes. *J. Membr. Sci.,* **1996**, *113*(2), 343-350.
[http://dx.doi.org/10.1016/0376-7388(95)00132-8]

[33]   Bottino, A.; Capannelli, G.; Comite, A. Preparation and characterization of novel porous $PVDF-ZrO_2$ composite membranes. *Desalination,* **2002**, *146*(1), 35-40.
[http://dx.doi.org/10.1016/S0011-9164(02)00469-1]

[34]   Arthanareeswaran, G.; Devi, T.S.; Raajenthiren, M. Effect of silica particles on cellulose acetate blend ultrafiltration membranes: part I. *Separ. Purif. Tech.,* **2008**, *64*(1), 38-47.
[http://dx.doi.org/10.1016/j.seppur.2008.08.010]

[35]   Arthanareeswaran, G.; Devi, T.S.; Mohan, D. Development, characterization and separation performance of organic–inorganic membranes: part II. Effect of additives. *Separ. Purif. Tech.,* **2009**, *67*(2), 271-281.
[http://dx.doi.org/10.1016/j.seppur.2009.03.037]

[36]   Arthanareeswaran, G.; Thanikaivelan, P. Fabrication of cellulose acetate–zirconia hybrid membranes for ultrafiltration applications: performance, structure and fouling analysis. *Separ. Purif. Tech.,* **2010**, *74*(1), 230-23.
[http://dx.doi.org/10.1016/j.seppur.2010.06.010]

[37]   Damodar, R.A.; You, S.J.; Chou, H.H. Study the self cleaning, antibacterial and photocatalytic properties of TiO_2 entrapped PVDF membranes. *J. Hazard. Mater.,* **2009**, *172*(2-3), 1321-1328.
[http://dx.doi.org/10.1016/j.jhazmat.2009.07.139] [PMID: 19729240]

[38]   Cao, X.; Ma, J.; Shi, X.; Ren, Z. Effect of TiO_2 nanoparticle size on the performance of PVDF membrane. *Appl. Surf. Sci.,* **2006**, *253*(2-3), 2003-2010.
[http://dx.doi.org/10.1016/j.apsusc.2006.03.090]

[39]   Méricq, J.P.; Mendret, J.; Brosillon, S.; Faur, C. High performance $PVDF-TiO_2$ membranes for water treatment. *Chem. Eng. Sci.,* **2015**, *1231*, 283-291.

[40]   Bae, T.H.; Tak, T.M. Effect of TiO_2 nanoparticles on fouling mitigation of ultrafiltration membranes for activated sludge filtration. *J. Membr. Sci.,* **2005**, *249*(1), 1-8.

[41]   Luo, M.l.; Tang, W.; Zhao, J.Q.; Pu, C.S. Hydrophilic modification of poly (ether sulfone) used TiO_2 nanoparticles by a sol-gel process. *J. Mater. Process. Technol.,* **2006**, *172*(2), 431-436.
[http://dx.doi.org/10.1016/j.jmatprotec.2005.11.004]

[42]   Rahimpour, A.; Madaeni, S.; Taheri, A.; Mansourpanah, Y. Coupling TiO_2 nanoparticles with UV irradiation for modification of polyethersulfone ultrafiltration membranes. *J. Membr. Sci.,* **2008**, *313*(1), 158-169.
[http://dx.doi.org/10.1016/j.memsci.2007.12.075]

[43]   Bae, T.H.; Tak, T.M. Preparation of TiO_2 self-assembled polymeric nanocomposite membranes and examination of their fouling mitigation effects in a membrane bioreactor system. *J. Membr. Sci.,* **2005**, *266*(1), 1-5.

[44]   Bae, T.H.; Kim, I.C.; Tak, T.M. Preparation and characterization of fouling-resistant TiO_2 self-assembled nanocomposite membranes. *J. Membr. Sci.,* **2006**, *275*(1), 1-5.
       [http://dx.doi.org/10.1016/j.memsci.2006.01.023]

[45]   Luo, M.L.; Zhao, J.Q.; Tang, W.; Pu, C.S. Hydrophilic modification of poly (ether sulfone) ultrafiltration membrane surface by self-assembly of TiO_2 nanoparticles. *Appl. Surf. Sci.,* **2005**, *249*(1), 76-84.
       [http://dx.doi.org/10.1016/j.apsusc.2004.11.054]

[46]   Rahimpour, A.; Madaeni, S.; Taheri, A.; Mansourpanah, Y. Coupling TiO_2 nanoparticles with UV irradiation for modification of polyethersulfone ultrafiltration membranes. *J. Membr. Sci.,* **2008**, *313*(1), 158-169.
       [http://dx.doi.org/10.1016/j.memsci.2007.12.075]

[47]   Li, J.H.; Xu, Y.Y.; Zhu, L.P.; Wang, J.H.; Du, C.H. Fabrication and characterization of a novel TiO_2 nanoparticle self-assembly membrane with improved fouling resistance. *J. Membr. Sci.,* **2009**, *326*(3), 659-666.
       [http://dx.doi.org/10.1016/j.memsci.2008.10.049]

[48]   Abdallah, H.; Moustafa, A.; AlAnezi, A.A.; El-Sayed, H. Performance of a newly developed titanium oxide nanotubes/polyethersulfone blend membrane for water desalination using vacuum membrane distillation. *Desalination,* **2014**, *346*(1), 30-36.
       [http://dx.doi.org/10.1016/j.desal.2014.05.003]

[49]   Shaban, M.; AbdAllah, H.; Said, L.; Hamdy, H.S.; Khalek, A.A. Titanium dioxide nanotubes embedded mixed matrix PES membranes characterization and membrane performance. *Chem. Eng. Res. Des.,* **2015**, *95*(2), 307-316.
       [http://dx.doi.org/10.1016/j.cherd.2014.11.008]

[50]   Rabiee, H.; Farahani, M.H.D.A.; Vatanpour, V. Preparation and characterization of emulsion poly(vinyl chloride) (EPVC)/TiO_2 nanocomposite ultrafiltration membrane. *J. Membr. Sci.,* **2014**, *472*(1), 185-193.
       [http://dx.doi.org/10.1016/j.memsci.2014.08.051]

[51]   Rahimpour, A.; Jahanshahi, M.; Mollahosseini, A.; Rajaeian, B. Structural and performance properties of UV-assisted TiO_2 deposited nano-composite PVDF/SPES membranes. *Desalination,* **2012**, *285*(1), 31-38.
       [http://dx.doi.org/10.1016/j.desal.2011.09.026]

[52]   Vatanpour, V.; Madaeni, S.S.; Khataee, A.R.; Salehi, E.; Zinadini, S.; Monfared, H.A. TiO_2 embedded mixed matrix PES nanocomposite membranes: influence of different sizes and types of nanoparticles on antifouling and performance. *Desalination,* **2012**, *292*(1), 19-29.
       [http://dx.doi.org/10.1016/j.desal.2012.02.006]

[53]   Wu, G.; Gan, S.; Cui, L.; Xu, Y. Preparation and characterization of PES/TiO_2 composite membranes. *Appl. Surf. Sci.,* **2008**, *254*(21), 7080-7086.
       [http://dx.doi.org/10.1016/j.apsusc.2008.05.221]

[54]   Li, J.F.; Xu, Z.L.; Yang, H.; Yu, L.Y.; Liu, M. Effect of TiO_2 nanoparticles on the surface morphology and performance of microporous PES membrane. *Appl. Surf. Sci.,* **2009**, *255*(20), 4725-4732.
       [http://dx.doi.org/10.1016/j.apsusc.2008.07.139]

[55]   Yu, Z.; Liu, X.; Zhao, F.; Liang, X.; Tian, Y. Fabrication of a low-cost nano-SiO_2/PVC composite ultrafiltration membrane and its antifouling performance. *J. Appl. Polym. Sci.,* **2015**, *132*(2), 41267.
       [http://dx.doi.org/10.1002/app.41267]

[56]   Rahimpour, A.; Jahanshahi, M.; Rajaeian, B.; Rahimnejad, M. TiO_2 entrapped nanocomposite

PVDF/SPES membranes: preparation, characterization, antifouling and antibacterial properties. *Desalination,* **2011**, *278*(2), 343-353.
[http://dx.doi.org/10.1016/j.desal.2011.05.049]

[57]   Pourjafar, S.; Rahimpour, A.; Jahanshahi, M. Synthesis and characterization of PVA/ PES thin film composite nanofiltration membrane modified with TiO$_2$ nanoparticles for better performance and surface properties. *J. Ind. Eng. Chem.,* **2012**, *18*(6), 1398-1405.
[http://dx.doi.org/10.1016/j.jiec.2012.01.041]

[58]   Kumar, R.; Isloor, A.M.; Ismail, A.F.; Rashid, S.A.; Al Ahmed, A. Permeation, antifouling and desalination performance of TiO$_2$ nanotube incorporated PSf/CS blend membranes. *Desalination,* **2013**, *316*(1), 76-84.
[http://dx.doi.org/10.1016/j.desal.2013.01.032]

[59]   Yang, Y.; Zhang, H.; Wang, P.; Zheng, Q.; Li, J. The influence of nano-sized TiO$_2$ fillers on the morphologies and properties of PSF UF membrane. *J. Membr. Sci.,* **2007**, *288*(1), 231-238.
[http://dx.doi.org/10.1016/j.memsci.2006.11.019]

[60]   Simone, S.; Galiano, F.; Faccini, M.; Boerrigter, M.E.; Chaumette, C.; Drioli, E.; Figoli, A. Preparation and Characterization of Polymeric-Hybrid PES/TiO$_2$ Hollow Fiber Membranes for Potential Applications in Water Treatment. *Fibres,* **2017**, *14*(5), 2-19.

[61]   Jeong, B.H.; Hoek, E.M.; Yan, Y.; Subramani, A.; Huang, X.; Hurwitz, G.; Ghosh, A.K.; Jawor, A. Interfacial polymerization of thin film nanocomposites: a new concept for reverse osmosis membranes. *J. Membr. Sci.,* **2007**, *294*(1), 1-7.
[http://dx.doi.org/10.1016/j.memsci.2007.02.025]

[62]   Dong, C.; He, G.; Li, H.; Zhao, R.; Han, Y.; Deng, Y. Antifouling enhancement of poly (vinylidene fluoride) microfiltration membrane by adding Mg(OH)$_2$ nanoparticles. *J. Membr. Sci.,* **2012**, *387*(1), 40-47.
[http://dx.doi.org/10.1016/j.memsci.2011.10.007]

[63]   Nair, A.K.; Isloor, A.M.; Kumar, R.; Ismail, A. Antifouling and performance enhancement of polysulfone ultrafiltration membranes using CaCO$_3$ nanoparticles. *Desalination,* **2013**, *322*(1), 69-75.
[http://dx.doi.org/10.1016/j.desal.2013.04.031]

[64]   Gohari, R.J.; Halakoo, E.; Lau, W.J.; Kassim, M.A.; Matsuura, T.; Ismail, A.F. Novel polyethersulfone (PES)/hydrous manganese dioxide (HMO) mixed matrix membranes with improved anti-fouling properties for oily wastewater treatment process. *RSC Advances,* **2014**, *4*, 17587-17596.
[http://dx.doi.org/10.1039/C4RA00032C]

[65]   Madaeni, S.; Ghaemi, N. Characterization of self-cleaning RO membranes coated with TiO$_2$ particles under UV irradiation. *J. Membr. Sci.,* **2007**, *303*(1), 221-233.
[http://dx.doi.org/10.1016/j.memsci.2007.07.017]

[66]   Damodar, R.A.; You, S.J.; Chou, H.H. Study the self cleaning, antibacterial and photocatalytic properties of TiO$_2$ entrapped PVDF membranes. *J. Hazard. Mater.,* **2009**, *172*(2-3), 1321-1328.
[http://dx.doi.org/10.1016/j.jhazmat.2009.07.139] [PMID: 19729240]

[67]   Miyauchi, M.; Nakajima, A.; Watanabe, T.; Hashimoto, K. Photocatalysis and photo induced hydrophilicity of various metal oxide thin films. *Chem. Mater.,* **2002**, *14*(6), 2812-2816.
[http://dx.doi.org/10.1021/cm020076p]

[68]   Zhang, Y.; Li, H.; Li, H.; Li, R.; Xiao, C. Preparation and characterization of modified polyvinyl alcohol ultrafiltration membranes. *Desalination,* **2006**, *192*(1), 214-223.
[http://dx.doi.org/10.1016/j.desal.2005.07.037]

[69]   Zhang, Q.; Zhang, S.; Dai, L.; Chen, X. Novel zwitterionic poly (arylene ether sulfone) s as antifouling membrane material. *J. Membr. Sci.,* **2010**, *349*(1), 217-224.

[70]   Rana, D.; Matsuura, T. Surface modifications for antifouling membranes. *Chem. Rev.,* **2010**, *110*(4), 2448-2471.

[http://dx.doi.org/10.1021/cr800208y] [PMID: 20095575]

[71]   Kim, J.; Van der Bruggen, B. The use of nanoparticles in polymeric and ceramic membrane structures: review of manufacturing procedures and performance improvement for water treatment. *Environ. Pollut.,* **2010**, *158*(7), 2335-2349.
[http://dx.doi.org/10.1016/j.envpol.2010.03.024] [PMID: 20430495]

[72]   Liang, S.; Xiao, K.; Mo, Y.; Huang, X. A novel ZnO nanoparticle blended polyvinylidene fluoride membrane for anti-irreversible fouling. *J. Membr. Sci.,* **2012**, *394*(1), 184-192.
[http://dx.doi.org/10.1016/j.memsci.2011.12.040]

[73]   Shen, J.N.; Ruan, H.M.; Wu, L.G.; Gao, C.J. Preparation and characterization of PES-SiO_2 organic-inorganic composite ultrafiltration membrane for raw water pre-treatment. *Chem. Eng. J.,* **2011**, *168*(7), 1272-1278.
[http://dx.doi.org/10.1016/j.cej.2011.02.039]

[74]   Yu, S.; Zuo, X.; Bao, R.; Xu, X.; Wang, J.; Xu, J. Effect of SiO_2 nanoparticle addition on the characteristics of a new organic–inorganic hybrid membrane. *Polymer (Guildf.),* **2009**, *50*(3), 553-559.
[http://dx.doi.org/10.1016/j.polymer.2008.11.012]

[75]   Yu, L.Y.; Xu, Z.L.; Shen, H-M.; Yang, H. Preparation and characterization of PVDF-SiO_2 composite hollow fiber UF membrane by sol-gel method. *J. Membr. Sci.,* **2009**, *337*(1-2), 257-265.
[http://dx.doi.org/10.1016/j.memsci.2009.03.054]

[76]   Jadav, G.L.; Singh, P.S. Synthesis of novel silica-polyamide nanocomposite membrane with enhanced properties. *J. Membr. Sci.,* **2009**, *328*(1-2), 257-267.
[http://dx.doi.org/10.1016/j.memsci.2008.12.014]

[77]   Ng, L.Y.; Leo, C.P.; Mohammad, A.W. Optimizing the incorporation of silica nanoparticles in polysulfone/poly (vinyl alcohol) membranes with response surface methodology. *J. Appl. Polym. Sci.,* **2011**, *121*(10), 1804-1814.
[http://dx.doi.org/10.1002/app.33628]

[78]   Xu, H.P.; Yu, Y.H.; Lang, W.Z.; Yan, X.; Guo, Y.J. Hydrophilic modification of polyvinylchloride hollow fiber membranes by silica with a weak *in situ* sol-gel method. *RSC Advances,* **2015**, *5*(18), 13733-13742.
[http://dx.doi.org/10.1039/C4RA15687K]

[79]   Li, J.H.; Shao, X.S.; Zhou, Q.; Li, M.Z.; Zhang, Q.Q. The double effects of silver nanoparticles on the PVDF membrane: surface hydrophilicity and antifouling performance. *Appl. Surf. Sci.,* **2013**, *265*(3), 663-670.
[http://dx.doi.org/10.1016/j.apsusc.2012.11.072]

[80]   Bhushan, B. *Handbook of Nanotechnology*; Springer Science & Business Media, **2010**.
[http://dx.doi.org/10.1007/978-3-642-02525-9]

[81]   Chaudhari, L.B.; Murthy, Z.V.P. Separation of Cd and Ni from multicomponent aqueous solutions by nanofiltration and characterization of membrane using IT model. *J. Hazard. Mater.,* **2010**, *180*(1-3), 309-315.
[http://dx.doi.org/10.1016/j.jhazmat.2010.04.032] [PMID: 20452729]

[82]   Balta, S.; Sotto, A.; Luis, P.; Benea, L.; Van der Bruggen, B.; Kim, J. A new outlook on membrane enhancement with nanoparticles: the alternative of ZnO. *J. Membr. Sci.,* **2012**, *389*(1), 155-161.
[http://dx.doi.org/10.1016/j.memsci.2011.10.025]

[83]   Hong, J.; He, Y. Effects of nano sized zinc oxide on the performance of PVDF microfiltration membranes. *Desalination,* **2012**, *302*(1), 71-79.
[http://dx.doi.org/10.1016/j.desal.2012.07.001]

[84]   Hong, J.; He, Y. Polyvinylidene fluoride ultrafiltration membrane blended with nano- ZnO particle for photo-catalysis self-cleaning. *Desalination,* **2014**, *332*(1), 67-75.
[http://dx.doi.org/10.1016/j.desal.2013.10.026]

[85] Leo, C.P.; Lee, W.C.; Ahmad, A.L.; Mohammad, A.W. Polysulfone membranes blended with ZnO nanoparticles for reducing fouling by oleic acid. *Separ. Purif. Tech.,* **2012**, *89*(1), 51-56.
[http://dx.doi.org/10.1016/j.seppur.2012.01.002]

[86] Zhang, X.; Wang, Y.; Liu, Y.; Xu, J.; Han, Y.; Xu, X. Preparation, performances of PVDF/ ZnO hybridmembranes and their applications in the removal of copper ions. *Appl. Surf. Sci.,* **2014**, *316*(2), 333-340.
[http://dx.doi.org/10.1016/j.apsusc.2014.08.004]

[87] Zhao, S.; Yan, W.; Shi, M.; Wang, Z.; Wang, J.; Wang, S. Improving permeability and antifouling performance of polyethersulfone ultrafiltration membrane by incorporation of ZnO-DMF dispersion containing nano-ZnO and polyvinylpyrrolidone. *J. Membr. Sci.,* **2015**, *478*(1), 105-116.
[http://dx.doi.org/10.1016/j.memsci.2014.12.050]

[88] Andrews, R.; Jacques, D.; Qian, D.; Rantell, T. Multiwall carbon nanotubes: synthesis and application. *Acc. Chem. Res.,* **2002**, *35*(12), 1008-1017.
[http://dx.doi.org/10.1021/ar010151m] [PMID: 12484788]

[89] Son, M.; Choi, H.G.; Liu, L.; Celik, E.; Park, H.; Choi, H. Efficacy of carbon nanotube positioning in the polyethersulfone support layer on the performance of thin-film composite membrane for desalination. *Chem. Eng. J.,* **2015**, *266*(2), 376-384.
[http://dx.doi.org/10.1016/j.cej.2014.12.108]

[90] Hummers, W.S., Jr; Offeman, R.E. Preparation of graphitic oxide. *J. Am. Chem. Soc.,* **1958**, *80*, 1339-1339.
[http://dx.doi.org/10.1021/ja01539a017]

[91] Marcano, D.C.; Kosynkin, D.V.; Berlin, J.M.; Sinitskii, A.; Sun, Z.; Slesarev, A.; Alemany, L.B.; Lu, W.; Tour, J.M. Improved synthesis of graphene oxide. *ACS Nano,* **2010**, *4*(8), 4806-4814.
[http://dx.doi.org/10.1021/nn1006368] [PMID: 20731455]

[92] Qiu, S.; Wu, L.; Pan, X.; Zhang, L.; Chen, H.; Gao, C. Preparation and properties of functionalized carbon nanotube/PSF blend ultrafiltration membranes. *J. Membr. Sci.,* **2009**, *342*(1), 165-172.
[http://dx.doi.org/10.1016/j.memsci.2009.06.041]

[93] Wu, H.; Tang, B.; Wu, P. Novel ultrafiltration membranes prepared from a multiwalled carbon nanotubes/polymer composite. *J. Membr. Sci.,* **2010**, *362*(2), 374-383.
[http://dx.doi.org/10.1016/j.memsci.2010.06.064]

[94] Roy, S.; Ntim, S.A.; Mitra, S.; Sirkar, K.K. Facile fabrication of superior nanofiltration membranes from interfacially polymerized CNT-polymer composites. *J. Membr. Sci.,* **2011**, *375*(1), 81-87.
[http://dx.doi.org/10.1016/j.memsci.2011.03.012]

[95] Vatanpour, V.; Madaeni, S.S.; Moradian, R.; Zinadini, S.; Astinchap, B. Fabrication and characterization of novel antifouling nanofiltration membrane prepared from oxidized multiwalled carbon nanotube/polyethersulfone nanocomposite. *J. Membr. Sci.,* **2011**, *375*(2), 284-294.
[http://dx.doi.org/10.1016/j.memsci.2011.03.055]

[96] Majeed, S.; Fierro, D.; Buhr, K.; Wind, J.; Du, B.; Boschetti-de-Fierro, A.; Abetz, V. Multiwalled carbon nanotubes (MWCNTs) mixed polyacrylonitrile (PAN) ultrafiltration membranes. *J. Membr. Sci.,* **2012**, *403*(1), 101-109.
[http://dx.doi.org/10.1016/j.memsci.2012.02.029]

[97] Murthy, Z.V.P.; Gaikwad, M.S. Preparation of chitosan-multiwalled carbon nanotubes blended membranes: characterization and performance in the separation of sodium and magnesium ions. *Nanoscale Microscale Thermophys. Eng.,* **2013**, *17*(1), 245-262.
[http://dx.doi.org/10.1080/15567265.2013.787571]

[98] Zinadini, S.; Zinatizadeh, A.A.; Rahimi, M.; Vatanpour, V.; Zangeneh, H. Preparation of a novel antifouling mixed matrix PES membrane by embedding graphene oxide nanoplates. *J. Membr. Sci.,* **2014**, *453*(2), 292-301.

[http://dx.doi.org/10.1016/j.memsci.2013.10.070]

[99] Wang, Z.; Yu, H.; Xia, J.; Zhang, F.; Li, F.; Xia, Y.; Li, Y. Novel GO-blended PVDF ultrafiltration membranes. *Desalination,* **2012**, *299*(1), 50-54.
[http://dx.doi.org/10.1016/j.desal.2012.05.015]

[100] Chang, X.; Wang, Z.; Quan, S.; Xu, Y.; Jiang, Z.; Shao, L. Exploring the synergetic effects of graphene oxide (GO) and polyvinylpyrrodione (PVP) on poly (vinylylidenefluoride) (PVDF) ultrafiltration membrane performance. *Appl. Surf. Sci.,* **2014**, *316*(4), 537-548.
[http://dx.doi.org/10.1016/j.apsusc.2014.07.202]

[101] Zhao, C.; Xu, X.; Chen, J.; Yang, F. Optimization of preparation conditions of poly(vinylidene fluoride)/graphene oxide microfiltration membranes by the Taguchi experimental design. *Desalination,* **2014**, *334*(1), 17-22.
[http://dx.doi.org/10.1016/j.desal.2013.07.011]

[102] Xia, S.; Ni, M. Preparation of poly (vinylidene fluoride) membranes with graphene oxide addition for natural organicmatter removal. *J. Membr. Sci.,* **2015**, *473*(1), 54-62.
[http://dx.doi.org/10.1016/j.memsci.2014.09.018]

[103] Zhang, J.; Xu, Z.; Shan, M.; Zhou, B.; Li, Y.; Li, B.; Niu, J.; Qian, X. Synergetic effects of oxidized carbon nanotubes and graphene oxide on fouling control and anti-fouling mechanism of polyvinylidene fluoride ultrafiltration membranes. *J. Membr. Sci.,* **2013**, *448*(1), 81-92.
[http://dx.doi.org/10.1016/j.memsci.2013.07.064]

[104] Zhao, Y.; Xu, Z.; Shan, M.; Min, C.; Zhou, B.; Li, Y.; Li, B.; Liu, L.; Qian, X. Effect of graphite oxide and multi-walled carbon nanotubes on the microstructure and performance of PVDF membranes. *Separ. Purif. Tech.,* **2013**, *103*(1), 78-83.
[http://dx.doi.org/10.1016/j.seppur.2012.10.012]

[105] Zinadini, S.; Zinatizadeh, A.A.; Rahimi, M.; Vatanpour, V.; Zangeneh, H. Preparation of a novel antifouling mixed matrix PES membrane by embedding graphene oxide nanoplates. *J. Membr. Sci.,* **2014**, *453*(2), 292-301.
[http://dx.doi.org/10.1016/j.memsci.2013.10.070]

[106] Xu, Z.; Zhang, J.; Shan, M.; Li, Y.; Li, B.; Niu, J.; Zhou, B.; Qian, X. Organosilane-functionalized graphene oxide for enhanced antifouling and mechanical properties of polyvinylidene fluoride ultrafiltration membranes. *J. Membr. Sci.,* **2014**, *458*(1), 1-13.

[107] Akin, I.; Zor, E.; Bingol, H.; Ersoz, M. Green synthesis of reduced graphene oxide/polyaniline composite and its application for salt rejection by polysulfone-based composite membranes. *J. Phys. Chem. B,* **2014**, *118*(21), 5707-5716.
[http://dx.doi.org/10.1021/jp5025894] [PMID: 24811756]

[108] Compton, O.C.; Nguyen, S.T. Graphene oxide, highly reduced graphene oxide, and graphene: versatile building blocks for carbon-based materials. *Small,* **2010**, *6*(6), 711-723.
[http://dx.doi.org/10.1002/smll.200901934] [PMID: 20225186]

[109] Mahmoud, K.A.; Mansoor, B.; Mansour, A.; Khraisheh, M. Functional graphene nanosheets: the next generation membranes for water desalination. *Desalination,* **2015**, *356*(1), 208-225.
[http://dx.doi.org/10.1016/j.desal.2014.10.022]

[110] Zhao, C.; Xu, X.; Chen, J.; Wang, G.; Yang, F. Highly effective antifouling performance of PVDF/graphene oxide composite membrane in membrane bioreactor (MBR) system. *Desalination,* **2014**, *340*(1), 59-66.
[http://dx.doi.org/10.1016/j.desal.2014.02.022]

[111] Haider, M.S.; Shao, G.N.; Imran, S.M.; Park, S.S.; Abbas, N.; Tahir, M.S.; Hussain, M.; Bae, W.; Kim, H.T. Aminated polyethersulfone-silver nanoparticles (AgNPs-APES) composite membranes with controlled silver ion release for antibacterial and water treatment applications. *Mater. Sci. Eng. C,* **2016**, *62*, 732-745.
[http://dx.doi.org/10.1016/j.msec.2016.02.025] [PMID: 26952479]

[112] Mo, J.; Son, S.H.; Jegal, J.; Kim, J.; Lee, Y.H. Preparation and characterization of polyamide nanofiltration composite membranes with TiO$_2$ layers chemically connected to the membrane surface. *J. Appl. Polym. Sci.,* **2007**, *105*(6), 1267-1274.
[http://dx.doi.org/10.1002/app.25767]

[113] Zhang, Y.; Shan, L.; Tu, Z.; Zhang, Y. Preparation and characterization of novel Ce doped nonstoichiometric nanosilica/polysulfone composite membranes. *Separ. Purif. Tech.,* **2008**, *63*(1), 207-212.
[http://dx.doi.org/10.1016/j.seppur.2008.05.015]

[114] Zhang, Y.; Cui, P.; Du, T.; Shan, L.; Wang, Y. Development of a sulfated Y-doped nonstoichiometric zirconia/polysulfone composite membrane for treatment of wastewater containing oil. *Separ. Purif. Tech.,* **2009**, *70*(1), 153-159.
[http://dx.doi.org/10.1016/j.seppur.2009.09.010]

[115] Zhang, Y.; Shan, X.; Jin, Z.; Wang, Y. Synthesis of sulfated Y-doped zirconia particles and effect on properties of polysulfone membranes for treatment of wastewater containing oil. *J. Hazard. Mater.,* **2011**, *192*(2), 559-567.
[http://dx.doi.org/10.1016/j.jhazmat.2011.05.058] [PMID: 21664050]

[116] Huang, J.; Arthanareeswaran, G.; Zhang, K. Effect of silver loaded sodium zirconium phosphate (nanoAgZ) nanoparticles incorporation on PES membrane performance. *Desalination,* **2012**, *285*(1), 100-107.
[http://dx.doi.org/10.1016/j.desal.2011.09.040]

[117] Kim, E.S.; Hwang, G.; Gamal El-Din, M.; Liu, Y. Development of nanosilver andmultiwalled carbon nanotubes thin-film nanocomposite membrane for enhanced watertreatment. *J. Membr. Sci.,* **2012**, *394-395*(1), 37-48.
[http://dx.doi.org/10.1016/j.memsci.2011.11.041]

[118] Gao, P.; Liu, Z.; Tai, M.; Sun, D.D.; Ng, W. Multifunctional graphene oxide-TiO$_2$ microsphere hierarchical membrane for clean water production. *Appl. Catal. B,* **2013**, *138*(1), 17-25.
[http://dx.doi.org/10.1016/j.apcatb.2013.01.014]

[119] Yu, H.; Zhang, Y.; Zhang, J.; Zhang, H.; Liu, J. Preparation and antibacterial property of SiO$_2$-Ag/PES hybrid ultrafiltration membranes. *Desalination Water Treat.,* **2013**, *51*(16), 3584-3590.
[http://dx.doi.org/10.1080/19443994.2012.752900]

[120] Huang, J.; Wang, H.; Zhang, K. Modification of PES membrane with Ag–SiO$_2$: reduction of biofouling and improvement of filtration performance. *Desalination,* **2014**, *336*(1), 8-17.
[http://dx.doi.org/10.1016/j.desal.2013.12.032]

[121] Gao, Y.; Hu, M.; Mi, B. Membrane surfacemodification with TiO$_2$–graphene oxide for enhanced photocatalytic performance. *J. Membr. Sci.,* **2014**, *455*(2), 349-356.
[http://dx.doi.org/10.1016/j.memsci.2014.01.011]

[122] Wu, H.; Tang, B.; Wu, P. Development of novel SiO$_2$–GO nanohybrid/Polysulfone membrane with enhanced performance. *J. Membr. Sci.,* **2014**, *451*(1), 94-102.
[http://dx.doi.org/10.1016/j.memsci.2013.09.018]

[123] Goei, R.; Lim, T.T. Ag-decorated TiO$_2$ photocatalytic membrane with hierarchical architecture: photocatalytic and anti-bacterial activities. *Water Res.,* **2014**, *59*(1), 207-218.
[http://dx.doi.org/10.1016/j.watres.2014.04.025] [PMID: 24805373]

[124] Safarpour, M.; Khataee, A.; Vatanpour, V. Preparation of a novel polyvinylidenefluoride (PVDF) ultrafiltration membrane modified with reduced graphene oxide/titanium dioxide (TiO$_2$) nanocomposite with enhanced hydrophilicity and antifouling properties. *Ind. Eng. Chem. Res.,* **2014**, *53*(20), 13370-13382.
[http://dx.doi.org/10.1021/ie502407g]

[125] Safarpour, M.; Khataee, A.; Vatanpour, V. Effect of reduced graphene oxide/TiO$_2$ nanocomposite with

different molar ratios on the performance of PVDF ultrafiltration membranes. *Separ. Purif. Tech.,* **2015**, *140*(1), 32-42.
[http://dx.doi.org/10.1016/j.seppur.2014.11.010]

[126] Pinnau, I.; Freeman, B. *Membrane formation and modification*; American Chemical Society: Washington, DC, **2000**.

[127] Ulbricht, M. () Advanced functional polymer membranes. *Polymer (Guildf.),* **2006**, *47*(8), 2217.
[http://dx.doi.org/10.1016/j.polymer.2006.01.084]

[128] Wang, L.K.; Chen, J.P.; Hung, Y.T.; Shammas, N.K. *Handbook of Environmental Engineering*; Springer Science and Business Media; LLC, **2011**.

[129] Loeb, S.; Sourirajan, S. Sea water demineralization by means of an osmotic membranes. *Adv. Chem. Ser.,* **1962**, *38*(1), 117.

[130] Kesting, R.E. Phase Inversion Membranes, American Chemical Society. *A.C.S Symposium Series,* **1984**, pp. 269-300.

[131] Ahmed, I. High Performance Ultrafiltration Polyethersulfone Membrane. *PhD Thesis, University Technology Malaysia: Johor Bahru,* **2008**.

[132] Wijmans, J.G.; Baaij, J.B.; Smolders, C.A. The Mechanism of formation of microporous or skinned membranes produced by immersion precipitation. *J. Membr. Sci.,* **1983**, *14*(2), 263-274.
[http://dx.doi.org/10.1016/0376-7388(83)80005-2]

[133] Mecham, S. Synthesis and Characterization of Phenylethynyl Terminated Poly (arylene ether sulfone) as Thermosetting Structural Adhesives and Composite Matrices. *PhD Thesis, Faculty of the Virginia Polytechnic Institute and State University,* **1997**.

[134] Basile, A.; Cassano, A.; Rastogi, N.K. *Advances in Membrane Technologies for Water Treatment Materials, Processes and Applications*; Woodhead Publishing. Elsevier, **2011**.

[135] Grulke, E.R. Solubility Parameter Values*Polymer handbook,* 4th ed.; Brandrup, J.; Immergut, E.H.; Grulke, E.A., Eds.; Wiley: New York, NY, **1999**, 7, pp. 675-714.

[136] Strathmann, H. Production of Microporous Media by Phase Inversion Processes *ACS Symposium Serial number 269. American Chemical Society Washington, DC,* **1985**, p. 165.
[http://dx.doi.org/10.1021/bk-1985-0269.ch008]

[137] Yun, Y.; Tian, Y.; Shi, G.; Li, J.; Chen, C. Preparation, morphologies and properties for flat sheet ppesk ultrafiltration membranes. *J. Membr. Sci.,* **2006**, *270*(1), 146-153.
[http://dx.doi.org/10.1016/j.memsci.2005.06.050]

[138] Garcia-Ivars, J.; Iborra-Clara, M.I.; Alcaina-Miranda, M.I.; der Bruggen, B.V. Comparison between hydrophilic and hydrophobic metal nanoparticles on the phase separation phenomena during formation of asymmetric polyethersulphone membranes. *J. Membr. Sci.,* **2015**, *493*(4), 709-722.
[http://dx.doi.org/10.1016/j.memsci.2015.07.009]

[139] Sadrzadeh, M.; Bhattacharjee, S. Rational design of phase inversion membranes by tailoring thermodynamics and kinetics of casting solution using polymer additives. *J. Membr. Sci.,* **2013**, *441*(1), 31-44.
[http://dx.doi.org/10.1016/j.memsci.2013.04.009]

[140] Ani, I.; Iqbal, A. A Production of polyethersulfone asymmetric membranes using mixture of two solvents and lithium chloride as additive. *Jurnal Teknologi.,* **2007**, *47*(1), 25-34.

[141] Nejati, S.; Boo, C.; Osuji, C.O. Menachem Elimelech, Engineering flat sheet microporous PVDF films for membrane distillation. *J. Membr. Sci.,* **2015**, *492*(3), 355-363.
[http://dx.doi.org/10.1016/j.memsci.2015.05.033]

[142] Fadhil, S.; Marino, T.; Makki, H.F.; Alsalhy, Q.F.; Blefari, S.; Macedonio, F.; Di Nicolò, E.; Giorno, L.; Drioli, E.; Figoli, A. Novel PVDF-HFP flat sheet membranes prepared by triethyl phosphate (TEP) solvent for direct contact membrane distillation, Chemical Engineering and Processing. *Process*

Intensification, **2016**, *102*(1), 16-26.
[http://dx.doi.org/10.1016/j.cep.2016.01.007]

[143] Yeowa, M.L.; Liub, Y.; Li, K. Preparation of porous PVDF hollow fibre membrane *via* a phase inversion method using lithium perchlorate (LiClO$_4$) as an additive. *J. Membr. Sci.,* **2005**, *258*(1), 16-22.
[http://dx.doi.org/10.1016/j.memsci.2005.01.015]

[144] Strathmann, H.; Scheible, P.; Baker, R.W. A Rationale for the Preparation of Loeb-Sourirajan-Type Cellulose Acetate Membranes. *J. Appl. Polym. Sci.,* **1971**, *15*(5), 811-828.
[http://dx.doi.org/10.1002/app.1971.070150404]

[145] Zhao, W.; Su, Y.; Li, C.; Shi, Q.; Ning, X.; Jiang, Z. Fabrication of antifouling polyethersulfone ultrafiltration membranes using Pluronic F127 as both surface modifier and pore-forming agent. *J. Membr. Sci.,* **2008**, *318*(3), 405-412.
[http://dx.doi.org/10.1016/j.memsci.2008.03.013]

[146] Saljoughi, E.; Sadrzadeh, M.; Mohammadi, T. Effect of preparation variables on morphology and pure water permeation flux through asymmetric cellulose acetate membranes. *J. Membr. Sci.,* **2009**, *326*(4), 627-634.
[http://dx.doi.org/10.1016/j.memsci.2008.10.044]

[147] Taurozzi, J.S.; Arul, H.; Bosak, V.Z.; Burban, A.F.; Voice, T.C.; Bruening, M.L.; Tarabara, V.V. Effect of filler incorporation route on the properties of polysulfone–silver nanocomposite membranes of different porosities. *J. Membr. Sci.,* **2008**, *325*(1), 58-68.
[http://dx.doi.org/10.1016/j.memsci.2008.07.010]

[148] Teoh, M.M.; Chung, T.S.; Yeo, Y.S. Dual-layer PVDF/PTFE composite hollow fibers with a thin macrovoid-free selective layer for water production *via* membrane distillation. *Chem. Eng. J.,* **2011**, *171*(5), 684-691.
[http://dx.doi.org/10.1016/j.cej.2011.05.020]

[149] Lin, J.; Ye, W.; Zhong, K.; Shen, J.; Jullok, N.; Sotto, A.; Van der Bruggen, B. Enhancement of polyethersulfone (PES) membrane doped by monodisperse Stöber silica for water treatment. *Chem. Eng. Process.,* **2016**, *107*, 194-205.
[http://dx.doi.org/10.1016/j.cep.2015.03.011]

[150] Pinnau, I.; Koros, W.J. A qualitative skin layer formation mechanism for membranes made by dry/wet phase inversion. *J. Polym. Sci., B, Polym. Phys.,* **1993**, *31*(2), 419-427.
[http://dx.doi.org/10.1002/polb.1993.090310406]

[151] Pesek, S.C.; Koros, W.J. Aqueous Quenched Asymmetric Polysulfone Hollow Fibers Prepared by Dry/wet Phase Separation. *J. Membr. Sci.,* **1994**, *88*(1), 1-19.
[http://dx.doi.org/10.1016/0376-7388(93)E0150-I]

[152] Pinnau, I.; Koros, W.J. Influence of quench medium on the structures and gas permeation properties of polysulfone membranes made by wet and dry/wet phase inversion. *J. Membr. Sci.,* **1991**, *71*(1-2), 81-96.
[http://dx.doi.org/10.1016/0376-7388(92)85008-7]

[153] Witte, P.V.; Dijkstra, P.J.; Berg, J.W.A.; Feijen, J. Phase Separation Processes in Polymer Solutions in Relation to Membrane Formation. *J. Membr. Sci.,* **1996**, *117*(1), 1-31.
[http://dx.doi.org/10.1016/0376-7388(96)00088-9]

[154] Cahn, J.W. Phase separation by spinodal decomposition in isotropic systems. *J. Chem. Phys.,* **1965**, *42*(1), 93-98.
[http://dx.doi.org/10.1063/1.1695731]

[155] Barzin, J.; Sadatnia, B. Theoretical phase diagram calculation and membrane morphology evaluation for water/solvent/polyethersulfone systems. *Polymer (Guildf.),* **2007**, *48*(10), 1620-1631.
[http://dx.doi.org/10.1016/j.polymer.2007.01.049]

[156] Aroon, M.A.; Ismail, A.F.; Montazer-Rahmati, M.M.; Matsuura, T. Morphology and permeation properties of polysulfone membranes for gas separation: effects of non-solvent additives and co-solvent. *Separ. Purif. Tech.,* **2010**, *72*(1), 194-202.
[http://dx.doi.org/10.1016/j.seppur.2010.02.009]

[157] Zhou, B.; Tang, Y.; Li, Q.; Lin, Y.; Yu, M.; Xiong, Y.; Wang, X. Preparation of polypropylene microfiltration membranes *via* thermally induced (solid-liquid or liquid-liquid) phase separation method. *J. Appl. Polym. Sci.,* **2015**, *132*, 42490.
[http://dx.doi.org/10.1002/app.42490]

CHAPTER 8

Nano-catalyst and Nano-catalysis: State of the Arts and Prospects

Shahid Ali Khan[1,2,3], Sher Bahadar Khan[1,2,*], Kalsoom Akhtar[4] and **Aliya Farooq[5]**

[1] *Center of Excellence for Advanced Materials Research, King Abdulaziz University, Jeddah 21589, Saudi Arabia*

[2] *Chemistry Department, Faculty of Science, King Abdulaziz University, Jeddah 21589, Saudi Arabia*

[3] *Department of Chemistry, University of Swabi, Anbar-23561, Khyber Pakhtunkhwa, Pakistan*

[4] *Division of Nano Sciences and Department of Chemistry, Ewha Womans University, Seoul 120-750, Korea,*

[5] *Department of Chemistry, Shaheed Benazir Bhutto Women University, Peshawar, Pakistan*

Abstract: Nanocatalysts have received much attention because of their vast applications in various organic pollutants removal as well as their selectivity, ease in separation, eradicating the expensive materials and hazardous chemicals. Nanocatalysts are the heart of scientific field because of their vast applications in multidisciplinary sciences. This chapter provided a brief knowledge using nanocatalyst for the removal of organic toxins, for instance nitrophenols and dyes, which are at alarming condition. The readers will be benefited from this chapter and will learn about photocatalysis and its application. The use of semiconductors is appropriate in photocatalysis, therefore, much information is available about semiconductor. An important morphological nanocatalyst is the layered double hydroxide; also worked as a solar catalyst for the removal of contaminants and various support are present for these material to enhance its catalytic performance. Supported materials in nanocatalysis are required to stabilize the nanoparticles (Nps) from being aggregated and also offer any easy route in separating the catalyst after the reaction. Such criteria are highly demanded at industrial level.

Keywords: Adsorption, Chemical degradation, Environmental pollution, Nanofiltration, Nanomaterials, Photodegradation.

---

* **Corresponding author Sher Bahadar Khan**: Chemistry Department and Center of Excellence for Advanced Materials Research (CEAMR), King Abdulaziz University, Jeddah 21589, Saudi Arabia; Tel/Fax: +966-59-3541984; Email: sbkhan@kau.edu.sa

INTRODUCTION

Water treatment is highly related to the health issue and quality of life. Waste water has a high degree of negative impact on the environment, therefore, clean and safe water is the necessity of every human being [1]. However, unfortunately, majority of the people have less access to the potable and safe drinking water. Contaminated water is the cause of numerous gastro-intestinal discomforts. Some organic pollutants are extremely carcinogenic and mutagenic. Prolonged deficiencies of clean and drinking water, and their safe treatment in economic point of views are the world prime concern. It was estimated that approximately 3,900 children die each year from various diseases, caused by polluted water, and approximately 1.2 billion people have little access to safe and clean water [1]. In order to overcome these worst conditions, different technologies has been developed for the cleansing of water including adsorption, reverse osmosis, electrodialysis, electrolysis, membrane, filtration, ion exchanger and advance oxidation process. Among all these stated methods, advance oxidation process is the most useful technique which uses semiconductors. Among various substances, metal oxide nanomaterial is the well known catalyst. Various nanomaterials as semiconductor have been evaluated for their catalytic potentials, for instance, TiO_2, ZnO, WO_3, V_2O_5, ZrO_2, CdS, CdO and many more.

NANOMATERIALS

The substances in which one dimension is less than 100 nanometers are called nanomaterials or nano-scale materials. Nano is one billionth 10^{-9} and is one millionth of a millimeter and nearly 100,000 times smaller than a human hair. The size of the nanomaterials is between the size of the atoms or molecules and a bulk structure of the molecules [2].

By decreasing the particle size to one millionth of a millimeter, the volume of atoms assembling the particle becomes much smaller as many as several hundreds or thousands times. At the Nano-scale level, some of the vital physical characteristics are significantly changed. For instance, a significant change in the melting point of materials and the sintering of ceramic materials at lower temperature. Also, as the particles get smaller than the wave length of visible light, they not only become transparent but also emit special light by plasma absorption. Although, the smaller particle size and larger particle size constitue the same things, however, smaller particles indicate absolutely diverse physicochemical or electromagnetic properties from their bulk counterparts [3].

The description of Nps varies depending on the materials, applications and areas in question. For instance, it could be as smaller as 10–20 nm, where the characteristics of most of the solid constituents changed drastically. However, particles ranging from 1 nm to 1 μm (1000 nm) may also consider as Nps [4].

Significant interest has been taken due to the unprecedented scale with diverse electrical, magnetic, electronics, optical, medicine, and catalytic properties. Nanomaterials are the milestones in nanotechnology and nanoscience. In the past few decades, nanostructure expertise has been the emerging and growing area in science and technology. It is an interdisciplinary science and a broad term which covers many fields. Nanomaterials are revolutionizing the synthesis of majority of the industrial products that can be synthesized with expensive routes. Nanomaterials have a high commercial impact, and are exponentially increases day by day [5].

Mostly, the nanomaterials are engineered with control morphology, texture and appropriate size for specific purpose. There are two main explanations about the distinct characteristics of nanomaterials. First, their high surface-to-volume ratio and second their quantum size effects [6]. Nanomaterials have much larger surface-to-volume area as compared to their bulk counterparts, providing them to high chemical reactivity and strength. Similarly, the high optical, magnetic, electrical and catalytic properties is due to their quantum effects [7]. The use of nanomaterial in practical applications largely depends on the particle size in powder-particles. The powder is usually constructed from diverse particles size, hence, it is essential to acquire the mean particle size as well as the size distribution. Various technologies have been developed for the investigation of particle size, particularly, analytical methods which, covering particle size of different size and shape with high repeatability. For instance, desire particle size can be obtained with laser scattering and diffraction approaches [8].

Nanomaterials have driven the complex and important organic scaffold, which have immense importance in industries and are economically at its highest level in the market. The use of nanomaterials in laboratories and industrial scales for the multiphase organic reactions finds immense inspiration on decreasing the costs of industrial process and minimizing or in some cases diminishing the waste disposal. As compared to heterogeneous catalyst, homogenous catalyst is practically not favorable due to its high costs, difficulty in separation and recycling. To cope with this dilemma, several technologies have been put forward for the separation of catalyst, improving recyclability and regeneration. Such as nanofiltration, liquid–liquid phase separation; comprising of fluorous phases, supercritical solvents, ionic liquids, and polymeric supports. Each of these methods has some flaws in one or other directions due to cost and secondary waste generation. Heterogeneous catalyst overwhelmed these problems by serving as support, or catalysts are supported on some solid media.

Nanomaterials are commercially applied from the past few decades. The availability of nanomaterials based commercial products are very vast and are largely used in electronics, sunscreens, including wrinkle-free textiles, cosmetics, stain-resistant, varnishes and paints.

Another fascinating and promising practical application of nanomaterials is the use of nanocomposites and nanocoatings in important products, like sports equipment, windows, automobiles and bicycles. Similarly, coatings of novel UV-blocking on glass bottles of beverages, protecting them from injurious radiation of sunlight, and longer-lasting tennis balls through butyl rubber/nano-clay composites. Nano-scale titanium dioxide is primarily used in sun-block creams, cosmetics, and self-cleaning windows, while the nano-scale silica is used as a filler in cosmetics and dental fillings [9].

NANO-CATALYSIS

Thanks to the advent of nanotechnology which provided novel protocol for the synthesis and development of complex solids particles with control size and shape [10]. Catalysis at the nano-scale level has attracted significant attention in the past two decades due to their unique properties. The ability of controlling the specific shapes or size of the particles, complex nanostructures of solid materials, may accomplish selectivity of the reactions. This relationship between nanotechnology and surface chemistry led to many exciting progression, and potentials for the development in chemical synthesis. Rational design of catalysts with exceptional morphological and textural characteristics have long been applied in catalysis [11]. To a high degree, the activity of a catalysts largely depends on their textural, morphological features and their chemical composition, effecting the number of active sites in a catalyst, approachability of the reactant on to the active sites of the catalytic surface, pores to the substrates and mass-transport of the reactant and product molecules [12].

The rapid progress in the field of nanoscience enables the construction and study of solid catalysts on the nanometre and sometimes even atomic or molecular scales. These advancements led to the development of nanocatalysis, which involves the use of nanostructured catalysts for a variety of catalytic applications, and the study of the chemical mechanisms of such nanostructure induced catalytic properties.

SUPPORTED NANOCATALYST ON SOLID MATRICES

In the last few decades, nanotechnology has become the emerging field in the scientific area, which deals with the synthesis, exploration and characterization of nanomaterials, where noteworthy nano-scale systems have been developed and applied [13]. Nanomaterials find many attractive applications in diverse fields of science, including chemistry, electronics, physics, biology, catalysis and medicine. Especially, nanomaterials is of precise awareness in the synthesis of chemicals, pharmaceutical, fine chemicals and drugs from raw materials [14]. Nps have large surface-to-volume metal ratio as compared to their bulk-counterparts. The Nps show less activity in the bulk because of the accumulation of misarranged atoms on the surface [15]. The Nps are used efficiently in the many electron transfer reactions, because the electron in nano-scale materials is restricted to small space like the width of few atoms which attributed to high fermi potential level and giving small quantum size effect [16]. Nps created much interest in both basic and applied fields of chemistry. They are tremendously important and reactive by virtue of their inherent electronic properties and high surface-to-volume ratio. The Nps are of small size and unstable which are agglomerated quickly and therefore, decreases their catalytic activity. The stability of these Nps against agglomeration can be accomplished by using solid porous matrices for heterogeneous catalysis and also act as stabilizers in solution-phase.

Recently, metal Nps immobilization onto solid medium got the attention of researcher due to their huge prominent in the field of nanotechnology [17]. This type of recognition played an important role in practical heterogeneous catalysis [18]. The loading of metal Nps in the solid matrices and their recapturing has got importance in the field of catalysis. A huge benefit can be taken from such type of immobilized Nps in solid materials, and can be used for a variety of chemical reactions. Such supported materials removed the phase transfer process of Nps. Moreover, controlling the composition, morphology and surface modification of metal Nps is extremely important for specific purpose. The use of matrix can provide an easy route for the separation of Nps after reaction which is indispensable in catalytic reactions [19].

The most widely reported supports used to adsorb and stabilized Nps are inorganic solids, such as alumina, charcoal, silica, and oxides such as mesoporous MgO or TiO_2 and so forth. Another procedure to graft and immobilized the Nps on the solid support by chemical methods are the use of polyacrylamide gel [20]. Latex and resins are some important stabilizer polymeric materials. Such type of finding, stabilization and immobilization will help the use of these Nps in sorbents, catalysis, sensors, bactericides, and reducing agents, and electrochemical devices [21].

In our group we synthesized and immobilized the mono and bimetallic metal Nps on supported materials [22]. Similarly, we synthesized the metal nanocomposite (metal oxide + polymer) and immobilized various transition metals, which are excellently used in the hydrogenation of nitrophenol and dyes. These catalyst have electron donating group originating from polymeric material and metal oxide, acting as ligands for the empty d-orbital of the transition metals. These metal ligands bonds immobilized the Nps and avoids agglomeration and aggregation. These supported metal Nps are used for the catalytic reduction of nitrophenol to amino phenol with a reducing agent like $NaBH_4$.

Nitrophenols is the most extensively used industrial nitro-aromatic compounds and often employed as an intermediates in the production of pharmaceuticals, synthetic dyes, herbicides, pesticides, explosives, pigments, dyes, wood preservatives and rubber chemicals [23]. However, it is a very toxic and carcinogenic compound [17]. Although, nitrophenols are useful intermediates in the fabrication of various aforementioned materials, however, they also act as common environmental pollutants because of their toxicity and resistance to microbial degradation [24]. For the reasons, nitrophenols are considered as a priority pollutant. The U. S. Environmental Protection Agency has registered 4-nitrophenol, 2-nitrophenol, and 2,4-dinitrophenol as 'Priority Pollutants' and make strict restriction for controlling their concentrations in natural waters to <10 ng/L (U.S. EPA, 1976). The agency has also set pre-treatment standards for industries to discharge nitrophenols after using or manufacturing these compounds (U.S. EPA, 1988). A large amount of nitrophenols have been discharged into wastewater due to their wide application in industrial and agriculture setup. They are injurious to the liver, kidney, central nervous system, and both human and animal blood. Hence, their removal from the environment is essential.

By the process of photochemical oxidation nitrophenols degraded in surface soil, surface water and become the part of water resources. However, going down to the depth of soil and groundwater, the degradation process becomes slow and then relies on biodegradation, therefore, nitrophenols possibly stay in these sites for a long time, and resistance in such area to microbial degradation [25]. Therefore, it is utmost to develop an efficient method to control nitrophenols from the effluents before entering into water resources. Various methods and technologies have been put forward for their removal, for instance, adsorption [26], photocatalytic degradation [27], electro-Fenton method [28], electrocoagulation [29], biodegradation [30], catalytic degradation [31], advanced oxidation processes [32], and electrochemical methods [33]. Moreover, many absorbents such as resin [34], carbon nanosphere [35], chemical reduction, activated carbon [26b] and zeolite [35], *etc.* have been used for nitrophenol removal. However, due to high cost, difficulty in separation, and generation of toxic intermediates limited the scope of these techniques.

Chemical reduction is the method of choice for nitrophenols removal from the effluent. In this process, the nitrophenols are converted to amino-phenol in the presence of an efficient catalyst and reducing agent. Amio-phenol itself is an important industrial chemical which finds applications in corrosion inhibitor, photographic drying agent, and films developer. Amino-phenol also employed as a raw material for the production of antipyretic and analgesic drugs [36].

PHOTOCATALYSIS

Heterogeneous photocatalysis has currently arose as an effective method for waste water treatment. Currently, water treatment is one of the major concerns for the mankind [37]. Clean and potable water is the fundamental need for human beings. Largely, earth is covered by water, however, clean water supply is limited to human beings. Unfortunately, our water resources are rapidly contaminated by improper sewage, fast growing industrial sectors, man-made activities and natural calamities. Due to the aforementioned issues, increasing safe water demand and shortages of safe and potable water resources are the prolong term and the world leading problems. The people in the developing countries have less or no access to the clean and potable water [38].

In a survey, it was reported that about 1.2 billion people have less or no access to safe drinking water, and 2.6 billion people have slight or no access to clean water. Similarly, millions of people die annually and 3,900 children die per day from various disease and discomfort conveyed from harmful and unsafe water or human excreta to water resources [38, 39]. Besides, various infectious water-borne diseases, such as diarrhea, fever, intestinal infection, constipation, abdominal pain and several other diseases caused the death of millions of lives [40]. Approximately, 7, 00 microbial, inorganic and organic contaminants have been reported from water, in which certain inorganic and organic toxins are carcinogenic. In fact, some organic pollutants are not biodegradable and therefore, retain in water resources for prolong time. Therefore, high risk health and environmental anxieties of these toxins propel researchers to sanitize the contaminated water [38].

Among these organic compounds, dyes are highly concerned due to its mutagenic and carcinogenic nature to aquatic life. Dyes in small concentration make the water colorful, and form a foam like layer on the surface of water, thus preventing the diffusion of oxygen and sunlight, and thereby affecting the biological phenomena of aquatic flora and fauna [1].

With the advancement in science and technologies, the manufacturing of newer chemicals at the industrial level is increased. With this high demand, the organic dyes emerged as a new chemical family and makes people life better due to their vast industrial applications. Dyes mostly used to color our clothes, food ingredients and beverages, even dyes are also usd to make some medicines colored [41]. Around 10,000 dyes are commercially available, and globally 7, 00,000 tons are manufactured each year and approximately 11% of these dyes are discharged into water resources during their synthesis and uses. Nearly, 10-15% of the dyes stuff are released into the environment which are esthetically not feasible [42]. Recently, dyes stuff are of prominent environmental distress, due to thier mutagenicity and carcinogenicity [43].

It was also reported that more than 90% of around 4000 dyes were practiced in an ETAD report (Ecological and Toxicological Association of the Dye stuffs) indicating more than 2×10^3 mg/kg LD_{50} values [42]. Similarly, the maximum toxicity was originated from diazo and basic dyes [3]. In September 1997, UK make strategies based on environmental protection, which stated that "zero synthetic chemical substances are to be released in the marine environment and guaranteed that the textile industries should treat their wastewater before dumping into the water resources" [42]. Similarly, European community and developed countries make policies which are more strict for controlling the dyes stuff from the industrial effluents [42].

It is well known that dyes industries play an important role in the progress and prosperity of country, but inappropriately, most of the industries dumped their dyes stuff into the water resources without any treatment [5]. Due to this, the underground water resources have become contaminated and thus creating a high risk to human health and other organisms. These dyes become an indispensable part of industrial effluent due to their remarkable uses. It is well documented that majority of these dyes possess potential carcinogenic and toxic nature, therefore, their eradication from the industrial effluents is of prime importance [43]. For this reason, various technologies have been designed to handle dyes removal from water; which included the coagulation, biodegradation, adsorption, membrane process and advanced oxidation process (AOP) [41, 44]. All these processes have some flaws in one or many ways.

Among the aforementioned processes, photocatalysis is the most effective and advance method. Fujishima and Honda discovered photocatalysis in 1972. After that, several scientists attempted to develop more proficient photo-catalysts for the treatment of water contamination [41, 45]. Photocatalysis is a part of advance oxidation method in which the catalyst worked under the light. This technology is appropriate as compared to other advance oxidation methods, due to the ease in availability of semiconductor, low cost and high efficiency.

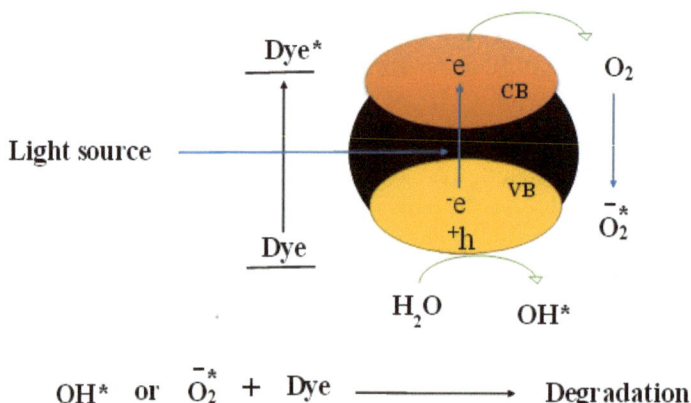

Fig. (1). Mechanistic pathway of dye degradation in the presence of light and catalyst.

When light of appropriate frequency strikes on a catalyst, the electron promoted from the valance band of the catalyst to the excited state. This high energy electron is captured by the O_2 or OH^{-1}, generated from water molecule in the system. The O_2 captures an electron and goes to higher energy state to form superoxide free radical anion $O_2^{-\cdot}$, while the $OH^{\cdot}$ is formed from the OH^{-1} after receiving an electron available in the high energy state. It could be possible that the dye itself absorbed light and excited to the higher energy state. These extremely high energy free radical species invade the dyes molecule and mineralized. It could also be possible that both $O_2^{-\cdot}$ and $OH^{\cdot}$ free radical transfer to the surface of the catalyst, where charge transfer takes place for the degradation of dye molecules [1].

Some efficient metal oxides, for instance, zinc oxide (ZnO), titanium dioxide (TiO_2), strontium titanate ($SrTiO_3$), tungsten oxide (WO_3), and hematite (Fe_2O_3) have been shown excellent catalytic activity in the UV range [4]. Majority of these metal oxides have 3.2 ev ($\lambda = 387$ nm) band gap in the UV region, and therefore they work well upon the irradiation of UV radiation. These catalysts worked well only in the presence of UV light, however, sunlight comprises merely 5–7% of UV light, while the remaining spectrum consists of 46% and 47% visible and infrared radiation, respectively [41].

The main problem related to the photocatalysis is the quick recombination of electron-hole pair. This can be solved by the used of scavenger. Molecular oxygen is a better option as a scavenger, however, many transition metals are used to prevent the recombination and enhanced charge transfer. This problem is solved by the doping techniques.

The process of photocatalysis is affected by the addition of some additives, for instance, some ions such as SO_4^{-2}, BrO^{3-}, CO_3^{-2}, HCO_3^-, Zn^{+2}, Fe^{+2}, Na^{+1} and Ag^{+1}. These ions slow down the rate of dye degradation, because they scavenge the $\overset{\cdot}{O}H$ and convert them to ^-OH, at the end of the day the ^-OH become the part of water molecule [46].

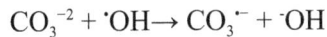

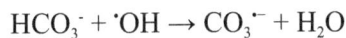

$$Fe^{+2} + \overset{\cdot}{\,}OH \rightarrow Fe^{+3} + \overset{-}{\,}OH$$

$$CO_3^{-2} + \overset{\cdot}{\,}OH \rightarrow CO_3^{\cdot-} + \overset{-}{\,}OH$$

$$HCO_3^- + \overset{\cdot}{\,}OH \rightarrow CO_3^{\cdot-} + H_2O$$

The above reactions show the $\overset{\cdot}{O}H$ scavenging mechanism by decreasing the rate of dye degradation.

Importance of Doping

The best way to avoid electron-hole pair recombination is the doping technology. Here we take the example of TiO_2 and other transition metal as dopants. When transitional metals (dopant) are added to TiO_2 catalyst it decreases the recombination of electrons-holes pairs and also improved its catalytic activity [47]. Doping brought a bathochromic shift (increase in wavelength) to the catalyst by decreasing the band gap or we can also say that doping phenomena introduce an intra-band gap states, which result in more visible light absorption. The influence of dopants on the photocatalytic activity is a complex problem. Doping changed the photocatalytic properties of TiO_2 and the overall effect of dopant with TiO_2 led to the enhanced light absorptivity and charge transfer. Further, the TiO_2 efficiency will be enhanced by using suitable dopants [87-90]. When a dopant is mixed with TiO_2, the 3d electron of Ti overlapped with the d-orbital of the used transition metal and thus caused a red shift, which enabled the synthesized catalyst to absorb visible light and enhanced the photocatalytic activity. In this way, the UV light-based catalysts are transformed to visible light-based catalyst of widened range depending on the dopant metal.

Various transition metals, such as Ni^{2+}, Fe^{3+}, Zn^{2+}, and Cr^{3+} can be simply integrated into the crystal lattice of TiO_2 due to their comparable same ionic radii. For instance, ($Zn^{2+} = 0.74$ Å, $Cr^{3+} = 0.76$ Å, $Ni^{2+} = 0.72$ Å, and $Fe^{3+} = 0.69$ Å), which are very close to the ionic radii of $Ti^{4+} = 0.75$ Å ions [91]. The process of dyes degradation is generally faster in doped material as compared to single material, due to the efficient consumption of photo-generated holes by the oxidation of dyes, thus attenuating electron-hole recombination [48].

Layered double hydroxide as photo-catalyst in dye degradation

Another strategy to make efficient catalyst under visible light is the engineering of layered double hydroxide morphology (LDH). The LDH works well in visible light. These types of morphologies are formed by stacking two-dimensional units linked to each other by weak forces. LDH is an anionic clay with common formula [Zn(1-x) M(III) (x)(OH)$_2$][Anx/n·mH$_2$O] (M(III) = Al, Cr, Ga; An- = CO_3^{2-}, Cl-, NO^{3-}, CH_3COO^-) [49]. The trivalent cations generated a complete positive charge which is counter-balanced by the intercalation of anions in the interlayers of brucite type structure [49]. ZnAl-LDH is considered as the member of LDH family due to its vast tunable ability of metal cations, and its anion exchange capability arises due to various host-guest associations of Zn^{2+}/Al^{3+} molar ratios. One can make a wide range of LDH morphology by changing the metal cations and enhancing their anion exchange capacity, making them smart contestants in the field of catalysis, absorbents, and ion exchangers [50]. Heating the LDH ingredients from 300–600 °C, the decomposition started and formed an extremely active mixed metal oxide nanocomposites with large thermal stability and high specific surface area [50a], these mixed metal oxide are used for the degradation of organic toxins, such as dyes and phenol. Different types of molecules can be introduced into the interlayered spaced between the two dimensional (2-D) layered nanostructures with flexible chemical and physical characteristics [51]. For instance, ZnAl-LDH are reported with high catalytic properties, such as nanocomposite hydrogel, energy storage, proficient luminescence source, chemical nanocontainer, absorbent, anion-exchanger and biological nanocarrier [50b]. Similarly, it is also used as a precursor for mixed metal oxide [50b]. LDHs can also be synthesized with morphologies comprising three or even more cations in the layers. This wide range and flexible composition permits LDH with versatile characteristics. A highly ordered and well organized arrangement has been observed on the atomic level through solid state multinuclear NMR spectroscopy. This is why, these materials have broad spectrum characteristics [52]

Supported Material for Photocatalysis

Indeed photocatalysis related to the quality of life is helpful in pollutants removal from wastewater. However, the semiconductor for instance, titanium oxide has two main flaws. Firstly, quick recombination of $\bar{e}/h^+$pair, and secondly, powdered TiO$_2$ are easily lost and difficult to recycle.

The first problem was solved by many researchers and much data are available in the literature where researchers doped the TiO$_2$ with other transition metals, thereby preventing the process of recombination. The second drawback was handled by immobilizing or combining with strong adsorbent which itself works as a co-oxidant, for example, alumina, silica, zeolites or clays and activated carbon [53]. The TiO_2 also supported on the glass plate, however, it is detached from the glass plate due to the week adhesion force between TiO_2 and glass surface and therefore, its efficiency is declined with time. Compared to the other sorbent the activated carbon showed superiority due to high capability as adsorbant as well as their low cost [54].

Dispersion of TiO_2 with material of high surface area and well powdered form in the reaction system can enhance the photocatalytic activity. For instance, two types of TiO_2 composites were synthesized with carbon; TiO_2-coated exfoliated graphite and carbon-coated TiO_2. The carbon-coated TiO_2 showed remarkable catalytic activity in the decomposition of polyvinyl alcohol in aqueous media under the exposure of UV light and was indicating good recyclability [55]. Matos and his co-workers (2001) examined the effect of various activated carbons with titania and study its effect on the photocatalytic-degradation of organic contaminants in aqueous media under the influence of UV light [56]. They investigated that, by adding activated carbon to Titania slurry, it could prompt a valuable results on the photocatalytic degradation of p-chlorophenol, methyl orange dye and herbicide in reliance of the properties of activated carbon under UV irradiation. Similarly, phenol degradation and other model toxins are efficiently degraded in the presence of TiO_2 and activated carbon [55]. These references make incredible acknowledgment in the removal of pollutants with significant kinetics rate. It was also proved that the degradation is faster in mixed system comprising TiO_2 and activated carbon. The so nominated synergy have been enlightened by forming common contact at the interface between the solid phases of different materials, for instance, activated carbon and TiO_2. In this combination, activated carbon act as an effective adsorbing trapping material for the organic molecules, and therefore, the molecules is transferred efficiently to the surface of TiO_2, where it is directly degraded photo-catalytically [55].

Activated carbon, also called active carbon or activated charcoal, is an amorphous carbon and solid porous material. Primarily, it is derived from carbonaceous substances, such as coal (bituminous, lignite) or plant based (lignocellulosic) material, such as wood, peat, or nutshells (*i.e.* oil palm, coconut shells) [57]. Usually, the activated carbons are prepared in two phases; carbonization and activation process [57]. In the process of carbonization the by-products, such as tars, hydrocarbons (*i.e.* volatile organic compounds) and gases produced are eliminated from the raw material by drying and heating. In this process, 400-600 °C temperature is provided along with inert atmosphere which makes the oxygen deficient environment to stop the oxidation of the synthesized carbonizing materials [57]. In the second phase, the so formed carbonized materials are then activated by treating them with an activating agent, such as CO_2 or steam at similar or elevated temperature as compared to the first phase [57]. The by-products formed in the first phase are subjected to the process of oxidation, which led to the production of three-dimensional graphite lattice structure with high porosity and exceptionally high surface area [58]. Usually, 1 g activated carbon has approximately 500-2000 m^2 surface area. This is normally recognized by using nitrogen gas or a simple liquid adsorption, calculating Langmuir isotherm for monolayer adsorption, while for multilayer adsorption isotherm the BET (Brunauer, Emmett and Teller) are employed. The high demand and versatile nature of these fascinating materials are much documented in the literature and approximately 182 scientific papers and 1341 patents (SCOPUS July 2007) are reported for multipurpose applications. Some chemical treatments are employed for activated carbon for specific purposes [55].

CONCLUSION

This chapter is design to explain various aspects of nano-scale material and their application in wastewater treatment. Nanocatalyst is the core of nanotechnology. The nano-scale size of these materials are tremendously important for their catalytic activity and vast application as compared to their bulk counterparts. Clean water are the prime source for all human beings, but it is mainly contaminated by human beings. In this chapter, we briefly explained various methodology for wastewater treatment. The role of layered double hydroxide in the process of degradation was explained. The stabilization of zero-valent metal Nps was explained in detailed for the degradation of organic pollutants.

CONSENT FOR PUBLICATION

Not applicable.

CONFLICT OF INTEREST

The author (editor) declares no conflict of interest, financial or otherwise.

ACKNOWLEDGEMENTS

The authors are grateful to the Center of Excellence for Advanced Materials Research (CEAMR), King Abdulaziz University, Saudi Arabia for providing research facilities.

REFERENCES

[1]   Khan, S.A.; Khan, S.B.; Kamal, T.; Asiri, A.M.; Akhtar, K. Recent Development of Chitosan Nanocomposites for Environmental Applications. *Recent Pat. Nanotechnol.,* **2016**, *10*(3), 181-188.
[http://dx.doi.org/10.2174/1872210510666160429145339] [PMID: 27136929]

[2]   Pumera, M. Graphene-based nanomaterials and their electrochemistry. *Chem. Soc. Rev.,* **2010**, *39*(11), 4146-4157.
[http://dx.doi.org/10.1039/c002690p] [PMID: 20623061]

[3]   Ng, L.Y.; Mohammad, A.W.; Leo, C.P.; Hilal, N. Polymeric membranes incorporated with metal/metal oxide nanoparticles: a comprehensive review. *Desalination,* **2013**, *308*, 15-33.
[http://dx.doi.org/10.1016/j.desal.2010.11.033]

[4]   Nogi, K.; Naito, M.; Yokoyama, T. *Nanoparticle technology handbook*; Elsevier, **2012**.

[5]   Stark, W.J.; Stoessel, P.R.; Wohlleben, W.; Hafner, A. Industrial applications of nanoparticles. *Chem. Soc. Rev.,* **2015**, *44*(16), 5793-5805.
[http://dx.doi.org/10.1039/C4CS00362D] [PMID: 25669838]

[6]   Rao, C.; Cheetham, A. Science and technology of nanomaterials: current status and future prospects. *J. Mater. Chem.,* **2001**, *11*(12), 2887-2894.
[http://dx.doi.org/10.1039/b105058n]

[7]   Wang, Z.L. Zinc oxide nanostructures: growth, properties and applications. *J. Phys. Condens. Matter,* **2004**, *16*(25), R829.
[http://dx.doi.org/10.1088/0953-8984/16/25/R01]

[8]   Ma, Z.; Merkus, H.G.; de Smet, J.G.; Heffels, C.; Scarlett, B. New developments in particle characterization by laser diffraction: size and shape. *Powder Technol.,* **2000**, *111*(1), 66-78.
[http://dx.doi.org/10.1016/S0032-5910(00)00242-4]

[9]   Alagarasi, A. *Introduction to nanomaterials*; National Center for Environmental Research, **2011**.

[10]   (a). Zaera, F. The new materials science of catalysis: Toward controlling selectivity by designing the structure of the active site. *J. Phys. Chem. Lett.,* **2010**, *1*(3), 621-627.
[http://dx.doi.org/10.1021/jz9002586]
(b). Lee, I.; Albiter, M.A.; Zhang, Q.; Ge, J.; Yin, Y.; Zaera, F. New nanostructured heterogeneous catalysts with increased selectivity and stability. *Phys. Chem. Chem. Phys.,* **2011**, *13*(7), 2449-2456.
[http://dx.doi.org/10.1039/C0CP01688H] [PMID: 21103527]

[11]   Schlögl, R.; Abd Hamid, S.B. Nanocatalysis: mature science revisited or something really new? *Angew. Chem. Int. Ed. Engl.,* **2004**, *43*(13), 1628-1637.
[http://dx.doi.org/10.1002/anie.200301684] [PMID: 15038028]

[12]   Khan, S.A.; Khan, S.B.; Asiri, A.M.; Ahmad, I. Zirconia-based catalyst for the one-pot synthesis of coumarin through Pechmann reaction. *Nanoscale Res. Lett.,* **2016**, *11*(1), 345.
[http://dx.doi.org/10.1186/s11671-016-1525-3] [PMID: 27460593]

[13]   Karakas, K.; Celebioglu, A.; Celebi, M.; Uyar, T.; Zahmakiran, M. Nickel nanoparticles decorated on electrospun polycaprolactone/chitosan nanofibers as flexible, highly active and reusable nanocatalyst in the reduction of nitrophenols under mild conditions. *Appl. Catal. B,* **2017**, *203*, 549-562.
[http://dx.doi.org/10.1016/j.apcatb.2016.10.020]

[14]   Zahmakıran, M.; Özkar, S. Metal nanoparticles in liquid phase catalysis; from recent advances to future goals. *Nanoscale,* **2011**, *3*(9), 3462-3481.
[http://dx.doi.org/10.1039/c1nr10201j] [PMID: 21833406]

[15]   Doyle, A.M.; Shaikhutdinov, S.K.; Jackson, S.D.; Freund, H.J. Hydrogenation on metal surfaces: why are nanoparticles more active than single crystals? *Angew. Chem. Int. Ed. Engl.,* **2003**, *42*(42), 5240-5243.
[http://dx.doi.org/10.1002/anie.200352124] [PMID: 14601183]

[16]   Pradhan, N.; Pal, A.; Pal, T. Catalytic reduction of aromatic nitro compounds by coinage metal nanoparticles. *Langmuir,* **2001**, *17*(5), 1800-1802.
[http://dx.doi.org/10.1021/la000862d]

[17]   Ling, Y.; Zeng, X.; Tan, W.; Luo, J.; Liu, S. Quaternized chitosan/rectorite/AgNP nanocomposite

catalyst for reduction of 4-nitrophenol. *J. Alloys Compd.*, **2015**, *647*, 463-470.
[http://dx.doi.org/10.1016/j.jallcom.2015.06.110]

[18] (a). Sun, L.; Crooks, R.M. Dendrimer-mediated immobilization of catalytic nanoparticles on flat, solid supports. *Langmuir*, **2002**, *18*(21), 8231-8236.
[http://dx.doi.org/10.1021/la020498d] (b). Praharaj, S.; Nath, S.; Ghosh, S.K.; Kundu, S.; Pal, T. Immobilization and recovery of au nanoparticles from anion exchange resin: resin-bound nanoparticle matrix as a catalyst for the reduction of 4-nitrophenol. *Langmuir*, **2004**, *20*(23), 9889-9892.
[http://dx.doi.org/10.1021/la0486281] [PMID: 15518467]

[19] Li, X-H.; Wang, X.; Antonietti, M. Mesoporous g-C$_3$N$_4$ nanorods as multifunctional supports of ultrafine metal nanoparticles: hydrogen generation from water and reduction of nitrophenol with tandem catalysis in one step. *Chem. Sci. (Camb.)*, **2012**, *3*(6), 2170-2174.
[http://dx.doi.org/10.1039/c2sc20289a]

[20] (a). Zhao, X.; Lv, L.; Pan, B.; Zhang, W.; Zhang, S.; Zhang, Q. Polymer-supported nanocomposites for environmental application: a review. *Chem. Eng. J.*, **2011**, *170*(2), 381-394.
[http://dx.doi.org/10.1016/j.cej.2011.02.071] (b). Harish, S.; Mathiyarasu, J.; Phani, K.; Yegnaraman, V. Synthesis of conducting polymer supported Pd nanoparticles in aqueous medium and catalytic activity towards 4-nitrophenol reduction. *Catal. Lett.*, **2009**, *128*(1-2), 197-202.
[http://dx.doi.org/10.1007/s10562-008-9732-x]

[21] (a). Lisha, K.; Pradeep, A.; Pradeep, T. Towards a practical solution for removing inorganic mercury from drinking water using gold nanoparticles. *Gold Bull.*, **2009**, *42*(2)
[http://dx.doi.org/10.1007/BF03214924]
(b). Astruc, D.; Lu, F.; Aranzaes, J.R. Nanoparticles as recyclable catalysts: the frontier between homogeneous and heterogeneous catalysis. *Angew. Chem. Int. Ed. Engl.*, **2005**, *44*(48), 7852-7872.
[http://dx.doi.org/10.1002/anie.200500766] [PMID: 16304662]
(c). Rao, C.; Govindaraj, A.; Gundiah, G.; Vivekchand, S. Nanotubes and nanowires. *Chem. Eng. Sci.*, **2004**, *59*(22), 4665-4671.
[http://dx.doi.org/10.1016/j.ces.2004.07.067]
(d). Miehr, R.; Tratnyek, P.G.; Bandstra, J.Z.; Scherer, M.M.; Alowitz, M.J.; Bylaska, E.J. Diversity of contaminant reduction reactions by zerovalent iron: role of the reductate. *Environ. Sci. Technol.*, **2004**, *38*(1), 139-147.
[http://dx.doi.org/10.1021/es034237h] [PMID: 14740729]
(e). Wang, L.; Wang, Z.; Zhang, X.; Shen, J.; Chi, L.; Fuchs, H. A new approach for the fabrication of an alternating multilayer film of poly (4-vinylpyridine) and poly (acrylic acid) based on hydrogen bonding. *Macromol. Rapid Commun.*, **1997**, *18*(6), 509-514.
[http://dx.doi.org/10.1002/marc.1997.030180609]

[22] (a). Kamal, T.; Ahmad, I.; Khan, S.B.; Asiri, A.M. Synthesis and catalytic properties of silver nanoparticles supported on porous cellulose acetate sheets and wet-spun fibers. *Carbohydr. Polym.*, **2017**, *157*, 294-302.
[http://dx.doi.org/10.1016/j.carbpol.2016.09.078] [PMID: 27987930]
(b). Ahmad, I.; Kamal, T.; Khan, S.B.; Asiri, A.M. An efficient and easily retrievable dip catalyst based on silver nanoparticles/chitosan-coated cellulose filter paper. *Cellulose*, **2016**, 1-12.
(c). Al-Mubaddel, F. S.; Haider, S.; Aijaz, M. O.; Haider, A.; Kamal, T.; Almasry, W. A.; Javid, M.; Khan, S. U.-D. Preparation of the chitosan/polyacrylonitrile semi-IPN hydrogel *via* glutaraldehyde vapors for the removal of Rhodamine B dye. *Polymer Bulletin*, **2017**, 1-17.
(d). Kamal, T.; Khan, S.B.; Asiri, A.M. Nickel nanoparticles-chitosan composite coated cellulose filter paper: An efficient and easily recoverable dip-catalyst for pollutants degradation. *Environ. Pollut.*, **2016**, *218*, 625-633.
[http://dx.doi.org/10.1016/j.envpol.2016.07.046] [PMID: 27481647]
(e). Haider, S.; Kamal, T.; Khan, S.B.; Omer, M.; Haider, A.; Khan, F.U.; Asiri, A.M. Natural polymers supported copper nanoparticles for pollutants degradation. *Appl. Surf. Sci.*, **2016**, *387*, 1154-1161.
[http://dx.doi.org/10.1016/j.apsusc.2016.06.133]
(f). Kamal, T.; Khan, S.B.; Asiri, A.M. Synthesis of zero-valent Cu nanoparticles in the chitosan

coating layer on cellulose microfibers: evaluation of azo dyes catalytic reduction. *Cellulose,* **2016**, *23*(3), 1911-1923.
[http://dx.doi.org/10.1007/s10570-016-0919-9]
(g). Bello, B.A.; Khan, S.A.; Khan, J.A.; Syed, F.Q.; Anwar, Y.; Khan, S.B. Antiproliferation and antibacterial effect of biosynthesized AgNps from leaves extract of Guiera senegalensis and its catalytic reduction on some persistent organic pollutants. *J. Photochem. Photobiol. B,* **2017**, *175*, 99-108.
[http://dx.doi.org/10.1016/j.jphotobiol.2017.07.031] [PMID: 28865320]
(h). Bakhsh, E.M.; Khan, S.A.; Marwani, H.M.; Danish, E.Y.; Asiri, A.M.; Khan, S.B. Performance of cellulose acetate-ferric oxide nanocomposite supported metal catalysts toward the reduction of environmental pollutants. *Int. J. Biol. Macromol.,* **2018**, *107*(Pt A), 668-677.
[http://dx.doi.org/10.1016/j.ijbiomac.2017.09.034] [PMID: 28919532]

[23]   Ng, S.W.; Kumar Das, V.; Holeček, J.; Lyčka, A.; Gielen, M.; Drew, M.G. Organostannate derivatives of dicyclohexylammonium hydrogen 2, 6-pyridinedicarboxylate: solution/solid-state 13C, 119Sn NMR and *in vitro* antitumour activity of bis (dicyclohexylammonium) bis (2, 6-pyridinedicarboxylato) dibutylstannate, and the crystal structure of its monohydrate. *Appl. Organomet. Chem.,* **1997**, *11*(1), 39-45.
[http://dx.doi.org/10.1002/(SICI)1099-0739(199701)11:1<39::AID-AOC539>3.0.CO;2-Q]

[24]   Unicef. , *Progress on drinking water and sanitation: 2012 update*; UNICEF: Nueva York, **2012**.

[25]   Uberoi, V.; Bhattacharya, S.K. Toxicity and degradability of nitrophenols in anaerobic systems. *Water Environ. Res.,* **1997**, *69*(2), 146-156.
[http://dx.doi.org/10.2175/106143097X125290]

[26]   (a). Wang, P.; Tang, L.; Wei, X.; Zeng, G.; Zhou, Y.; Deng, Y.; Wang, J.; Xie, Z.; Fang, W. Synthesis and application of iron and zinc doped biochar for removal of p-nitrophenol in wastewater and assessment of the influence of co-existed Pb (II). *Appl. Surf. Sci.,* **2017**, *392*, 391-401.
[http://dx.doi.org/10.1016/j.apsusc.2016.09.052]
(b). Tang, D.; Zheng, Z.; Lin, K.; Luan, J.; Zhang, J. Adsorption of p-nitrophenol from aqueous solutions onto activated carbon fiber. *J. Hazard. Mater.,* **2007**, *143*(1-2), 49-56.
[http://dx.doi.org/10.1016/j.jhazmat.2006.08.066] [PMID: 17030422]

[27]   Di Paola, A.; Augugliaro, V.; Palmisano, L.; Pantaleo, G.; Savinov, E. Heterogeneous photocatalytic degradation of nitrophenols. *J. Photochem. Photobiol. Chem.,* **2003**, *155*(1), 207-214.
[http://dx.doi.org/10.1016/S1010-6030(02)00390-8]

[28]   Yuan, S.; Tian, M.; Cui, Y.; Lin, L.; Lu, X. Treatment of nitrophenols by cathode reduction and electro-Fenton methods. *J. Hazard. Mater.,* **2006**, *137*(1), 573-580.
[http://dx.doi.org/10.1016/j.jhazmat.2006.02.069] [PMID: 16650528]

[29]   Kumar, S.; Singh, S.; Srivastava, V.C. Electro-oxidation of nitrophenol by ruthenium oxide coated titanium electrode: parametric, kinetic and mechanistic study. *Chem. Eng. J.,* **2015**, *263*, 135-143.
[http://dx.doi.org/10.1016/j.cej.2014.11.051]

[30]   Yi, S.; Zhuang, W-Q.; Wu, B.; Tay, S.T-L.; Tay, J-H. Biodegradation of p-nitrophenol by aerobic granules in a sequencing batch reactor. *Environ. Sci. Technol.,* **2006**, *40*(7), 2396-2401.
[http://dx.doi.org/10.1021/es0517771] [PMID: 16646480]

[31]   Lai, B.; Zhang, Y.; Chen, Z.; Yang, P.; Zhou, Y.; Wang, J. Removal of p-nitrophenol (PNP) in aqueous solution by the micron-scale iron–copper (Fe/Cu) bimetallic particles. *Appl. Catal. B,* **2014**, *144*, 816-830.
[http://dx.doi.org/10.1016/j.apcatb.2013.08.020]

[32]   Gimeno, O.; Carbajo, M.; Beltrán, F.J.; Rivas, F.J. Phenol and substituted phenols AOPs remediation. *J. Hazard. Mater.,* **2005**, *119*(1-3), 99-108.
[http://dx.doi.org/10.1016/j.jhazmat.2004.11.024] [PMID: 15752854]

[33]   Bakheet, B.; Qiu, C.; Yuan, S.; Wang, Y.; Yu, G.; Deng, S.; Huang, J.; Wang, B. Inhibition of polymer formation in electrochemical degradation of p-nitrophenol by combining electrolysis with ozonation.

Chem. Eng. J., **2014**, *252*, 17-21.
[http://dx.doi.org/10.1016/j.cej.2014.04.103]

[34]   Chen, T.; Liu, F.; Ling, C.; Gao, J.; Xu, C.; Li, L.; Li, A. Insight into highly efficient coremoval of copper and p-nitrophenol by a newly synthesized polyamine chelating resin from aqueous media: competition and enhancement effect upon site recognition. *Environ. Sci. Technol.,* **2013**, *47*(23), 13652-13660.
[http://dx.doi.org/10.1021/es4028875] [PMID: 24164273]

[35]   Lazo-Cannata, J.C.; Nieto-Márquez, A.; Jacoby, A.; Paredes-Doig, A.L.; Romero, A.; Sun-Kou, M.R.; Valverde, J.L. Adsorption of phenol and nitrophenols by carbon nanospheres: Effect of pH and ionic strength. *Separ. Purif. Tech.,* **2011**, *80*(2), 217-224.
[http://dx.doi.org/10.1016/j.seppur.2011.04.029]

[36]   Woo, Y.T.; Lai, D.Y. *Aromatic amino and nitro–amino compounds and their halogenated derivatives*; Patty's Toxicology, **2001**.

[37]   Nair, R.G.; Roy, J.K.; Samdarshi, S.; Mukherjee, A. Mixed phase V doped titania shows high photoactivity for disinfection of Escherichia coli and detoxification of phenol. *Sol. Energy Mater. Sol. Cells,* **2012**, *105*, 103-108.
[http://dx.doi.org/10.1016/j.solmat.2012.05.008]

[38]   Khan, S.A.; Khan, S.B.; Kamal, T.; Asiri, A.M.; Akhtar, K. Recent development of chitosan nanocomposites for environmental applications. *Recent Pat. Nanotechnol.,* **2016**, *10*(3), 181-188.
[http://dx.doi.org/10.2174/1872210510666160429145339] [PMID: 27136929]

[39]   Malato, S.; Fernández-Ibáñez, P.; Maldonado, M.; Blanco, J.; Gernjak, W. Decontamination and disinfection of water by solar photocatalysis: recent overview and trends. *Catal. Today,* **2009**, *147*(1), 1-59.
[http://dx.doi.org/10.1016/j.cattod.2009.06.018]

[40]   John De Zuane, P. Handbook of drinking water quality standards and control. *Van Nostrand Reinhold, New York. Kudryavtseva LP (1999). Assessment of drinking water quality in the city of Apatit. Water Res.,* **1990**, *26*(2), 659-665.

[41]   Rauf, M.; Ashraf, S.S. Fundamental principles and application of heterogeneous photocatalytic degradation of dyes in solution. *Chem. Eng. J.,* **2009**, *151*(1), 10-18.
[http://dx.doi.org/10.1016/j.cej.2009.02.026]

[42]   Khan, S.A.; Khan, S.B.; Abdullah, M. Layered double hydroxide of Cd-Al/C for the Mineralization and De-coloration of Dyes in Solar and Visible Light Exposure. *Scientific Reports,* **2016**. accepted manuscript.

[43]   Padhi, B. Pollution due to synthetic dyes toxicity & carcinogenicity studies and remediation. *Int. J. Environ. Sci.,* **2012**, *3*(3), 940.

[44]   (a). Mo, J.H.; Lee, Y.H.; Kim, J.; Jeong, J.Y.; Jegal, J. Treatment of dye aqueous solutions using nanofiltration polyamide composite membranes for the dye wastewater reuse. *Dyes Pigments,* **2008**, *76*(2), 429-434.
[http://dx.doi.org/10.1016/j.dyepig.2006.09.007]
(b). Arslan, I.; Balcioğlu, I.A.; Bahnemann, D.W. Advanced chemical oxidation of reactive dyes in simulated dyehouse effluents by ferrioxalate-Fenton/UV-A and TiO_2/UV-A processes. *Dyes Pigments,* **2000**, *47*(3), 207-218.
[http://dx.doi.org/10.1016/S0143-7208(00)00082-6]

[45]   Arshad, T.; Khan, S.A.; Faisal, M.; Shah, Z.; Akhtar, K.; Asiri, A.M.; Ismail, A.A.; Alhogbi, B.G.; Khan, S.B. Cerium based photocatalysts for the degradation of acridine orange in visible light. *J. Mol. Liq.,* **2017**, *241*, 20-26.
[http://dx.doi.org/10.1016/j.molliq.2017.05.079]

[46]   (a). Behnajady, M.A.; Modirshahla, N.; Shokri, M. Photodestruction of Acid Orange 7 (AO7) in aqueous solutions by UV/H2O2: influence of operational parameters. *Chemosphere,* **2004**, *55*(1), 129-

134.
[http://dx.doi.org/10.1016/j.chemosphere.2003.10.054] [PMID: 14720555]
(b). Ashraf, S.S.; Rauf, M.A.; Alhadrami, S. Degradation of Methyl Red using Fenton's reagent and the effect of various salts. *Dyes Pigments,* **2006**, *69*(1), 74-78.
[http://dx.doi.org/10.1016/j.dyepig.2005.02.009]

[47]   (a). Alaton, I.A.; Balcioglu, I.A. Photochemical and heterogeneous photocatalytic degradation of waste vinylsulphone dyes: a case study with hydrolyzed Reactive Black 5. *J. Photochem. Photobiol. Chem.,* **2001**, *141*(2), 247-254.
[http://dx.doi.org/10.1016/S1010-6030(01)00440-3]
(b). Zielińska, B.; Grzechulska, J.; Grzmil, B.; Morawski, A.W. Photocatalytic degradation of reactive Black 5: a comparison between TiO_2-Tytanpol A11 and TiO_2-Degussa P25 photocatalysts. *Appl. Catal. B,* **2001**, *35*(1), L1-L7.
[http://dx.doi.org/10.1016/S0926-3373(01)00230-2]

[48]   (a). Qamar, M.; Saquib, M.; Muneer, M. Semiconductor-mediated photocatalytic degradation of anazo dye, chrysoidine Y in aqueous suspensions. *Desalination,* **2005**, *171*(2), 185-193.
[http://dx.doi.org/10.1016/j.desal.2004.04.005]
(b). Habibi, M.H.; Hassanzadeh, A.; Mahdavi, S. The effect of operational parameters on the photocatalytic degradation of three textile azo dyes in aqueous TiO_2 suspensions. *J. Photochem. Photobiol. Chem.,* **2005**, *172*(1), 89-96.
[http://dx.doi.org/10.1016/j.jphotochem.2004.11.009]

[49]   Khan, S.B.; Khan, S.A.; Asiri, A.M. A fascinating combination of Co, Ni and Al nanomaterial for oxygen evolution reaction. *Appl. Surf. Sci.,* **2016**, *370*, 445-451.
[http://dx.doi.org/10.1016/j.apsusc.2016.02.062]

[50]   (a). Shao, M.; Han, J.; Wei, M.; Evans, D.G.; Duan, X. The synthesis of hierarchical Zn–Ti layered double hydroxide for efficient visible-light photocatalysis. *Chem. Eng. J.,* **2011**, *168*(2), 519-524.
[http://dx.doi.org/10.1016/j.cej.2011.01.016]
(b). Baek, S-H.; Nam, G-H.; Park, I-K. Morphology controlled growth of ZnAl-layered double hydroxide and ZnO nanorod hybrid nanostructures by solution method. *RSC Advances,* **2015**, *5*(74), 59823-59829.
[http://dx.doi.org/10.1039/C5RA10374F]

[51]   Xu, K.; Chen, G.; Shen, J. Facile synthesis of submicron-scale layered double hydroxides and their direct decarbonation. *RSC Advances,* **2014**, *4*(17), 8686-8691.
[http://dx.doi.org/10.1039/c3ra45602a]

[52]   Sideris, P.J.; Nielsen, U.G.; Gan, Z.; Grey, C.P. Mg/Al ordering in layered double hydroxides revealed by multinuclear NMR spectroscopy. *Science,* **2008**, *321*(5885), 113-117.
[http://dx.doi.org/10.1126/science.1157581] [PMID: 18599785]

[53]   Matos, J.; Laine, J.; Herrmann, J-M. Synergy effect in the photocatalytic degradation of phenol on a suspended mixture of titania and activated carbon. *Appl. Catal. B,* **1998**, *18*(3), 281-291.
[http://dx.doi.org/10.1016/S0926-3373(98)00051-4]

[54]   Sun, J-h.; Wang, Y-k.; Sun, R-x.; Dong, S-y. Photodegradation of azo dye Congo Red from aqueous solution by the WO_3 TiO_2/activated carbon (AC) photocatalyst under the UV irradiation. *Mater. Chem. Phys.,* **2009**, *115*(1), 303-308.
[http://dx.doi.org/10.1016/j.matchemphys.2008.12.008]

[55]   Li, Y.; Li, X.; Li, J.; Yin, J. Photocatalytic degradation of methyl orange by TiO_2-coated activated carbon and kinetic study. *Water Res.,* **2006**, *40*(6), 1119-1126.
[http://dx.doi.org/10.1016/j.watres.2005.12.042] [PMID: 16503343]

[56]   Matos, J.; Laine, J.; Herrmann, J-M. Effect of the type of activated carbons on the photocatalytic degradation of aqueous organic pollutants by UV-irradiated titania. *J. Catal.,* **2001**, *200*(1), 10-20.
[http://dx.doi.org/10.1006/jcat.2001.3191]

[57]   Li Puma, G.; Bono, A.; Krishnaiah, D.; Collin, J.G. Preparation of titanium dioxide photocatalyst

loaded onto activated carbon support using chemical vapor deposition: a review paper. *J. Hazard. Mater.,* **2008**, *157*(2-3), 209-219.
[http://dx.doi.org/10.1016/j.jhazmat.2008.01.040] [PMID: 18313842]

[58]   Oberlin, A. High-resolution TEM studies of carbonization and graphitization. *Chem. Phys. Carbon,* **1989**, *22*(1)

SUBJECT INDEX

A

Acids 183, 264, 266
  humic 264, 266
  salicylic 183
Acids to form salts of iron 232
Activated carbon 9, 11, 12, 13, 176, 179, 182,
   269, 295, 301, 302
Active ingredients (AIs) 51
Adsorbents 1, 2, 3, 4, 5, 6, 7, 9, 11, 12, 13, 14,
   15, 17, 67, 68, 69, 70, 167, 168, 176,
   178, 179, 181, 182, 183, 184, 203
Adsorbents and catalysts 1, 2, 5
Adsorption 1, 2, 7, 8, 11, 13, 15, 20, 23, 24,
   27, 68, 71, 167, 176, 177, 179, 180, 181,
   182, 184, 195, 205, 239, 264, 271, 272,
   290, 291, 295, 297
  vapor 271, 272
Adsorption capacity 11, 12, 178, 179, 195
Adsorption efficiency 12, 177, 178, 179
Adsorption mechanism 181
Adsorption of dichlorobenzene 12
Adsorption sites, high 13
Adsorptive removal 68, 70, 72
Advanced oxidation process (AOP) 295, 297
Alkali metals 195, 232
Alkalis 232, 234, 239, 246
Anatase 124, 125, 128
Antimicrobial properties 27, 83
Applications of environmental
   nanotechnology 43, 44
Applications of photocatalysis 145
Aquatic environments 175, 176
Aqueous environment 166, 176, 178, 179,
   183, 184
Aqueous solutions 8, 12, 14, 26, 38, 68, 71,
   74, 90, 122, 144, 178, 182, 239
Aqueous system 175, 176, 193
Arsenophosphate 198, 205
Asymmetric membranes 269, 273, 274, 275
  skinned 274, 275
Atomic force microscopy (AFM) 174, 227

Automobiles 64, 65, 167, 293

B

Band, conduction 16, 120, 128, 132, 136, 137,
   148
Band gap energy 15, 120, 132, 135
Benefits of environmental monitoring 86
Bentonite 177, 179, 180, 182, 183, 203
  modified 177, 179, 180, 183
Biodegradation 176, 295, 297
Bismuth 133, 134, 197
BSA solution 266
Bulk materials 6, 38, 39, 40, 222

C

Cadmium 15, 55, 74, 175, 176
Carbon black (CB) 11, 13, 19, 64, 66, 120,
   140
Carbon materials 13, 19, 118
Carbon nanodots 139
Carbon nanomaterials 11, 13, 58, 77, 265
Carbon nanostructures 69, 136
Carbon nanotubes 6, 11, 12, 13, 17, 19, 27, 45,
   137, 168, 264
Casting solution 271, 272
Catalyst concentration 124
Catalysts 1, 2, 3, 4, 5, 8, 14, 15, 17, 18, 20, 21,
   27, 47, 48, 49, 73, 75, 76, 77, 78, 120,
   122, 124, 148, 237, 246, 247, 290, 291,
   292, 293, 294, 297, 298, 299
  conventional 47, 48
  heterogeneous 73, 292
Catalyst substrate 122
Catalyst surface 122, 124, 134, 140
Catalytic properties 18, 19, 20, 67, 87, 292
Catalytic reaction 18, 21, 22, 294
Cation-exchangers 191, 196, 198, 202
Chemical degradation 1, 15, 27, 184, 290
Chemical properties 39, 68, 77, 79, 125, 196,
   207, 230, 232, 235, 264

www.ingramcontent.com/pod-product-compliance
Lightning Source LLC
Chambersburg PA
CBHW041725210326
41598CB00008B/778